Public Transportation: Planning, Operations and Management

Public Transportation: Planning, Operations and Management

Editor: Jacob Rowe

NY RESEARCH PRESS

New York

Published by NY Research Press
118-35 Queens Blvd., Suite 400,
Forest Hills, NY 11375, USA
www.nyresearchpress.com

Public Transportation: Planning, Operations and Management
Edited by Jacob Rowe

International Standard Book Number: 978-1-63238-845-2 (Hardback)

Trademark Notice: Registered trademark of products or corporate names are used only for explanation and identification without intent to infringe.

Cataloging-in-Publication Data

Public transportation : planning, operations and management / edited by Jacob Rowe.
    p. cm.
Includes bibliographical references and index.
ISBN 978-1-63238-845-2
1. Transportation. 2. Transportation engineering. 3. Transportation--Planning.
4. Transportation--Management. 5. Traffic engineering. I. Rowe, Jacob.
TA1145 .P83 2022
629.04--dc23

Contents

Preface

The purpose of the book is to provide a glimpse into the dynamics and to present opinions and studies of some of the scientists engaged in the development of new ideas in the field from very different standpoints. This book will prove useful to students and researchers owing to its high content quality.

Public transportation is the transport of passengers using group travel systems operating on established routes managed on a schedule for a fee. The public transport used globally includes passenger trains, city buses, trams, rapid transit, trolley buses and ferries. Between cities, travel can be accommodated by intercity rail, airlines and coaches. Public transport systems are characterized by set embarkation/disembarkation points and prearranged time tables. Public transport infrastructure is vital for the smooth transportation of people. This can be exclusive or shared with freight and private transport. This book provides comprehensive insights into the planning, operations and management of public transportation. It discusses the fundamentals as well as modern approaches of this area of study. For all readers who are interested in public transportation, the case studies included in this book will serve as an excellent guide to develop a comprehensive understanding.

At the end, I would like to appreciate all the efforts made by the authors in completing their chapters professionally. I express my deepest gratitude to all of them for contributing to this book by sharing their valuable works. A special thanks to my family and friends for their constant support in this journey.

Editor

1

Shared Autonomous Taxi System and Utilization of Collected Travel-Time Information

Zhiguang Liu [ID],[1] **Tomio Miwa** [ID],[2] **Weiliang Zeng** [ID],[3]
Michael G. H. Bell,[4] **and Takayuki Morikawa**[5]

[1]*Department of Civil Engineering, Nagoya University, Nagoya 464-8603, Japan*
[2]*Institute of Materials and Systems for Sustainability, Nagoya University, Nagoya 464-8603, Japan*
[3]*School of Automation, Guangdong University of Technology, Guangzhou, Guangdong 510006, China*
[4]*Institute of Transport and Logistics Studies, Business School, The University of Sydney, Sydney, NSW, Australia*
[5]*Institute of Innovation for Future Society, Nagoya University, Nagoya 464-8603, Japan*

Correspondence should be addressed to Weiliang Zeng; weiliangzeng@gdut.edu.cn

Academic Editor: Vinayak Dixit

Shared autonomous taxi systems (SATS) are being regarded as a promising means of improving travel flexibility. Each shared autonomous taxi (SAT) requires very precise traffic information to independently and accurately select its route. In this study, taxis were replaced with ride-sharing autonomous vehicles, and the potential benefits of utilizing collected travel-time information for path finding in the new taxi system examined. Specifically, four categories of available SATs for every taxi request were considered: currently empty, expected-empty, currently sharable, and expected-sharable. Two simulation scenarios—one based on historical traffic information and the other based on real-time traffic information—were developed to examine the performance of information use in a SATS. Interestingly, in the historical traffic information-based scenario, the mean travel time for taxi requests and private vehicle users decreased significantly in the first several simulation days and then remained stable as the number of simulation days increased. Conversely, in the real-time information-based scenario, the mean travel time was constant. As the SAT fleet size increased, the total travel time for taxi requests significantly decreased, and convergence occurred earlier in the historical information-based scenario. The results demonstrate that historical traffic information is better than real-time traffic information for path finding in SATS.

1. Introduction

Autonomous vehicles (AVs) have undergone rapid development in recent years. Many automobile manufacturers and IT companies around the world, including Google, Uber, Tesla, and Toyota, are testing their AV products on real road networks [1]. In some countries, AVs are allowed to enter select areas, but not the entire public road network. A Singaporean technology company called "nuTonomy" uses AVs as taxis within an area of 2.5 square miles [2]. On December 2, 2017, four self-driving buses were tested on public roads in Shenzhen, China, which is believed to be the first live test of autonomous buses in the world [3]. There is considerable evidence to indicate that the use of AVs will become more widespread in the near future.

AVs can be designed to connect with each other and can exchange traffic information related to the road network [4]. The AVs start when orders are transmitted to the controller inside the vehicles, and they drive automatically based on the instructions from the controller. Owing to these advantages, AVs are expected to make transportation more efficient and comfortable and reduce cost, environmental impact, and congestion [5].

In a traditional taxi operation system, a driver can serve only a single customer or a group of customers in a point-to-point journey. For example, the average occupancy rate of a taxi in New York City is only 1.2 passengers per trip [6]. Empty taxis usually cruise along urban roads looking for customers. This is economically burdensome because they increase the traffic volume, which may lead to traffic congestion on the urban road network. Overall, the operational efficiency is relatively low. The main reason why customers decide to hire taxis is that they have limited time, but heavy traffic prevents customers from reaching their destinations on time [7]. Meanwhile, the current system is unable to match the entire taxi demand during peak hours. Measures to solve such problems are urgently needed.

Accordingly, this study was conducted with the objective of developing a shared autonomous taxi system (SATS) and investigating the efficiency of the system through simulation experiments. In the studied system, customers are assumed to be willing to share a taxi with other customers and can request a taxi through a smartphone. Shared autonomous taxis (SATs) remain parked when unoccupied to reduce the road burden. A taxi is automatically assigned by the SATS to pick a customer up after considering both unoccupied and occupied taxis. By taking advantage of unoccupied seats, the SATS can reduce the total number of taxis required on the road network.

AVs are equipped with superior technological sensors [8], which are used to accurately perceive the surrounding environment, and collect traffic information. In this study, as probe vehicles, SATs were deemed capable of collecting valuable traffic information, including data pertaining to the vehicle location and link travel time. The collected traffic information can be used for path finding and categorized as historical and real-time information. The effectiveness of using the two types of information was investigated in this study.

The remainder of this paper is organized as follows. In Section 2, the literature on taxi sharing and the performance of shared AVs is reviewed. The SATS considered in this study is explained in Section 3. Section 4 presents the traffic information used for path finding in the SATS. The simulation design and results are described in Section 5. Finally, conclusions and a discussion of future work are presented in Section 6.

2. Literature Review

Autonomous taxi systems were proposed prior to the year 2000, e.g., the autonomous dial-a-ride taxi system [9], even though the technology for navigation and positioning was not sufficiently advanced at that time. With the development of AVs, a customer can request an SAT and ride it to his/her destination. The sharable taxis can be assigned to a single person or shared with other customers. The deployment of shared AVs (SAVs) can lead to a reduction in the total number of private cars on urban road networks. Fagnant and Kockelman [10] designed an agent-based model for SAV operations in which four strategies are used to relocate AVs with the aim of minimizing waiting times for future travelers. A 5-min interval is chosen as the iteration period. At the beginning of every 5-min interval, the travel demand in every zone is predetermined. A parameter called "block balance" is proposed, which represents the difference between the expected demand and supply for SAVs in the upcoming 5 min. By comparing the average number of trips served by an SAV with that by a private car, they concluded that

one SAV can replace approximately 11 conventional cars. Fagnant et al. [11] provided more details with regard to link-level travel times. Their proposed model comprises four submodules: SAV location and trip assignment, SAV fleet generation, SAV movement, and SAV relocation. An SAV is assigned first to the traveler who has been waiting for the longest time. Fagnant and Kockelman [12] developed SAV simulations for clients with different origins and destinations by considering dynamic ride sharing (DRS). Five conditions are considered to judge whether a ride could be shared: the total travel time and the increase in the remaining journey time for riding passengers, the increase in the total travel time for a new passenger, the possibility of the new passenger being picked up in the next 5 min, and the total travel time for the two passengers. Their experimental results suggested that DRS has the potential to reduce the total service time (which includes the waiting time), travel time, and cost for users. The authors also discussed the optimal SAV fleet size from an economic viewpoint. However, they did not provide delivery rules for the customers in a shared taxi. This is important because these rules determine the remaining time and travel time for each individual customer. Burghout et al. [13] replaced private vehicles in Stockholm with the SAVs in a simulation study. The sharing schemes included passengers with the same origin and destination, the same origin but different destinations, and different origins but the same destination. Their results indicated that only 5% of the current number of private cars would be needed to transport commuters, but the travel time increased by 13% on average. Lioris et al. [14] suggested that if the detour time incurred by serving a potential customer exceeds a maximal detour time, that customer should be rejected. Levin et al. [15] reported that, when choosing between an occupied SAV and an unoccupied SAV, the one that is able to arrive at the customer location first should be assigned to the customer. They further proposed that SAVs would increase congestion because of the additional trips made to reach each customer's origin. It was found that the difference in the vehicle miles traveled between SAV scenarios and non-SAV scenarios was primarily due to the repositioning trips required to pick the next passenger up.

To optimize conventional taxi systems in terms of customer convenience and mitigation of traffic congestion, a major strategy has been to investigate dial-a-ride and shared taxi systems. Studies on this strategy can be found in the literature. Teal [16] found that commuters could be the major users of carpool ride sharing, which can reduce travel costs. Even though ride sharing is generally applied to private cars, commuters can call a taxi with the intention of sharing it. Dial [9] reported on an autonomous dial-a-ride taxi system in which customers request a taxi via telephone and only the customer is involved in the process of requesting a ride, assigning the trip, scheduling the arrival, and routing the vehicle. The task of a driver is to simply follow instructions provided by the vehicle's computers. The author also investigated an ideal autonomous taxi system that has the ability to assign a taxi to a customer in the shortest possible time. Tao [17] used each customer's choice for the maximum acceptable number of sharing customers and acceptable gender as inputs

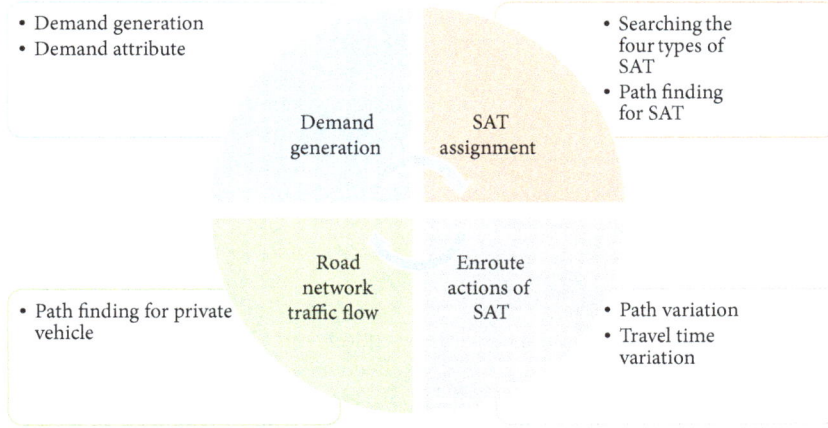

- Demand generation
- Demand attribute

- Searching the four types of SAT
- Path finding for SAT

Demand generation

SAT assignment

Road network traffic flow

Enroute actions of SAT

- Path finding for private vehicle

- Path variation
- Travel time variation

FIGURE 1: Shared autonomous taxi system (SATS) integrated into traffic flow.

to an algorithm, which then selects the taxi that is able to reach the customer's location most rapidly to provide the service. Orey et al. [18] proposed that a customer request be sent to all taxis or a subset of taxis and each taxi would respond with its cost associated with the trip to the customer. The customer would then determine the acceptable lowest cost for taxi sharing and choose a taxi to use. However, a customer finds it difficult to select one taxi from potential taxis. Ota et al. [19] studied a data-driven taxi ride-sharing simulation in which taxis cruise the road network even when empty. A trip is assigned to the taxi that provides the lowest additional cost if the passenger is assigned. Ma et al. [20] noted that a ride request, made through a smartphone app, should be assigned to whichever taxi that minimizes the increase in the travel distance resulting from the request while meeting the arrival time, capacity, and monetary constraints of both the new passenger and the existing passenger(s). Nourinejad and Roorda [21] verified the performance of ride sharing by maximizing the savings on the total vehicle kilometers traveled and maximizing the matching rate. Najmi et al. [22] investigated the effect of different dynamic sharing methods on the performance of the ride-sharing system.

Many previous studies have investigated SAV and taxi sharing. To correctly plan SAT paths, traffic information is essential in the SATS. However, providing sufficient information to enable accurate path selection is difficult. In this study, the link travel-time information collected by SATs was analyzed and the information applied to path finding for both SATs and private vehicles.

3. Shared Autonomous Taxi System

This section describes the components and implementation of an SATS. This system is an improvement of the system proposed in a previous study [23]. The SATS is built on two events: SAT assignment and enroute actions of the SAT, and the types of responses these actions warrant. The proposed system assumes that all SATs can be shared by two requests and these requests can only be made through a smartphone. It should be noted that a request can contain more than one customer.

This SATS operates on a network $G = (N, A, V, D)$, where N is the set of nodes and A is the set of links. The network has a set of SATs V that provides service to demand D. The integration of this system with road network traffic flow is illustrated in Figure 1. The implementation steps are grouped into four modules: (1) demand generation, (2) SAT assignment, (3) enroute actions of SAT, and (4) road network traffic flow. The remainder of this section describes these modules in detail.

3.1. Demand Generation. The demand generation module introduces customers to the proposed SATS. In each time step τ, this module outputs a set of new customers who request an SAT. The new customers and waiting customers are provided with the service during this time step, and the waiting customers are provided with this service before the new customers.

In this study, we assumed that demand can be separated into single requests. The origin and destination of each request $d \in D$ are denoted by O_d and D_d, respectively. For ride-sharing problems, departure time and arrival time are key factors in improving matching rate [24]; therefore, it was assumed that each request has an earliest departure time denoted by EDT_d, a latest arrival time denoted by LAT_d, and travel-time flexibility denoted by TTF_d [25]. Request d should be served in the time window (EDT_d, LAT_d) [26]. In this study, EDT was assumed to be the same as the SAT request time. Travel-time flexibility is the extra time that is acceptable to requests. If the minimum travel time from O_d to D_d is $T_{(O_d, D_d)}$, the travel-time flexibility of request d is $TTF_d = LAT_d - EDT_d - T_{(O_d, D_d)} \geq 0$. Once a request appears at τ, the SAT assignment event will be triggered.

3.2. SAT Assignment. The SAT assignment operates and responds to the appearance of new demand. This module describes the specific logic used to assign SATs to demand in the SATS.

The SATS assumes the existence of a virtual central control system that knows the status of all SATs and requests; this system can also assign an SAT to each request and select the path for the SAT. The output of the SAT assignment is

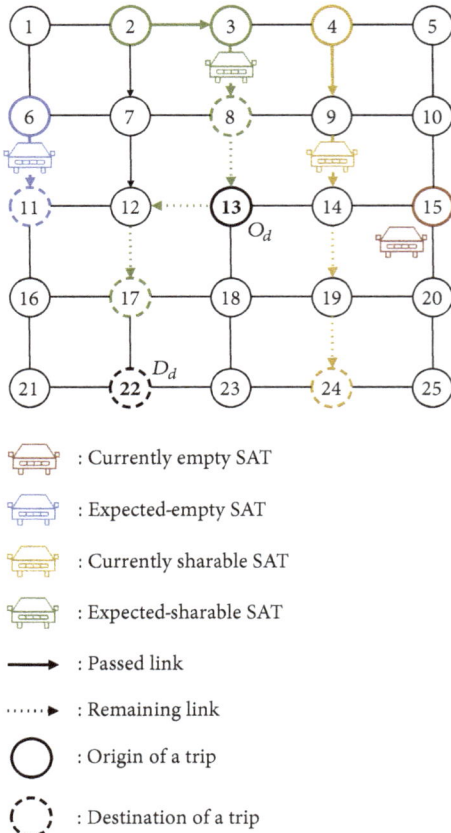

FIGURE 2: Four types of available SATs.

an update of the SAT status, including unoccupied, single-rider occupied, shared, and change-of-request status (i.e., in-service or waiting to be served). In this study, a rider corresponds to a request. Each SAT is either parked at a node or serving a customer at any given time; therefore, in this study, it was assumed that nodes have (infinite) parking space.

3.2.1. Four Types of Available SATs. In the SATS, once a new request is generated, the virtual central control system searches the available SATs for the request. All the searched SATs should be located within the distance D_{max} from the request, where D_{max} is the radius of the search area. There are four types of available SATs according to their status and possibility of sharing: currently empty, expected-empty, currently sharable, and expected-sharable SATs.

A currently empty SAT is an unoccupied SAT. As shown in Figure 2, when request d occurs at node O_d, the virtual central control system first tries to check whether a currently empty SAT is available. In this figure, an unoccupied SAT is parked at node 15.

An expected-empty SAT is a taxi that is occupied by other customer(s) presently, and the rider will get off within the maximum waiting time T_{max}. Next, the SAT can travel to d. As shown in Figure 2, the origin and destination of rider in an expected-empty SAT are node 6 and node 11, respectively, and the taxi is traveling on the planned path. In this case, this

SAT can first proceed to node 11 to drop the rider off and then travel to O_d to pick d up.

A currently sharable SAT is defined as a single-rider occupied taxi that can be shared with another request if the requirements for sharing can be met, which are threefold. The first requirement is that the origin of the new request O_d should be on a path in the path set of the rider c. The path set of each OD pair is predetermined using a k-shortest path algorithm [27, 28]. The k-shortest path algorithm can work in this study when the size of a network is not large. In addition, its application can reduce the time for processing every request since candidate SATs for sharing can be screened effectively. The second requirement is that the destination D_c of rider c should be on a path in the path set of the new request, or the destination of the new request D_d should be on a path in the path set of rider c. The third requirement is that both rider c and request d can arrive at their destinations within LAT_c and LAT_d, respectively. This is called sharing check in this SATS. As shown in Figure 2, the origin and destination of the rider are node 4 and node 24, respectively. The taxi may change its route from the original path (4 $\longrightarrow$ 9 $\longrightarrow$ 14 $\longrightarrow$ 19 $\longrightarrow$ 24) at node 14 to pick request d up, and the two trips will share this SAT on the link from 13 to 24.

An expected-sharable SAT is an SAT that is presently occupied by two riders and can be shared by request d and the remaining rider after one rider gets off the vehicle. For this type of SAT, the requirements for sharing are also threefold. The first two requirements are the same as the first two requirements for currently sharable SAT, and rider c in the sharing requirements is the rider who can arrive at their destination later than the other rider. The last requirement is that both the rider and request d can arrive at their destinations within their respective latest arrival time. As shown in Figure 2, an expected-sharable SAT is currently occupied by two riders. After one rider arrives at their destination (node 8), the SAT can be shared by the remaining rider and request d.

In this SATS, a currently sharable SAT is a single-rider occupied SAT, and a single-rider occupied SAT can also possibly be an expected-empty SAT. A two-rider occupied SAT can be an expected-empty SAT or an expected-sharable SAT. For searched occupied SAT, the virtual central control system checks the possibility of this SAT being a currently sharable or an expected-sharable SAT because the path set for each OD pair is found using the k-shortest path algorithm, and customers are prohibited from traveling on paths beyond the path set.

3.2.2. Path Finding for SAT. Before SAT assignment, the virtual central control system searches for a potential path for each of the four types of SATs using the shortest travel-time path algorithm. When an SAT is finally assigned, the SAT transports customers on the potential path. The path of a traveling SAT cannot be changed until customers arrive at their destination or the SAT is shared with a new request. For SATs that are not parked at O_d at the moment that path finding begins, to correctly find the shortest travel-time path, the mean travel time for each possible path should include the remaining part of the ongoing path. Figure 3 takes the

	Links						
	Path o			Path p		Path q	
	1	**2**	**3**	**4**	**5**	...	**m**
h	$\hat{\mu}_{1,h}$	$\hat{\mu}_{2,h}$	$\hat{\mu}_{3,h}$	$\hat{\mu}_{4,h}$	$\hat{\mu}_{5,h}$	⋮	$\hat{\mu}_{m,h}$
$h+\Delta h$		$h_0 + h_1 \geq h+\Delta h$	$\hat{\mu}_{3,h+\Delta h}$	$\hat{\mu}_{4,h+\Delta h}$	$\hat{\mu}_{5,h+\Delta h}$	⋮	$\hat{\mu}_{m,h+\Delta h}$
$h+2\Delta h$	$\hat{\mu}_{1,h+2\Delta h}$	$h_0 + h_2 \geq h+2\Delta h$		$\hat{\mu}_{4,h+2\Delta h}$	$\hat{\mu}_{5,h+2\Delta h}$	⋮	$\hat{\mu}_{m,h+2\Delta h}$
$h+3\Delta h$	$\hat{\mu}_{1,h+3\Delta h}$	$h_0 + h_3 \geq h+3\Delta h$			$\hat{\mu}_{5,h+3\Delta h}$	⋮	$\hat{\mu}_{m,h+3\Delta h}$
$h+4\Delta h$	$\hat{\mu}_{1,h+4\Delta h}$	$h_0 + h_4 \geq h+4\Delta h$				⋮	$\hat{\mu}_{m,h+4\Delta h}$
⋮	⋮	⋮	$h_0 + h_{m-2} \geq h+5\Delta h$			⋮	⋮

T

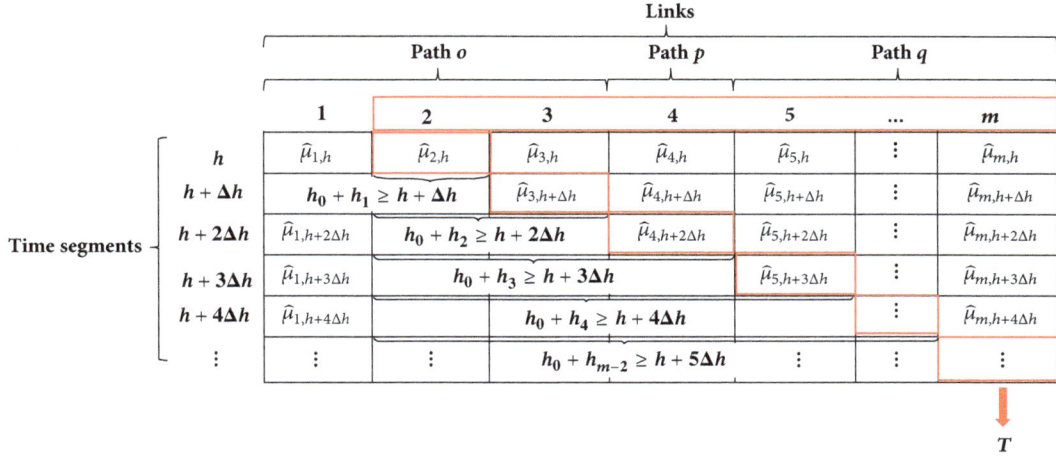

FIGURE 3: Path travel time of expected-empty taxi.

expected-empty SAT as an example. Information in this table is the mean travel time of each link in each time segment. In the figure, suppose that path o is the ongoing path of an expected-empty SAT and links 2 and 3 are the remaining links. The SAT travels to O_d through path p. Path q is one path from the origin of d to the destination. The travel time for this new path is the sum of the travel time for the remaining links of paths o, p, and q. The consideration of the transition in the traffic condition is explained in Section 4.

3.2.3. SAT Assignment and Status Update. To reduce the total travel time, a request is served by the SAT with the minimum travel time to its destination. The virtual central control system executes the following steps sequentially.

Step 1. The central control system searches for currently empty SATs within D_{max} from the origin of request d and then searches for the minimum travel-time paths for each searched SAT. The available path set of every currently empty SAT is the combination of two path sets: the path set from the SAT's location to O_d and the path set from O_d to D_d. The consideration of the combination of two path sets is reasonable because our aim is to ensure that the request arrives at the destination in the minimum travel time when time transition is considered, which is described in Section 4. The minimum travel times with each searched currently empty SAT can be recorded as $\{T_{ce,1}, T_{ce,2}, \ldots, T_{ce,n1}\}$, where $n1$ is the number of searched currently empty SATs. The minimum travel time is recorded as $T_{ce,min}$.

Step 2. The system searches for expected-empty SATs. The shortest travel-time path for each expected-empty SAT is searched as described in Section 3.2.2. The minimum travel times of each expected-empty SAT are $\{T_{ee,1}, T_{ee,2}, \ldots, T_{ee,n2}\}$, where $n2$ is the number of expected-empty taxis. The minimum travel time is recorded as $T_{ee,min}$.

Step 3. The system searches for currently sharable and expected-sharable SATs. The central control system first checks whether the searched SATs meet the requirements

for sharing. The searched SATs are feasible SATs if the requirements are met, and a potential path is assigned to each feasible SAT. The minimum travel times for the new request can be recorded as $\{T_{cs,1}, T_{cs,2}, \ldots, T_{cs,n3}\}$ with all feasible currently sharable SATs and as $\{T_{es,1}, T_{es,2}, \ldots, T_{es,n4}\}$ with all feasible expected-sharable SATs, where $n3$ and $n4$ are the number of feasible currently sharable and expected-sharable SATs. The minimum travel time with currently sharable SAT and that with expected-sharable SAT are $T_{cs,min}$ and $T_{es,min}$, respectively.

Step 4. The SAT with the minimum travel time, T_{min}, is assigned to the request eventually.

Step 5. If the virtual central control system cannot find T_{min}, which means that no SAT is available, the request, d, is added to the waiting list.

The SAT assignment procedure is shown in Figure 4. After the assignment of the SAT, the status of both the request and SAT will be updated. The completion of this module triggers the third module: enroute actions of the SAT.

3.3. Enroute Actions of SAT. When an SAT departs from the origin of a request, the module of enroute actions is triggered. In this process, the SAT takes the customer to the destination. The predicted travel time of the SAT varies with changing traffic conditions. The status and planned path of the SAT also change when the SAT proceeds to the destination or when the SAT is shared with a new request.

3.3.1. Status Variation of SAT and Its Causes. It is possible that the status of an SAT changes to an expected-empty SAT from a single-rider occupied SAT or shared SAT. This is because the SAT is selected to serve the next request. It is possible that the pick-up time for the next request is delayed or early according to variations in traffic conditions. In this study, we assume that once an SAT is assigned to a request, the request cannot refuse the assigned SAT. The SAT travels to the destination of the last rider and parks at the destination if it is not selected

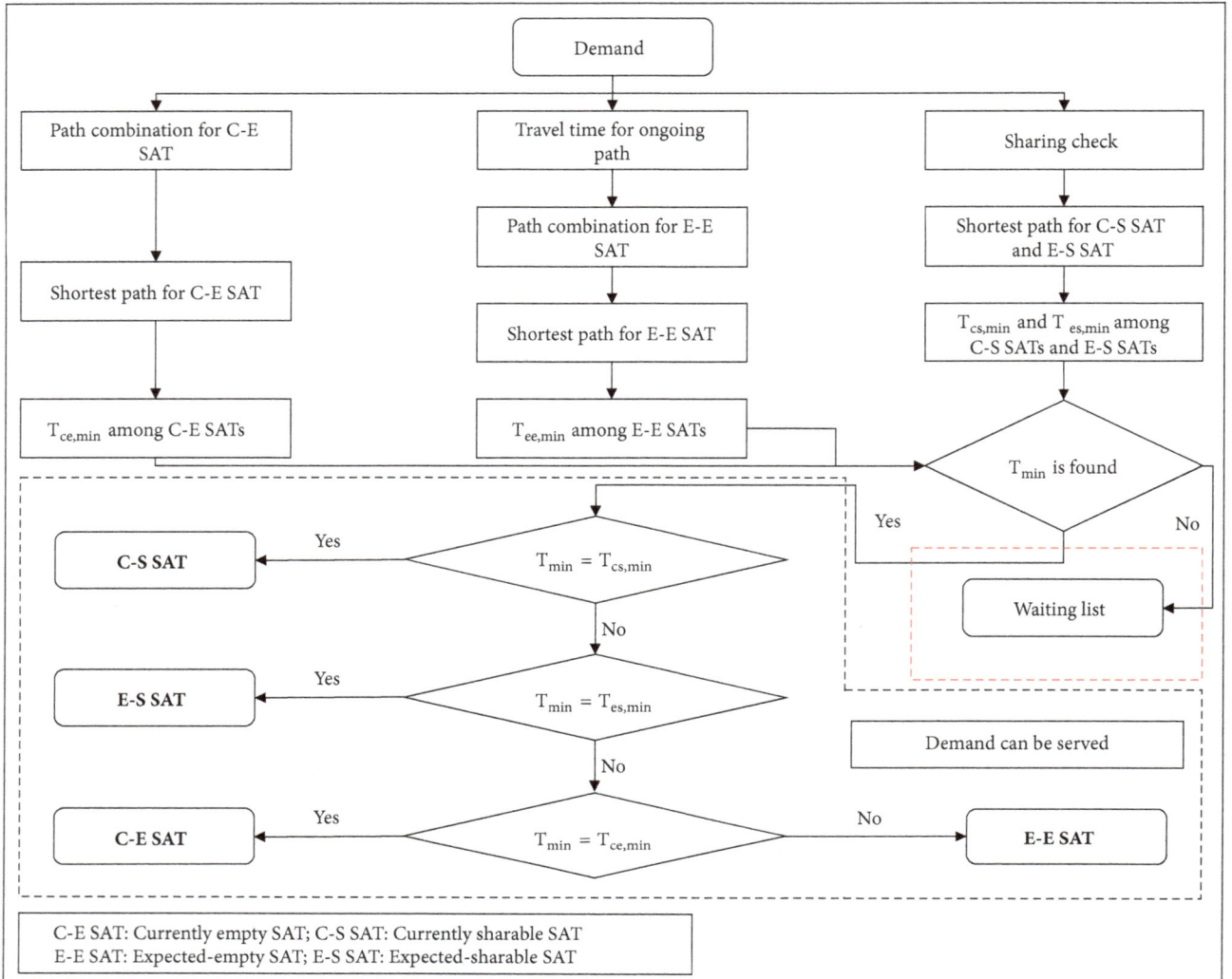

FIGURE 4: SAT assignment procedure.

to serve other requests soon. The status of the SAT is updated to unoccupied from single-rider occupied or shared.

The departed taxi is a single-rider occupied SAT or shared SAT. On the way to the destination, the status of the taxi can change from one of the two above statuses to another status. For a shared SAT, the status changes to single-rider occupied when one rider reaches the destination. The SAT can be shared with following requests, which results in SAT trip chaining [29]. The status of the SAT is still shared if it picks a new request up right at the node where one rider exits. If the departed SAT is only occupied by one rider, it can be shared with the following request on the way to the destination. In this study, two types of sharing are considered; these types determine the delivery sequence of the two riders and are described in Section 3.3.2. A new path is also necessary to be searched for picking up and dropping off customers; this aspect is described in Section 3.3.3.

3.3.2. Sharing Types and Customer-Delivery Rule. Based on the location of the origin and destination of the rider and those of the new request, sharing is divided into two types: extended-path sharing (EP sharing) and in-path sharing (IP sharing). EP sharing means that sharing extends the planned path of the taxi because the destination of the new request is not on any path in the path set of the rider, as shown in Figures 5(a) and 5(b). In IP sharing, both the origin and destination of the new request are on the *k*-shortest paths of the rider, as shown in Figures 5(c), 5(d), 5(e), and 5(f). Even if the origin and destination of the new request are located on different paths of the path set of the rider, the case can be considered provided there is a common node where SATs can detour to a new path to pick up or drop off a rider from the planned path.

In EP sharing, the first-in-first-out (FIFO) order is used as the customer-delivery rule; that is, the rider is delivered first.

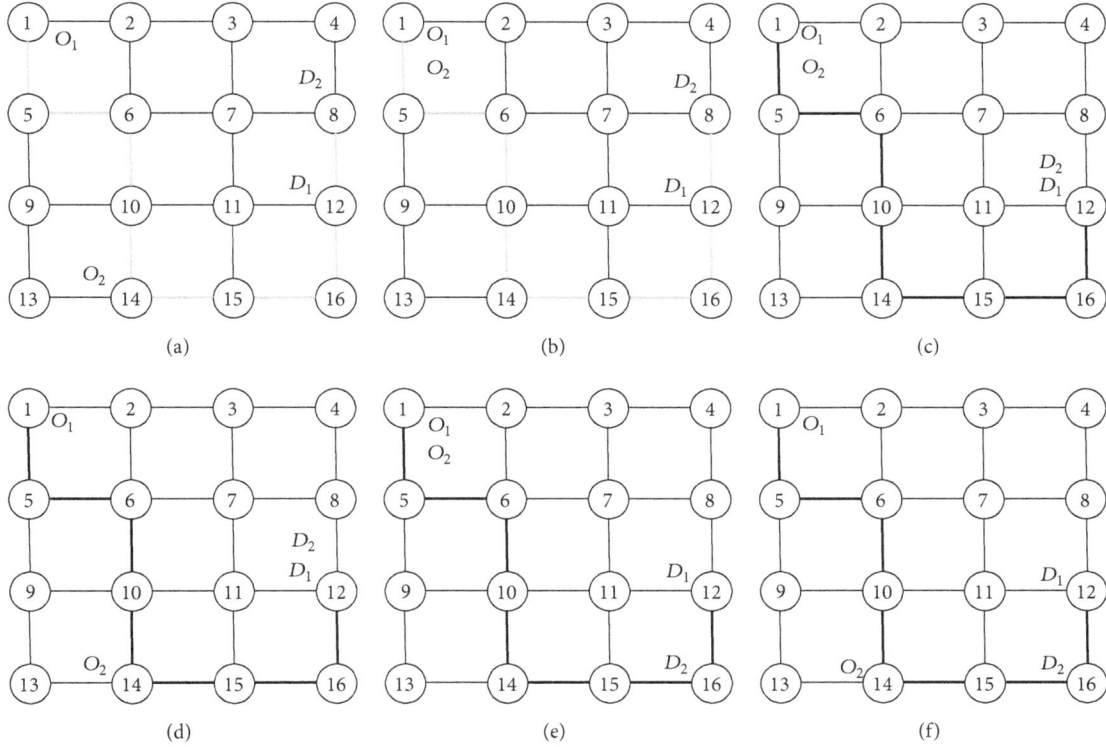

FIGURE 5: Sharing types in SATS.

In IP sharing, the new request can arrive at the destination earlier than or at the same time as the rider. It is necessary to determine the sharing type and delivery rule because it determines the minimum total travel time for the two trips.

3.3.3. Shared Path for Currently Sharable SAT and Expected-Sharable SAT. If an SAT meets the first two sharing requirements described in Section 3.2.3, the central control system finds a new path for each feasible sharable SAT as the shared path. The path set of this SAT is the combination of the path sets of the rider c and the new request d. The method mentioned in Section 3.2.2 is used to select paths that can make the rider c arrive at the destination within LAT_c. The travel times for these paths for the new request are $\{T_{s,s1}, T_{s,s2}, \ldots, T_{s,sn5}\}$, and the minimum time is $T_{s,smin}$. In this case, the path with $T_{s,smin}$ is assigned as the shared path to the taxi. The SAT travels on the shared path if it is finally selected to provide service to the new request.

The above three modules constitute the implementation of the SATS, which is integrated to the traffic system.

3.4. Road Network Traffic Flow. In the simulation, it was assumed that road network traffic flow consists of taxis and private vehicles. The path of a private vehicle was selected statistically. Considering the diverse preferences of private vehicle drivers, the path choice probability was calculated by using the path size logit model [30–32]. The model is formulated as follows [32]:

$$P\left(\frac{i}{C_n}\right) = \frac{PS_{in}e^{V_{in}}}{\sum_{j\in C_n} PS_{jn}e^{V_{jn}}} \tag{1}$$

$$PS_{in} = \sum_{a\in\Gamma_i} \left(\frac{l_a}{L_i}\right) \frac{1}{\sum_{j\in C_n}\left(L_i/L_j\right)\delta_{aj}} \tag{2}$$

where V_{in} is the systematic term of utility of path i for user n, C_n is the path set, PS_{in} is the size of path i, l_a is the length of link a, L_i is the length of path i, δ_{aj} is 1 if link a is in path j and 0 otherwise, and Γ_i is the set of links of path i.

4. Travel-Time Information

Like probe vehicles, traveling SATs can also collect traffic information, including data pertaining to the vehicle location and link travel time. The collected travel-time information can be used for path finding by both SATs and private vehicles. Link travel velocities are assumed to be calculated as $v_a = 16.1 * \ln(215/k_a)$ [33], where 215 vehicles/mile are the density of the traffic jam and k_a is the density of link a. The link density is updated every minute. The travel time for an SAT or a private vehicle on link a is assumed to be random and is normally distributed with $t_a \sim N(\mu_a, \sigma_a^2)$, where $\mu_a = (l_a/v_a)$ and $\sigma_a^2 = \alpha(l_a/v_a)^2$, in which μ_a and σ_a^2 are the population mean value and variance of the link

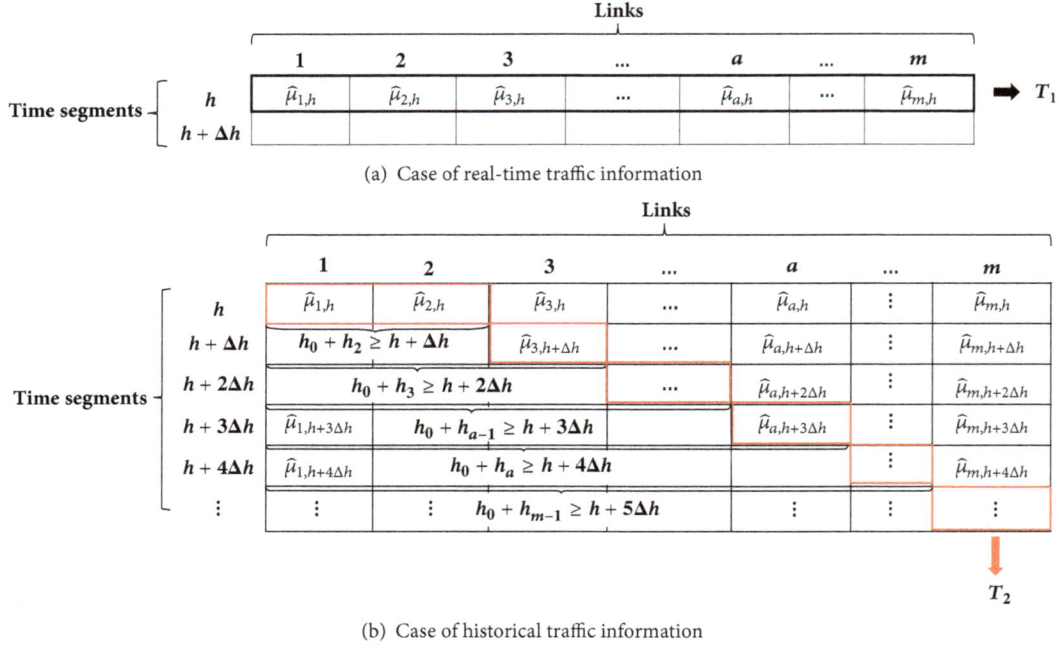

(a) Case of real-time traffic information

(b) Case of historical traffic information

FIGURE 6: Use of two types of traffic information in SATS.

travel time on link a, respectively [34]. In the proposed SATS, it is assumed that the link travel times are independent. The mean value of the path travel time can be expressed by $\widehat{\mu}_p = \sum_{a \in p} \widehat{\mu}_a$, where $\widehat{\mu}_a$ is the expected value of the travel time (t_a) experienced by SATs on link a. Considering the online information paradox [35], which means that the provision of incomplete or inaccurate information could deteriorate the transportation system, it is urgent and necessary to provide accurate information and use the information in an appropriate manner. In this study, two methodologies for using the collected information, real-time information, and historical information were investigated. However, this study has the limitation that no exogenous disturbance was considered. The reason for this lack of consideration is that the traffic environment is expected to become more stable and traffic safety can be improved in the autonomous vehicular transportation system.

In the case of real-time traffic information, the information collected by SATs at time segment h is used for path finding at $h + \Delta h$. That is, the link travel-time information for link a at $h + \Delta h$ is $\widehat{\mu}_{a,h}$. The information update interval is Δh. As shown in Figure 6(a), if the path of an SAT at $h + \Delta h$ contains m links, the travel-time information for the path will be the sum of the $\widehat{\mu}_{a,h}$ for the m links. This is the same for the current travel-time system.

In the case of historical traffic information, a traffic information database is first established, as shown in Figure 6(b). The information in the database is the mean value of the link travel time experienced by SATs in every time segment until the previous day. Through traffic information collection and accumulation, information in the database is updated each day. This information is utilized for path finding in the same

time segment in the following days. In this methodology, the transition of traffic condition is also considered [36]. As shown in Figure 6(b), the departure time of an SAT is h_0, which is greater than h and less than $h + \Delta h$. The total travel time for the first a links is h_a. The traffic information for link 3 at $h + \Delta h$ is used for path finding because the total travel time for the first two links exceeds $h + \Delta h - h_0$ and is less than $h + 2\Delta h - h_0$.

5. Simulation Design and Results

To determine the effect of traffic information and how different SAT fleet sizes perform in the SATS, in this study, several sets of simulation experiments were conducted in MATLAB on an Intel Core CPU running at 3.00 GHz.

The experiments were performed on the Sioux Falls network, as shown in Figure 7. The network had 24 nodes, 76 links, and 552 OD pairs. The metric used in the k-shortest path algorithm was distance, and k was set 10. The free-flow speed of all the links was set to 40 km/h. The travel time for a link was calculated using its length and the speed obtained using the equation in Section 4. In this study, α was set to 0.21 [37, 38].

The trip demand by taxi requests and private vehicle users is shown in Figure 8. The demand was generated from a Poisson distribution every minute and spread over a 24-h period based on the temporal distribution of US NHTS trip-start rates in 2009 [39]. The origin and destination of each trip were generated randomly.

Each experiment was simulated for 80 days. Initially, SATs were distributed evenly on every node in the network. We assumed that all SATs could be relocated at 0:00 a.m. to

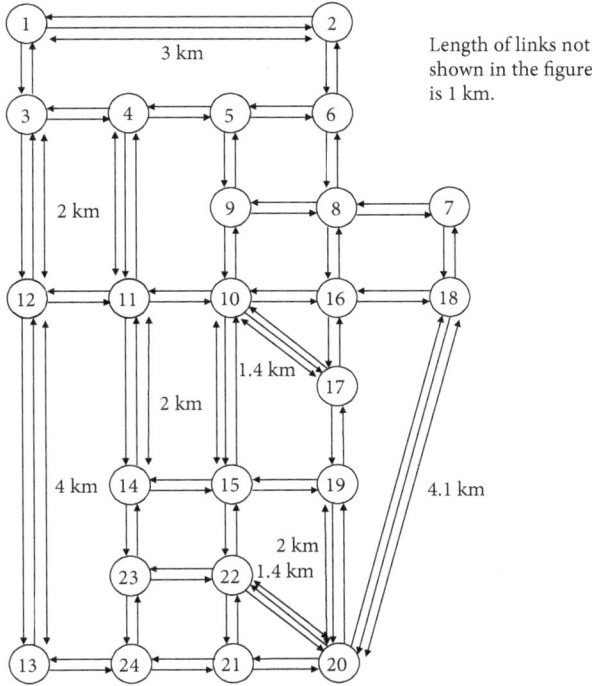

FIGURE 7: Sioux Falls network.

handle the demand of the next day. D_{max} was set as 10 km, and T_{max} was set as 10 min. The latest arrival time of a request was the sum of the travel time from its origin to the destination at the average travel speed (20 km/h) and an acceptable time that follows $U(0, 10)$. Δh was set as 5 min in all the scenarios. Both on the first simulation day of the historical traffic information-based scenario and in the first time segment of the real-time traffic information-based scenario, the travel-time information used for path finding was considered to have been obtained under free flow. The simulation results are shown in Figures 9–11.

5.1. Effect of Traffic Information. Figure 9(a) presents the day-to-day change in the mean travel time for PV users and SAT requests with real-time traffic information, Figure 9(b) illustrates the travel times with historical traffic information, and Figure 9(c) shows the mean waiting times for SAT requests in the case where the SAT fleet size was 30% of the fixed peak hour demand of SAT requests.

All the travel times and the waiting time in the historical traffic information-based scenario decreased gradually as the number of simulation days increased, and the times converged on approximately the 30th simulation day. The travel times and waiting time in the real-time information-based scenario were stable at a larger time level during the entire simulation. All the travel times at the stable level in the historical traffic information-based scenario had less fluctuation than those in the real-time traffic information-based scenario. In the historical information-based scenario, the times decreased because with the collection and accumulation of traffic information and the consideration of time transition, the central control center can find a more accurate

shortest travel-time path for each SAT compared to the past simulation days and the real-time traffic information-based scenario. These results confirm that historical traffic information is better than real-time traffic information for path finding in the proposed SATS. This is consistent with the conclusions in the literature [38].

In both scenarios, the total travel times for taxi requests are greater than those for private vehicle users. This is reasonable because the total travel time includes the waiting time until the SAT's arrival. The in-vehicle travel times for taxi requests are smaller than those for private vehicle users because the probabilistic path choice behavior of private vehicles is considered and some private vehicles do not select the shortest travel time path.

5.2. Effect of SAT Fleet Size. Figures 10 and 11 present the day-to-day changes in the total travel times and waiting times for the two traffic information source cases. The SAT fleet size was set to 30%, 40%, and 50% of the fixed peak hour demand of SAT requests.

The figures clearly show that the total travel time for the taxi requests in both scenarios decreases as the taxi fleet size increases. The difference among the different fleet sizes is due to the difference in the waiting time. The in-vehicle travel time does not vary significantly among different fleet sizes. That is, as the supply of SATs increases, the number of taxi requests who have to wait for taxis decreases. In particular, the difference between the cases with 30% and 40% of fleet sizes is considerably greater than the difference between the cases with 40% and 50% of fleet sizes. These results demonstrate that an insufficient fleet size in the SATS would nonlinearly worsen the service level of the SATS.

In addition, Figure 11(b) indicates that as the number of taxis increases, convergence occurs earlier in the historical information-based scenario. This is because in the case with a larger fleet size, the statistical reliability of the travel-time information accumulated in the historical information database can become higher because of the larger amount of collected data. Accordingly, the accuracy for searching for available SATs and the accuracy of path findings can increase.

6. Conclusions and Future Work

This paper proposed a framework for an SATS that utilizes collected travel-time information to reduce taxi demand and increase the possibility of reducing the travel time for taxi customers. In the proposed SATS framework, in which there are no drivers, SATs utilize collected travel-time information for path finding. The use of two types of information was investigated: historical information and real-time information. The results of the simulation experiments conducted in this study verify that using the two types of information is effective.

More specifically, the simulations demonstrated the effect of using historical travel-time information for path finding and reducing the travel time in the SATS. In particular, as the number of simulation days increased, the customers in the SATS with the historical traffic information obtained increasingly smaller travel times until the travel time converged

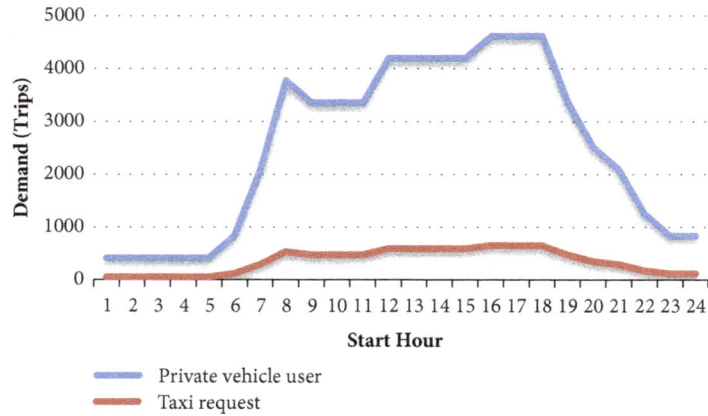

FIGURE 8: Trip demand by taxi requests and private vehicle users.

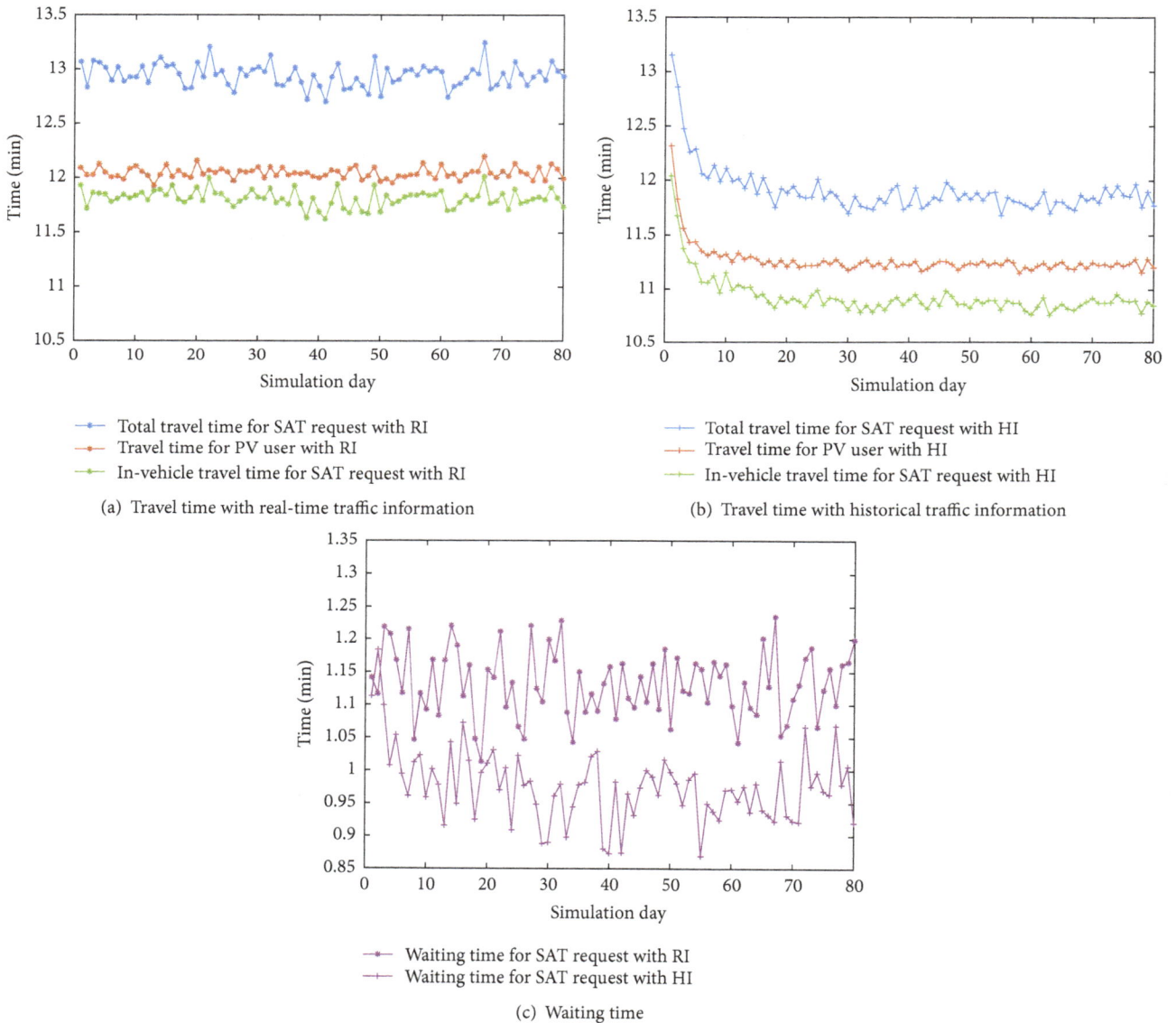

(a) Travel time with real-time traffic information

(b) Travel time with historical traffic information

(c) Waiting time

FIGURE 9: Day-to-day changes in (a) travel time with RI, (b) travel time with HI, and (c) waiting time for SAT request and PV user (RI: real-time information; HI: historical information).

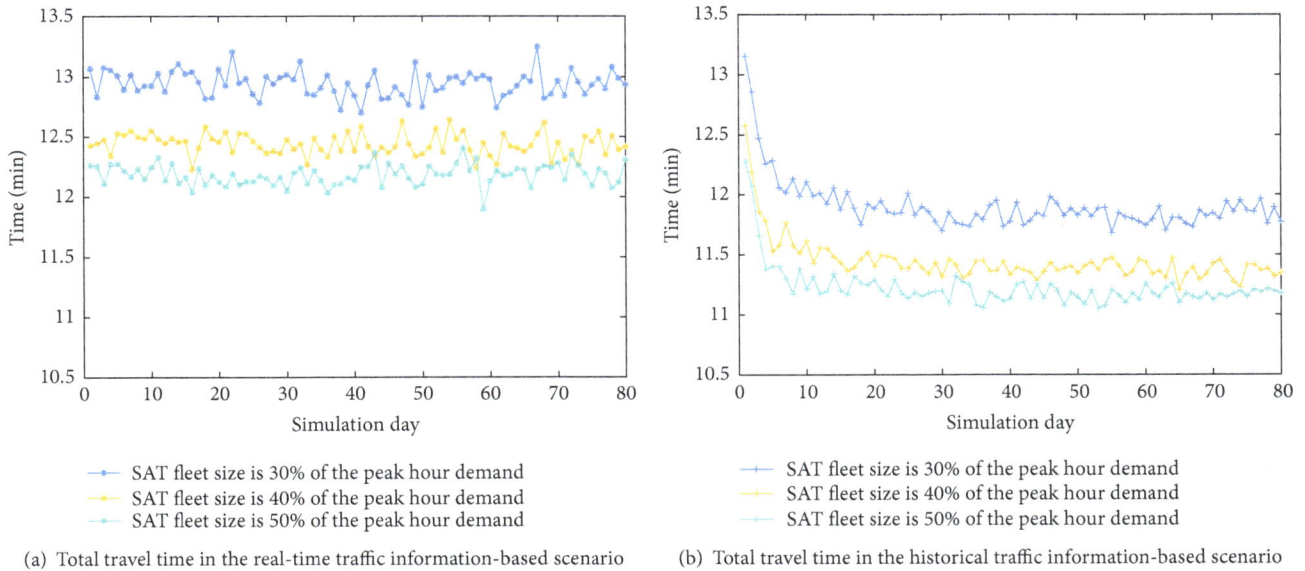

(a) Total travel time in the real-time traffic information-based scenario

(b) Total travel time in the historical traffic information-based scenario

FIGURE 10: Day-to-day changes in total travel time for taxi requests in SATS with several SAT fleet sizes.

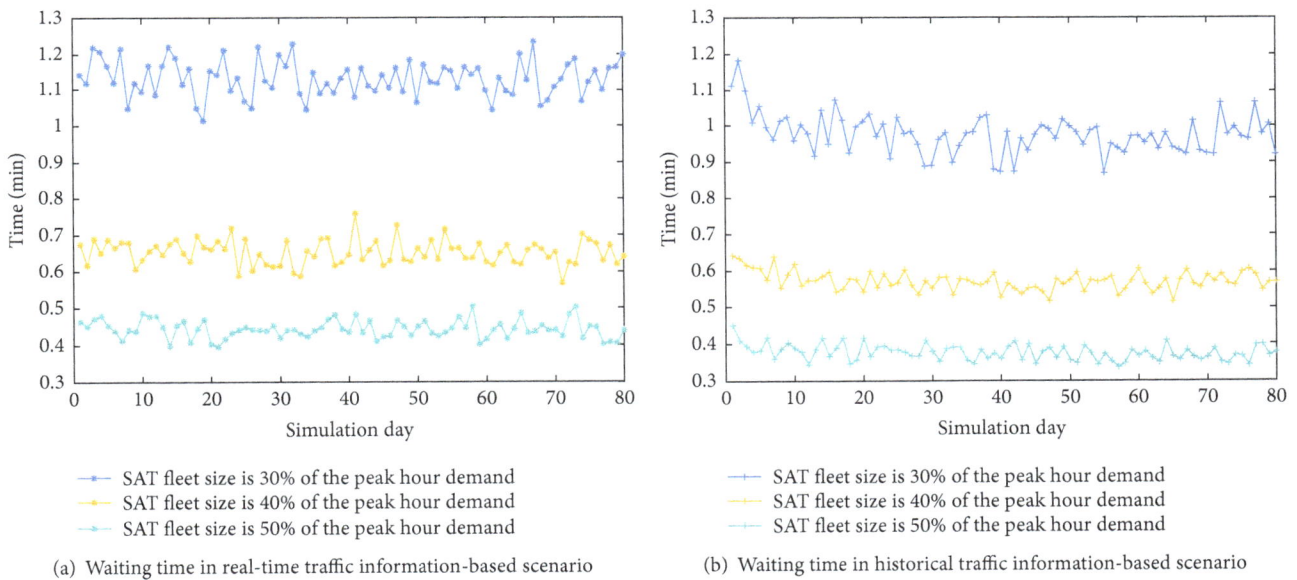

(a) Waiting time in real-time traffic information-based scenario

(b) Waiting time in historical traffic information-based scenario

FIGURE 11: Day-to-day changes in waiting time for taxi requests in SATS with several SAT fleet sizes.

to a steady level. The stable travel times were also smaller than those in the real-time traffic information-based system, where the travel times fluctuated slightly around the constant values. The study results also indicate that insufficient fleet size in the SATS would nonlinearly worsen the service level of the SATS and that as the number of SATs increases, convergence occurs earlier in the historical information-based scenario.

Future study on this topic can focus on the relocation process of SATs. Routing strategies for SATs and information use under exogenous disturbances also need to be studied. The variation in SAT demand to the different service level should be considered in our future research. This consideration will provide more realistic process until the convergence of traffic states. Another area of future research in order to offer a better level of service is examination of the effect of reassigning a new taxi to a request when the pick-up time of the assigned taxi increases owing to variations in traffic conditions.

Conflicts of Interest

The authors declare that there are no conflicts of interest regarding the publication of this paper.

Acknowledgments

This work was supported by JSPS KAKENHI Grant nos. 16H02367 and 16K12825. The first author would like to thank the China Scholarship Council (CSC) for financial support.

References

[1] S. Horl, F. Ciari, and K. W. Axhausen, *Recent Perspectives on The Impact of Autonomous Vehicles*, Institute for Transport Planning and Systems, 2016.

[2] J. Patel, "nuTonomy beats Uber to launch first self-driving taxi," 2016, http://www.autoblog.com.

[3] H. Chai, "4 Self-Driving Buses Tested in Shenzhen," http://www.chinadaily.com.cn/china/201.

[4] R. Krueger, T. H. Rashidi, and J. M. Rose, "Preferences for shared autonomous vehicles," *Transportation Research Part C: Emerging Technologies*, vol. 69, pp. 343–355, 2016.

[5] P. Bansal, K. M. Kockelman, and A. Singh, "Assessing public opinions of and interest in new vehicle technologies: An Austin perspective," *Transportation Research Part C: Emerging Technologies*, vol. 67, pp. 1–14, 2016.

[6] M. Bloomberg and D. Yassky, "Taxicab Fact Book," http://www.nyc.gov/html/.

[7] W. Wu, W. S. Ng, S. Krishnaswamy, and A. Sinha, "To taxi or not to taxi? - Enabling personalised and real-time transportation decisions for mobile users," in *Proceedings of the 2012 IEEE 13th International Conference on Mobile Data Management, MDM 2012*, pp. 320–323, India, July 2012.

[8] Z. Jia, A. Balasuriya, and S. Challa, "Sensor fusion-based visual target tracking for autonomous vehicles with the out-of-sequence measurements solution," *Robotics and Autonomous Systems*, vol. 56, no. 2, pp. 157–176, 2008.

[9] R. B. Dial, "Autonomous dial-a-ride transit introductory overview," *Transportation Research Part C: Emerging Technologies*, vol. 3, no. 5, pp. 261–275, 1995.

[10] D. J. Fagnant and K. M. Kockelman, "The travel and environmental implications of shared autonomous vehicles, using agent-based model scenarios," *Transportation Research Part C: Emerging Technologies*, vol. 40, pp. 1–13, 2014.

[11] D. J. Fagnant, K. M. Kockelman, and P. Bansal, "Operations of shared autonomous vehicle fleet for Austin, Texas, market," *Transportation Research Record*, vol. 2536, pp. 98–106, 2015.

[12] D. J. Fagnant and K. M. Kockelman, "Dynamic ride-sharing and fleet sizing for a system of shared autonomous vehicles in Austin, Texas," *Transportation*, pp. 1–16, 2016.

[13] W. Burghout, P. J. Rigole, and I. Andreasson, "Impacts of shared autonomous taxis in a metropolitan area," in *Proceedings of the 94th Annual Meeting of the Transportation Research Board*, Washington, wash, USA, 2015.

[14] E. Lioris, G. Cohen, and A. de La Fortelle, "Evaluation of Collective Taxi Systems by Discrete-Event Simulation," in *Proceedings of the 2010 Second International Conference on Advances in System Simulation (SIMUL)*, pp. 34–39, Nice, France, August 2010.

[15] M. W. Levin, T. Li, and S. D. Boyles, "A general framework for modeling shared autonomous vehicles," in *Proceedings of the 94th Annual Meeting of the Transportation Research Board Washington D*, 2016.

[16] R. F. Teal, "Carpooling: who, how and why," *Transportation Research Part A: General*, vol. 21, no. 3, pp. 203–214, 1987.

[17] C.-C. Tao, "Dynamic taxi-sharing service using intelligent transportation system technologies," in *Proceedings of the International Conference on Wireless Communications, Networking and Mobile Computing (WiCOM '07)*, pp. 3204–3207, September 2007.

[18] P. M. D'Orey, R. Fernandes, and M. Ferreira, "Empirical evaluation of a dynamic and distributed taxi-sharing system," in *Proceedings of the 2012 15th International IEEE Conference on Intelligent Transportation Systems, ITSC 2012*, pp. 140–146, USA, September 2012.

[19] M. Ota, H. Vo, C. Silva, and J. Freire, "A scalable approach for data-driven taxi ride-sharing simulation," in *Proceedings of the 3rd IEEE International Conference on Big Data, IEEE Big Data 2015*, pp. 888–897, USA, November 2015.

[20] S. Ma, Y. Zheng, and O. Wolfson, "Real-Time City-Scale Taxi Ridesharing," *IEEE Transactions on Knowledge and Data Engineering*, vol. 27, no. 7, pp. 1782–1795, 2015.

[21] M. Nourinejad and M. J. Roorda, "Agent based model for dynamic ridesharing," *Transportation Research Part C: Emerging Technologies*, vol. 64, pp. 117–132, 2016.

[22] A. Najmi, D. Rey, and T. H. Rashidi, "Novel dynamic formulations for real-time ride-sharing systems," *Transportation Research Part E: Logistics and Transportation Review*, vol. 108, pp. 122–140, 2017.

[23] Z. Liu, T. Miwa, W. Zeng, and T. Morikawa, "An agent-based simulation model for shared autonomous taxi system," *Asian Transport Studies*, vol. 5, no. 1, pp. 1–13, 2018.

[24] M. Stiglic, N. Agatz, M. Savelsbergh, and M. Gradisar, "Making dynamic ride-sharing work: The impact of driver and rider flexibility," *Transportation Research Part E: Logistics and Transportation Review*, vol. 91, pp. 190–207, 2016.

[25] H. Hosni, J. Naoum-Sawaya, and H. Artail, "The shared-taxi problem: Formulation and solution methods," *Transportation Research Part B: Methodological*, vol. 70, pp. 303–318, 2014.

[26] J. Y. Yen, "Finding the K shortest loopless paths in a network," *Management Science*, vol. 17, pp. 712–716, 1970/7119.

[27] Y. Molenbruch, K. Braekers, and A. Caris, "Typology and literature review for dial-a-ride problems," *Annals of Operations Research*, vol. 259, no. 1-2, pp. 295–325, 2017.

[28] W. Zeng, T. Miwa, and T. Morikawa, "Application of the support vector machine and heuristic k-shortest path algorithm to determine the most eco-friendly path with a travel time constraint," *Transportation Research Part D: Transport and Environment*, vol. 57, pp. 458–473, 2017.

[29] M. W. Levin, K. M. Kockelman, S. D. Boyles, and T. Li, "A general framework for modeling shared autonomous vehicles with dynamic network-loading and dynamic ride-sharing application," *Computers, Environment and Urban Systems*, vol. 64, pp. 373–383, 2017.

[30] E. Frejinger, M. Bierlaire, and M. Ben-Akiva, "Sampling of alternatives for route choice modeling," *Transportation Research Part B: Methodological*, vol. 43, no. 10, pp. 984–994, 2009.

[31] D. Li, T. Miwa, and T. Morikawa, "Dynamic route choice behavior analysis considering en route learning and choices," *Transportation Research Record*, no. 2383, pp. 1–9, 2013.

[32] S. Bekhor, M. Ben-Akiva, and M. S. Ramming, "Adaptation of Logit Kernel to route choice situation," *Transportation Research Record*, vol. 1805, no. 1, pp. 78–85, 2002.

[33] H. Greenberg, "An analysis of traffic flow," *Operations Research*, vol. 7, pp. 79–85, 1959.

[34] R. Seshadri and K. K. Srinivasan, "Algorithm for determining most reliable travel time path on network with normally distributed and correlated link travel times," *Transportation Research Record*, no. 2196, pp. 83–92, 2010.

[35] K. P. Wijayaratna, V. V. Dixit, D.-B. Laurent, and S. T. Waller, "An experimental study of the Online Information Paradox: Does en-route information improve road network performance?" *PLoS ONE*, vol. 12, no. 9, Article ID e0184191, 2017.

[36] T. Morikawa and T. Miwa, "Preliminary analysis on dynamic route choice behavior: Using probe-vehicle data," *Journal of Advanced Transportation*, vol. 40, no. 2, pp. 141–163, 2006.

[37] T. Yamamoto, T. Miwa, T. Takeshita, and T. Morikawa, "Updating dynamic origin-destination matrices using observed link travel speed by probe vehicles," *Transportation and Traffic Theory*, pp. 723–738, 2009.

[38] T. Miwa and M. G. Bell, "Efficiency of routing and scheduling system for small and medium size enterprises utilizing vehicle location data," *Journal of Intelligent Transportation Systems: Technology, Planning, and Operations*, vol. 21, no. 3, pp. 239–250, 2016.

[39] Federal Highway Administration, *National Household Travel Survey*, U.S. Department of Transportation, Washington, wash, USA, 2009, http://nhts.ornl.gov/2009/pub/stt.pdf.

An Association Rule Based Method to Integrate Metro-Public Bicycle Smart Card Data for Trip Chain Analysis

De Zhao [iD],[1,2] Wei Wang,[3] Ghim Ping Ong,[4] and Yanjie Ji [iD][3]

[1]Jiangsu Key Laboratory of Urban ITS, Southeast University, Si Pai Lou No. 2, Nanjing 210096, China
[2]Department of Civil and Environmental Engineering, National University of Singapore, Engineering Drive 2, E1A 08-20, Singapore 117576
[3]School of Transportation, Southeast University, Si Pai Lou No. 2, Nanjing 210096, China
[4]Department of Civil and Environmental Engineering, National University of Singapore, Engineering Drive 2, E1A 07-03, Singapore 117576

Correspondence should be addressed to De Zhao; zhaode_0@aliyun.com

Academic Editor: Lele Zhang

Smart card data provide valuable insights and massive samples for enhancing the understanding of transfer behavior between metro and public bicycle. However, smart cards for metro and public bicycle are often issued and managed by independent companies and this results in the same commuter having different identity tags in the metro and public bicycle smart card systems. The primary objective of this study is to develop a data fusion methodology for matching metro and public bicycle smart cards for the same commuter using historical smart card data. A novel method with association rules to match the data derived from the two systems is proposed and validation was performed. The results showed that our proposed method successfully matched 573 pairs of smart cards with an accuracy of 100%. We also validated the association rules method through visualization of individual metro and public bicycle trips. Based on the matched cards, interesting findings of metro-bicycle transfer have been derived, including the spatial pattern of the public bicycle as first/last mile solution as well as the duration of a metro trip chain.

1. Introduction

Public bicycle usage for metro access provides new opportunities for sustainable transportation, helping to address the "first-mile" and "last-mile" problems [1]. To understand the effect of integrating public bike and metro systems, transportation planners and researchers have been striving to evaluate the transfer efficiency and behavior through personal travel profiling [2], social-demographic information [1, 3], or public bicycle historical trips [4]. Previous attempts to understand metro and public bicycle transfer are limited, partly due to the difficulty in data collection. Conventional household travel surveys or diaries are time-consuming and laborious to carry out while convenient access to large travel datasets and integration across different data platforms are yet to be found.

In China, both metro and public bicycle transactions are made via Automatic Fare Collection (AFC) system, also known as smart card (SC). SC data, with a massive sample size, can provide valuable insights into the understanding of metro-public bicycle transfer behavior [5, 6]. Compared with conventional surveys, SC data collection, being a by-product of revenue collection, is a convenient method for retrieving travel patterns of commuters. Therefore, a great deal of research studies have emerged in terms of SC data mining [6–12].

However, the smart cards for metro and public bicycle systems are issued and managed by independent companies and in most cases, commuters need to hold two different smart cards to complete a public bicycle-metro trip chain. As such, the metro SC dataset and public bicycle SC dataset are saved independently without a common and unique identifier (ID) for a single commuter. This makes it difficult for researchers and transit agencies to leverage big SC data to investigate metro-bike transfer behavior effectively and efficiently, unless there exists a method to match the unique card IDs within each system. It is not possible to directly and accurately match the smart cards from different datasets, as accessing personal information may offend the privacy. Nevertheless, the detailed individual travel pattern hidden in the SC data makes it possible to match card IDs of the same commuter.

Therefore, the primary objective of this study is to develop a data fusion methodology for matching metro and public bicycle smart cards of commuter identity in an integrated metro-public bicycle network. To achieve the aim, this study provides a novel approach using association rules (AR), a concept in the machine learning domain. The smart card data from Nanjing metro and public bicycle in China is used to demonstrate and validate our developed method. The remainder of the paper is structured as follows. In the literature review section, previous studies on smart card data for trip chain, multisource data fusion, and association rules are reviewed. The methodology and data source section formulates the association rules to match smart card travel and identifier data and presents the data source preprocessing procedures and validation approach. The results and validation section applies the proposed data fusion method to empirical analysis, calibrates the key parameters, and validates the proposed method. The section also presents possible application after matching metro and public bicycle card IDs. The conclusions and recommendations section concludes the paper and gives the research limitations as well as recommendations for future study.

2. Literature Review

Transit smart card data records transit riders' detailed trip log, which can be used to analyze the transit riders' trip chain. A huge body of literature has grown with regard to SC data analysis [7, 11, 14–17]. Means of mining SC data are various, e.g., data fusion and machine learning. Three streams of research are relevant to this study: (1) smart card data for trip chain; (2) multisource data fusion; (3) association rules.

2.1. Smart Card Data for Trip Chain. Past research studies in the literature have analyzed historical SC data to estimate transit origin location [18], destination location [6, 19], and total daily or monthly transit trip chain pattern [8, 10]. Furthermore, long-term year-to-year changes in transit users' trip habits could also be tracked and analyzed [12]. As for public bicycle, the research just started in recent years [20]. Notably, most of the research used bicycle trip data [21] rather than true SC data, since the trip data is easier to obtain. Trip data is usually open to the public in cities of United States or Europe, where the bicycle rental is accomplished via credit card or cell phone app. In general, compared with SC data, bicycle trip data lacks card ID and thus cannot be used to model users' travel behavior. By far, public bicycle SC data has been used to investigate public bicycle users' travel patterns [22] as well as bicycle trip chains for men and women [14, 23]. Public bicycle SC data could also help to classify different types of behaviors and compare the trip disparity [24].

2.2. Multisource Data Fusion. SC data provides much detailed information about each trip, but not the information about trip purpose, user assessment, and ultimate destination. When integrated with other data sources, SC data can play a greater role in mining transit riders' behavior and validating previous research approaches. By integrating both SC and Global Positioning System (GPS) data, Munizaga and Palma [9] estimated the OD of multimodal transit systems and validated the results against metro OD surveys in Santiago, Chile [16]. Ma and Wang [25] built a data-driven platform by integrating SC and GPS data to monitor transit performance in Beijing. Researchers can also examine the spatial-temporal dynamics of bus passengers and estimate the trip purposes when matching SC data, respectively, with General Transit Feed Specification (GTFS) data [15] and person trip survey data [17]. Yet, very little research attempted to integrate metro SC data with public bicycle SC data for investigating metro-bicycle transfer.

2.3. Association Rules. AR was first introduced by Agrawal et al. [26], and they applied this model to the supermarket transaction data to find out what items people would buy together. They also proposed algorithms for finding the AR. Shortly after that, the method was applied to other fields as a popular machine learning technique, including transportation. AR was firstly used in the transportation area by Keuleers et al. [27] to learn the travel patterns of multiday activity diaries. Soon after that, Keuleers et al. [28] tried to recognize temporal effects that may exist in the same data. AR showed high efficiency and convenience in rules mining. Later, Kusumastuti et al. [29] explored individuals' thoughts about leisure-shopping travel decisions by means of AR. Diana [30] used AR analysis to explore travel patterns of different modes based on 2009 US National Household Travel Survey and found the substitution effect between private modes and public transit. In particular, Chu and Chapleau [31] used AR to mine behavior rules of SC users and found some potential regularities with a high level of confidence.

Among all the relevant studies presented in this section, there still remains lack of a data fusion methodology to match smart cards from different sources. As AR is capable of identifying potential relationships between items, this paper attempts to develop AR-based algorithm to match metro and public bicycle SC data of the same person within an integrated bus-public bicycle network. We convert metro SC data and public bicycle SC data into transaction datasets and follow the method of Agrawal et al. [26] to match the card IDs. In our paper, we also propose an approach to validate the developed method.

3. Methodology and Data Source

3.1. Association Rules. Let $I = \{i_1, i_2, \ldots, i_k\}$ be a set of items. A transaction d_i is defined as a group of items, namely, a subset of I. $D = \{d_1, d_2, \ldots, d_n\}$ is a set of all transactions called the transaction database. Each transaction d_i in D has a unique transaction number. An association rule is used to describe potential relations of several items in the transaction

---

(1) L_1 = {large 1 – itemsets};
(2) for $(k = 2; L_{k-1} \neq \phi; k + +)$ do begin
(3) C_k = apriori-gen(L_{k-1}); //New candidates
(4)   for all transactions $t \in D$ do begin
(5) C_t = subset(C_k, t); //Candidates contained in t
(6)   for all candidates $c \in C_t$ do
(7) $c.count + +$;
(8)   end
(9) $L_k = \{c \in C_k | c.count \geq$ min support$\}$
(10) end
(11) All frequent sets $= \bigcup_k L_k$

ALGORITHM 1: Apriori algorithm [13].

---

database D and is expressed as $X=>Y$, where $X, Y \subseteq I$ and $X \cap Y = \emptyset$. X is called antecedent or left-hand side (LHS), and Y is called consequent or right-hand side (RHS). For example, in supermarket sales data mining, the rule {butter, bread}=>{milk} means if a customer buys both "butter" and "bread", he is also likely to buy "milk". In this research, we set all metro card IDs and public bicycle card IDs as items I. An association rule would indicate that there is potential association between card IDs in LHS and RHS. To better match two smart cards by AR, we should try our best to cluster two cards of the same person into one transaction d_i.

In association rules, there are three key parameters: support, confidence, and lift. The corresponding definitions are listed below. The support value of X, represented as supp (X), means the probability that the item-set X appears in the database D, defined as the proportion of transactions that includes the item-set X in D, as in (1). Accordingly, supp$(X \Rightarrow Y)$ can be expressed by (2). The generalized expression $X \cup Y$ in association rules means the union of the items in X and Y rather than either X or Y:

$$\text{supp}(X) = P(X) = \frac{\text{num}(X)}{\text{num}(D)}, \qquad (1)$$

$$\text{supp}(X \Longrightarrow Y) = P(X \cup Y) = \frac{\text{num}(X \cup Y)}{\text{num}(D)}. \qquad (2)$$

The confidence value of a rule, expressed as conf$(X=>Y)$, indicates the proportion of the transactions containing both X and Y in those that contain X:

$$\text{conf}(X \Longrightarrow Y) = P(Y \mid X) = \frac{\text{supp}(X \cup Y)}{\text{supp}(X)}. \qquad (3)$$

The lift value is used to describe the effectiveness of the association rule, as defined in (4). A lift of 1.0 implies that the occurrence of X has nothing to do with that of Y. That is to say, no association rule can be found between X and Y when the lift = 1.0. When the lift is more than 1.0, $X=>Y$ is an effective association rule and greater value of lift indicates stronger association rule:

$$\text{lift}(X \Longrightarrow Y) = \frac{P(Y \mid X)}{P(Y)} = \frac{\text{supp}(X \cup Y)}{\text{supp}(X) \times \text{supp}(Y)}. \qquad (4)$$

The Minimum Support (MS) value and the Minimum Confidence (MC) value are set as a constraint on measure of significance to ensure that the rules under consideration are sufficiently significant. MS is used to search the most frequent item-sets and MC is used to form rules based on these frequent item-sets. The former process is computationally intensive. To accomplish the former process, the commonly used Apriori algorithm is applied, as shown in Algorithm 1. As the Apriori algorithm can be found in previous research [13, 27, 28], we do not expatiate in this study.

3.2. Data Source and Preprocessing. Two major datasets were used in this study: metro SC data and public bicycle SC data, as shown in Figures 1(a) and 1(b). The datasets were recorded from November 1, 2015, to November 24, 2015, obtained from Nanjing Smart Card Company and Public Bicycle Company, respectively. The metro SC data contains the card ID, departure station, tap-in time, arrival station, and tap-out time. The original public bicycle data contains card ID, rent station, rent time, return station, and return time. Based on the location of the bicycle rent/return station, we added its nearby metro station in the 300 m buffer to the data frame. A buffer of 300 m radius was used as the walkable distance for public bicycle trips [32], because the planning standard promulgated by Nanjing government has suggested the walkable distance for community public facility is 300 m (5-min walk) [33].

The 24-day metro SC data contains over 34 million rows of trips and the 24-day public bicycle SC data includes nearly 1.2 million rows of trips. To mine AR between two databases, we need to merge and convert the SC data into transaction data. As mentioned above, we need to cluster two cards of the same person into one transaction to the greatest extent possible. The most likely transaction is metro-bicycle transfer or bicycle-metro transfer.

Firstly, we divide metro SC data and public bicycle SC data into subsets of regular time slot (TS) based on tap-in time and return time, respectively. Then, we merged the metro SC subset and the public bicycle SC subset with the same time slot and metro station (departure station and return nearby metro station, respectively) as one transaction. For bicycle-metro transfer, two cards of the same person may probably appear in the same transaction. Similarly, we created the metro-bicycle transfer transactions by cutting and merging

(a) Metro SC Data Samples on Nov 1, 2015

Card ID	Departure station	Tap-in time	Arrival station	Tap-out time
0804322DA2612F8000	56	05:56:47	48	06:36:53
08040A344AC5208000	56	05:58:34	16	06:45:46
080415307AB5238000	56	06:01:16	25	06:37:33
0804321D4294248000	50	05:21:56	56	06:14:58
0804373F4954258000	58	06:12:50	56	06:25:07
...	...	...	...	...

+

(b) Expanded Public Bicycle SC Data Samples on Nov 1, 2015

Card ID	Rent station	Rent Nearby Metro station	Rent time	Return station	Return Nearby Metro station	Return time
8879	12027	/	05:47:47	12133	56	05:52:54
18702	12051	/	05:48:49	12018	56	05:58:22
56068	12039	/	05:39:00	12133	56	06:01:23
28871	12028	56	06:27:59	12029	/	06:43:56
98457	12031	56	06:28:27	12020	/	07:41:42
...	...	...	...	...	...	...

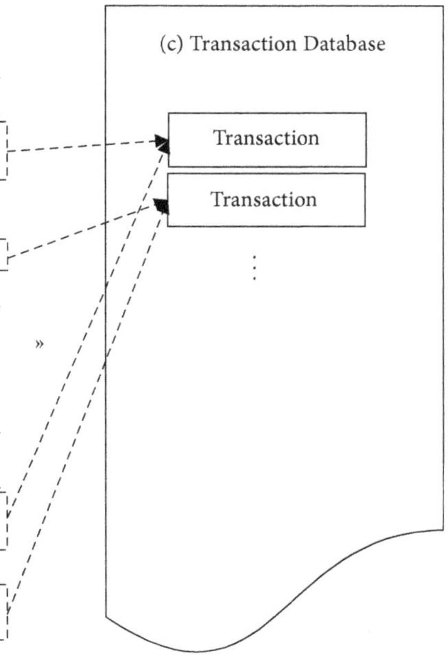

(c) Transaction Database

FIGURE 1: Smart card data and transaction dataset generation.

the datasets based on tap-out time, rent time, arrival station, and rent nearby metro station. All the transactions added up to create the transaction database D, as shown in Figure 1(c). In order to reduce the computational burden, we removed transactions with only metro card IDs or only public bicycle card IDs, because such transactions only contribute to finding internal relationship among metro cards or among public bicycle cards.

3.3. *Validation Approach.* Unfortunately, we cannot directly know whether two cards belong to the same person, because the only way to identify one person between different databases is to obtain nonanonymous personal information, which may violate individual privacy. Instead, we put forward a surrogated approach to validate the results: using the data in the first 20 days (train data) to train the association rules and the data in the last 4 days (test data) to validate the results. We assume the matched two card IDs do belong to the same person if they meet all the following three conditions.

(1) A transfer behavior of the two card IDs is also observed in the test data. A transfer behavior is defined as renting a public bicycle within 10 minutes after exiting the metro station or entering the metro station within 10 minutes after returning a public bicycle. We use 10 minutes as the maximum value of metro-bicycle (or bicycle-metro) transfer time. As mentioned above, the walkable distance between metro stations and public bicycle stations is 300 m, which is also equivalent to 5-min walk for an average person [33]. We set the maximum value of the transfer time as twice of 5 minutes.

(2) No time-overlap was observed between the matched two card IDs in 24-day datasets. In other words, one cannot

take the metro while renting the public bike or rent a public bike during the metro ride.

(3) One metro SC ID only matches with one public bicycle SC ID. We assume each person usually owns one metro SC or public bicycle SC. Therefore, it makes sense only if one item (card ID) associated with only one other item.

Meeting all above three conditions by chance is quite a small probability event. Because there are over 1.8 million unique card IDs in the metro SC database and over 0.1 million unique card IDs in the public bicycle SC database. Based on our data, we randomly choose one card ID from the metro SC database and one card ID from the public bicycle SC database to check if they can meet the three conditions at the same time. We repeated 10,000 times of the selections; there are only 16 pairs of card IDs meeting all the validation conditions. Hence, the probability of meeting all above three conditions by chance is 0.16%.

4. Results and Validation

After cutting the SC datasets into thousands of transactions by the proposed method and learning the data with AR, associations between SC IDs are retrieved. The Apriori algorithm is used to identify the most frequent item-sets (metro SC and public bicycle SC). The associated rules learning in this research was performed using R 3.4.0. Given the three model parameters and the database of Nanjing (totally 1,149,335 items and 38,014 transactions), the average calculation time of associated rules is 18 seconds with the help of Apriori algorithm on a PC with Intel i7-6700 3.4 GHz and 16 GB DDR4 RAM.

To capture the meaningful results, we only selected rules with one metro SC ID as LHS and one public bicycle SC

TABLE 1: Extracted association rules.

No.	Association rules (LHS $\Rightarrow$ RHS)	Support	Confidence	Lift
1	{0804322DA2612F8000} $\Rightarrow$ {00008879}	5.78×10^{-4}	0.79	853.37
2	{08046A39EA02218000} $\Rightarrow$ {00008429}	7.36×10^{-4}	0.97	1223.44
3	{0804142E2ABA268000} $\Rightarrow$ {00196771}	5.52×10^{-4}	0.72	1019.53
4	{080E4711130A4D3C00} $\Rightarrow$ {00168715}	5.52×10^{-4}	0.70	917.58
5	{0804293EE2382C8000} $\Rightarrow$ {00151192}	7.36×10^{-4}	0.93	1182.66
6	{08042E156A85368000} $\Rightarrow$ {00163152}	8.15×10^{-4}	0.91	936.75
...	...	...	...	...

ID as RHS. Because many-to-one rules, metro-metro rules, and bicycle-bicycle rules are all invalid, metro-metro rules probably mean that two or more cardholders often take the metro together. It is also true for bicycle-bicycle rules. These rules are not concerned in this research and thus removed from the association rules list. The extracted association rules are shown in Table 1. The metro card ID is represented as a string of 18 hex digits, while the public bicycle card ID is a string of 8 decimal digits. One AR indicates one metro SC ID matched with one public bicycle SC ID. All the extracted rules have a very high "lift" value, indicating the associations between cards are significantly strong. The support value scattered between 0 and 1.2×10^{-3}, indicating the parameter range for MS calibration.

There are three parameters MS, MC, and TS in our proposed model. They jointly determined the number of ARs (the number of matched IDs) and the accuracy of results. We need to obtain the optimal combination of three parameters in order to derive more ARs as well as better accuracy. Accuracy is defined as the ratio between the number of ARs meeting all the three validation conditions and the total number of ARs.

Figure 2 shows how the three parameters influenced the results. We set TS, respectively, as 2 min, 5 min, 10 min, and 20 min, as shown in Figures 2(a), 2(b), 2(c), and 2(d). In general, the accuracies of ARs under various parameter combinations are all very high, with most of them being over 90%. However, with increase of the number of ARs, the accuracy of results decreased. In other words, we cannot achieve the optimal levels of both number and accuracy simultaneously. To reach an accuracy of 100%, our proposed approach could at most identify 573 ARs (matched 573 pairs of SC IDs), with TS = 10 min, MS = 0.00055, and MC = 0.4 as marked in Figure 2(c). A too small a value of TS (e.g., 2 min) will reduce the maximum number of ARs because the metro card and public bicycle card of the same transfer could be divided into two transactions with high probability. On the other hand, too big TS (e.g., 20 min) could greatly improve the maximum number of ARs but may introduce many invalid ARs, thus decreasing the accuracy.

As mentioned above, there are totally 573 metro cards matched with bicycle cards under the accuracy of 100%. All the 573 paired cards have satisfied the 3 validation conditions. To ensure that the matched cards do belong to one person, we derived each individuals' trip log and visualized six of them as Figure 3 shows. The blue solid line segment indicates metro trip, while the red dashed line segment indicates public

bicycle trip. All the metro trips and public bicycle trips of 24 days are displayed in this plot. The metro-bicycle transfer behavior of an individual can be easily identified by adjacent metro and public bicycle trip.

Apparently, Individuals 1, 2, 5, and 6 have a regular trip pattern. They used public bicycle as a daily first-/last-mile connection to metro. The majority of their metro trips are connected with public bicycle trips, showing specific symmetries between morning trip chain and afternoon/evening one. Individuals 3 and 4 are probably not traditional office workers with routine commutes, but they still take public bicycle as a good way to address the first-/last-mile problem.

After matching two cards of the same person, interesting findings of metro-bicycle transfer could be obtained. Around 2/3 of the transfers between metro and public bicycle occurred in the peak hours. Figure 4 shows the spatial analysis of public bicycle trips that connected with metro trips by peak hours of the day. The plot provides us with a visual impression of public bicycle trips as a first-/last-mile connection.

We can find that there are usually several public bicycle stops within the vicinity of metro stations and the trip demand is not shared equally among them. Users prefer to rent bicycles from only one or two of these stations, which further exacerbates the burden of rebalancing. In zones far away from the metro station, public bikes even take on longer-distance connections, for example, in the northern part of the research area. The maximum straight-line distance is over 4 km.

Another interesting finding from Figure 4 is that the first-mile trips during morning peak (7:00~9:00) have the same spatial pattern with last-mile trips during evening peak (17:00~19:00). It is also true for last-mile trips during the morning peak and first-mile trips during the evening peak. This makes sense since people who ride a public bicycle from home to metro station (first mile) in the morning tend to ride one from metro station (last mile) in the evening, and people who ride a public bicycle from metro station to workplace (last mile) in the morning tend to ride one back to metro station (first mile) in the evening.

The results of duration of the metro-bicycle trip chain are presented in Figure 5, based on the 573 matched cards. Outliers have been removed since some overlong trips will overestimate the average trip duration. The outlier is defined as any data point that is over 1.5 interquartile ranges below the first quartile or above the third quartile.

(a) 2-minute time slot

(b) 5-minute time slot

(c) 10-minute time slot

(d) 20-minute time slot

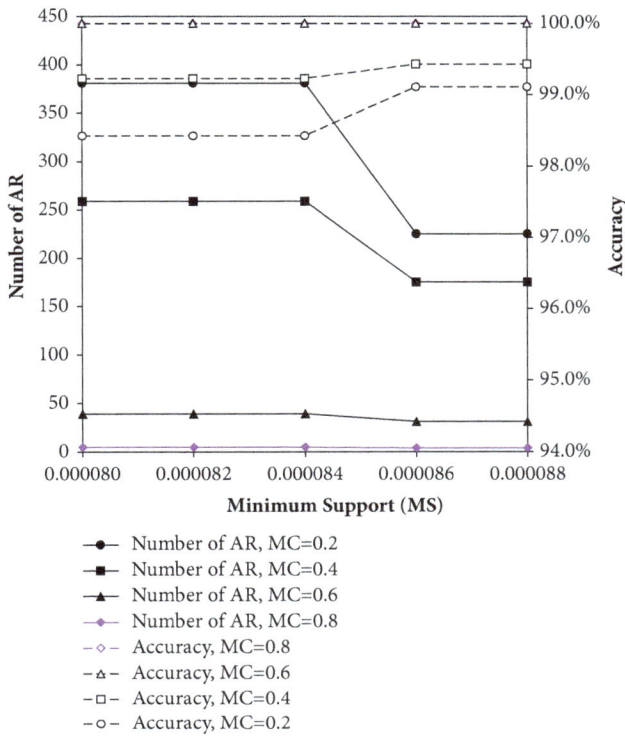

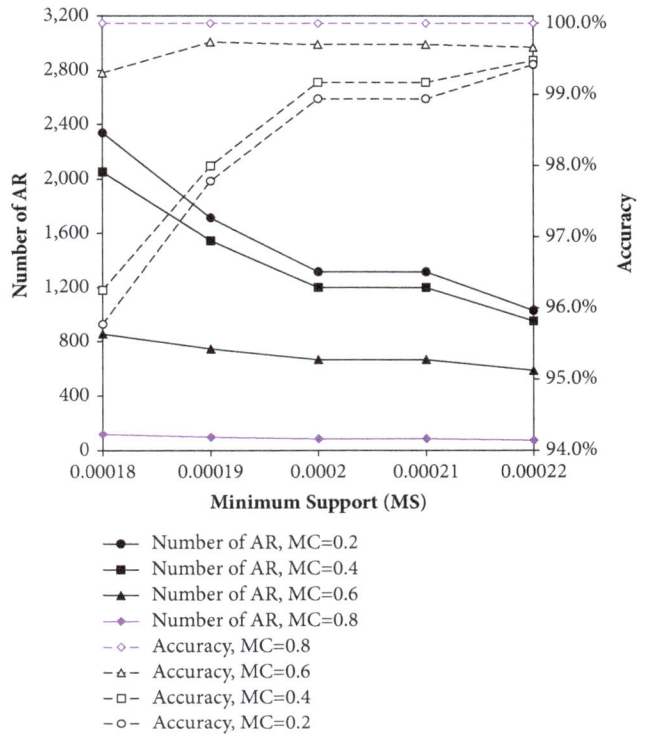

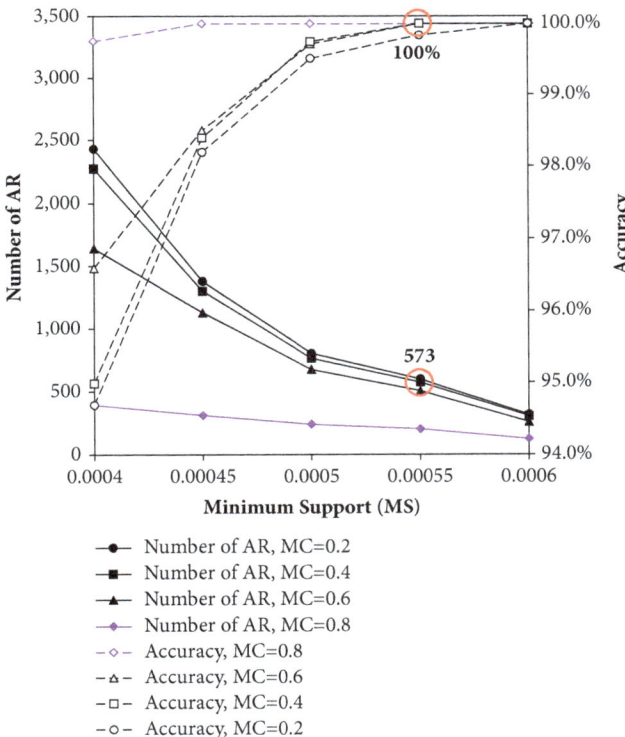

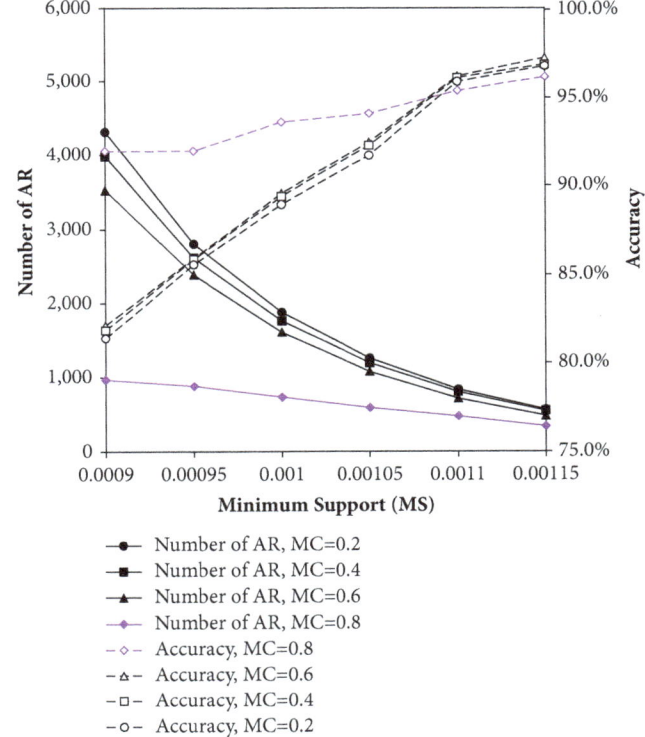

FIGURE 2: Parameter calibration.

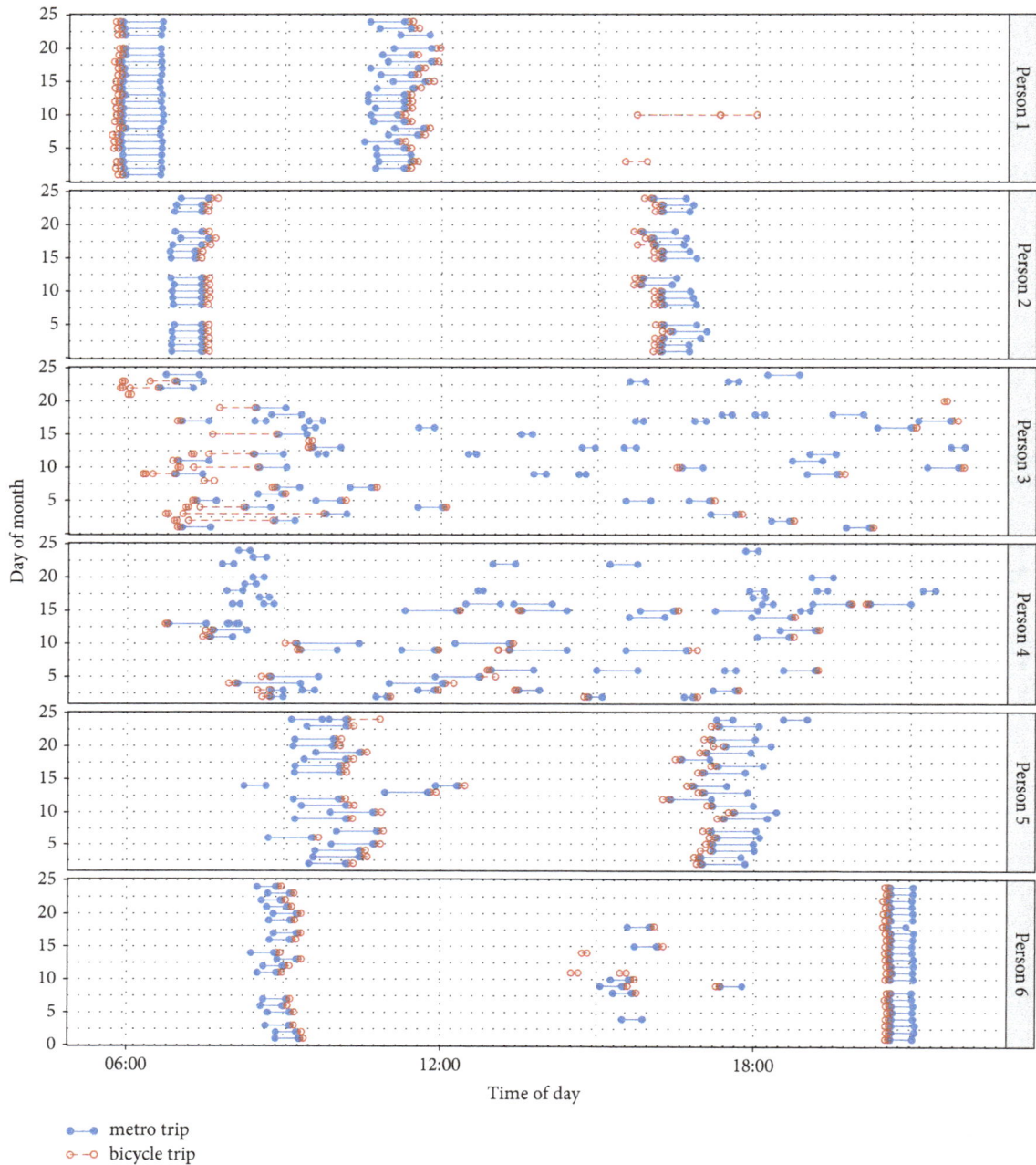

FIGURE 3: Individual trip log based on matched cards.

We divided metro-bicycle trip chain into metro trip, transfer, and public bicycle trip. For metro trip with last mile, the average metro trip time, transfer time and public bicycle time are respectively 23.59 min, 1.71 min, and 7.01 min. For metro trip with first mile, public bicycle time, transfer time and the average metro trip time are respectively 6.80 min, 1.75 min, and 23.50 min. The transfer time is very short, indicating that the walking distance between the metro station and bicycle stop is within a reasonable range. The connection time (public bicycle time and transfer time) takes around 27% of the total travel time.

When public bicycle was used as a first-mile mode, three parts of the trip chain in the evening peak are all longer than

morning peak or even nonpeak hours. This makes sense since people have to hurry to work in the morning, but they can take their time back home in the evening. Counterintuitively, for both first-mile and last-mile mode, the public bicycle trip in nonpeak hours is shorter than morning peak. This is probably because, in rush hours, people are more likely to spend much time on finding an unoccupied slot to return the bicycle.

Travel patterns comparison between matched public bicycle SC and unmatched public bicycle SC are shown in Figure 6. Outliers have also been removed to prevent overestimating the average trip duration or trip distance. The results showed that the average trip duration for matched

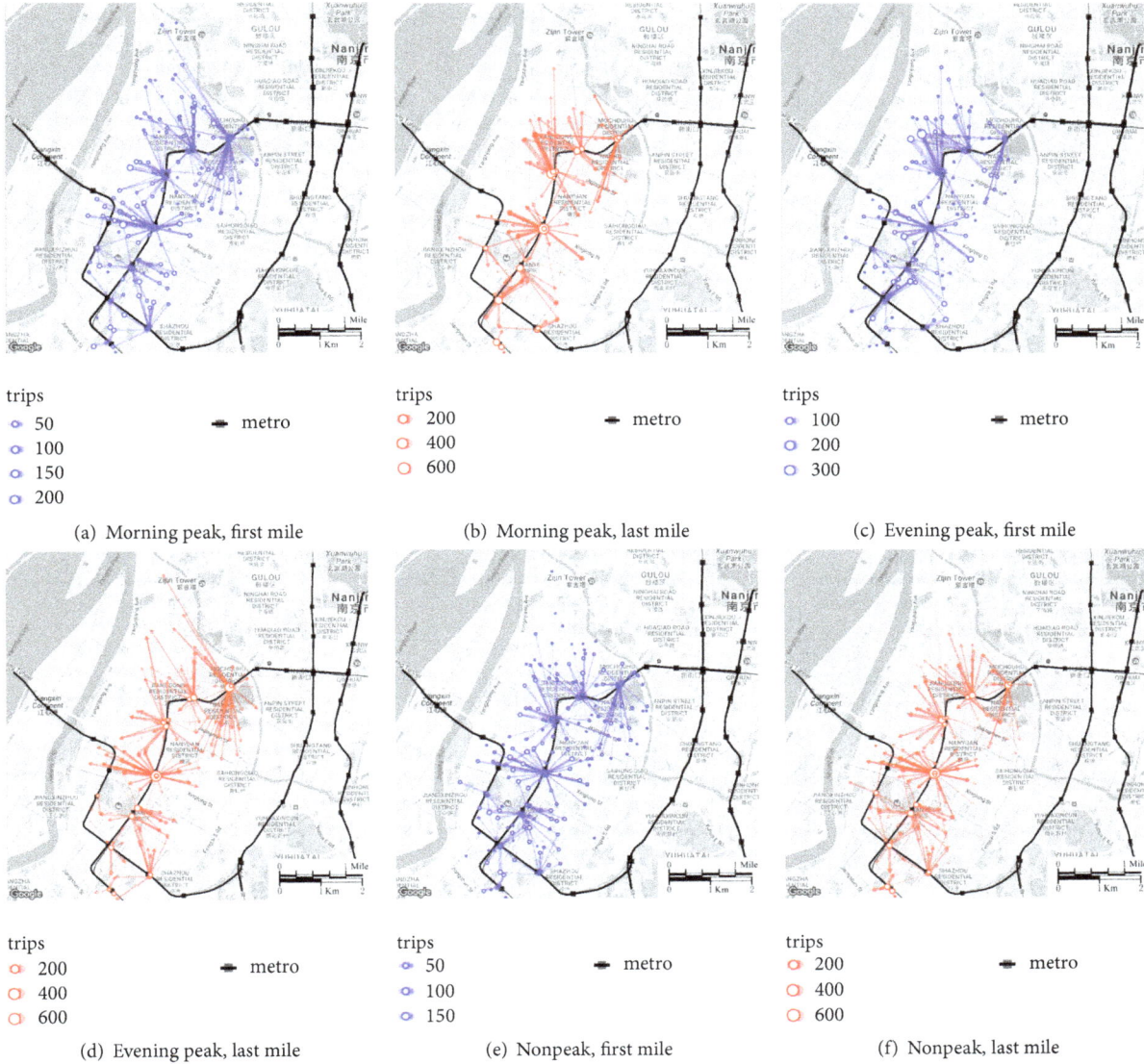

(a) Morning peak, first mile

(b) Morning peak, last mile

(c) Evening peak, first mile

(d) Evening peak, last mile

(e) Nonpeak, first mile

(f) Nonpeak, last mile

FIGURE 4: Spatial pattern of public bicycle as first-/last-mile solution.

public bicycle SC data is 6.80 min, while the average trip duration for unmatched public bicycle SC data is 10.02 min with 95% of the trip duration being less than 30 min, which is consistent with previous research studies by Zhao et al. [14]. The average trip distances for matched and unmatched public bicycle SC data are 0.95 km and 1.03 km, respectively, indicating the distance of first-mile or last-mile public bicycle trips is shorter than that of the other public bicycle trips. In contrast, the average trip distance is 0.99 km in Santander, Spain [24]. Notably, unmatched public bicycle SC data also contains first-mile or last-mile trips to connect metro.

5. Conclusions and Recommendations

This research has put forward a novel data fusion method using association rules mining to match metro and public bicycle smart cards of the same commuter. We attempt

to match SC IDs from different sources and propose a validation approach. We calibrated the three key parameters MS, MC, and TS by demonstrating how they influenced the number of ARs and the accuracy of results. The validation process showed that, with increase of the number of ARs, the accuracy of results decreased. The individual trip log has also been derived to validate the association rules method by visualizing individual metro and public bicycle trip of each day. Based on the matched cards of the same person, interesting findings of metro-bicycle transfer have been found, including spatial pattern of public bicycle as first-/last-mile solution as well as duration of metro trip chain.

Our paper contributes to the state of knowledge by taking advantage of linked-SC data to analyze metro-bicycle transfers. We demonstrated that it is possible to match two cards of the same person based on historical SC data. Our proposed method successfully matched 573 pairs of smart cards with an accuracy of 100%, when setting three key

(a) Metro trip chain with last mile

(b) Metro trip chain with first mile

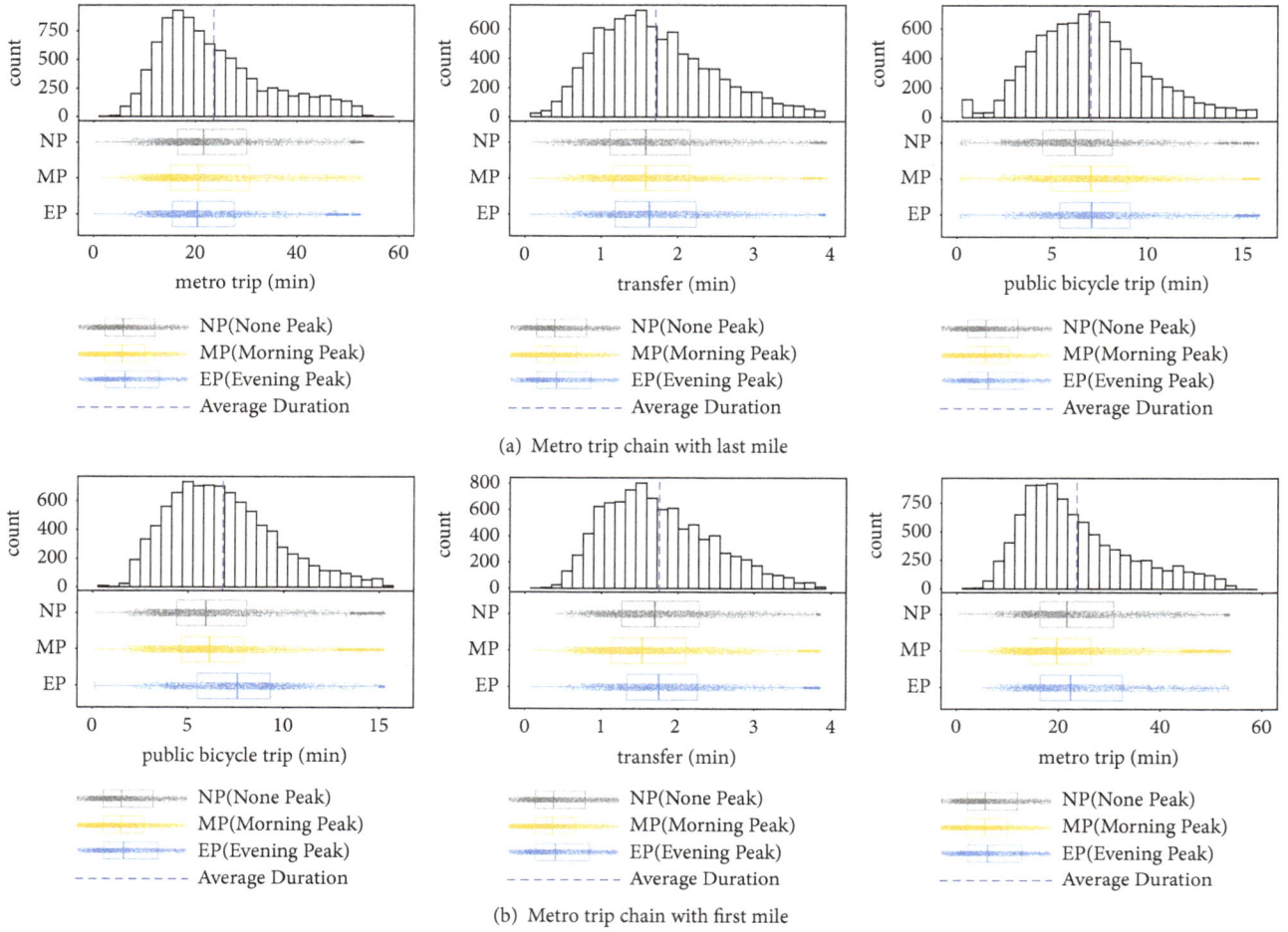

FIGURE 5: Duration of metro trip chain.

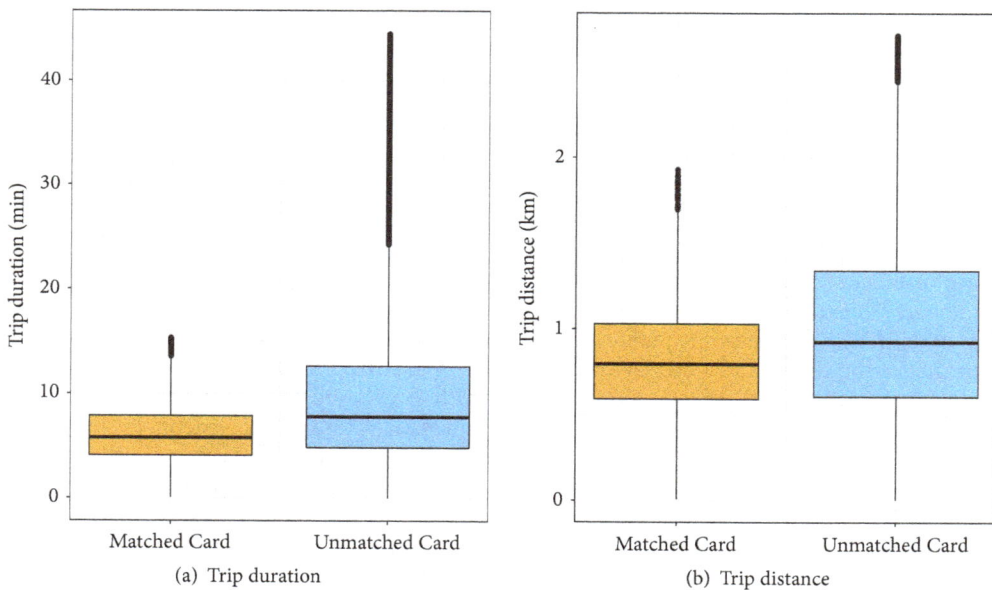

(a) Trip duration

(b) Trip distance

FIGURE 6: Travel patterns comparison between matched public bicycle SC and unmatched public bicycle SC.

parameters as TS = 10 min, MS = 0.00055, and MC = 0.4. TS determined the quality of conversion from SC data to transaction data, because TS is highly related to the metro-bicycle transfer time. However, even when transfer time is shorter than time slot (TS), the start time and end time of a transfer may still fall in two different transactions. Therefore, properly increasing the TS will help to reduce the rate of wrong grouping. Although the average transfer time is less than 1.8 min, the optimal TS of the proposed method is around 10 min.

Findings from this study suggest that around 2/3 of the transfers between metro and public bicycle in Nanjing occurred during peak hours. Although there are usually several public bicycle stops around metro station, users prefer to rent bicycles from only one or two of them, which of course exacerbates the burden of rebalancing. This paper also sheds light on the duration of the entire metro-bicycle trip chain. The connection time (public bicycle trip time and transfer time) takes around 27% of the total metro trip chain in Nanjing. When public bicycle was used as a first-mile mode, the trip duration in the evening peak is longer than morning peak or even nonpeak hours. This finding is consistent with common sense that people have to hurry to work in the morning, but they can take their time back home or dinner in the evening, while for both first-mile and last-mile mode, the public bicycle trip in nonpeak hours is even shorter than morning peak. This is probably because of the difficulty to find an unoccupied slot to return the bicycle during rush hours.

The proposed approach could be applied to other cities, where people also use different smart cards for different transportation modes. However, the parameters (MS, MC, and TS) in the model should be recalibrated based on actual SC data of the city. Because the city scale, transportation modes, and travel behaviors in different cities may greatly influence the standard of the parameters. There are several limitations of this study. Firstly, the total number of correctly matched cards were limited by the historical data. The dataset used in this research is not huge (only 24 days) due to the difficulty of obtaining both metro SC data and public bicycle SC data within the same period. Secondly, the proposed validation conditions are somewhat strict. This is because at least one metro-bicycle transfer (or bicycle-metro transfer) should occur in the remaining 4 days. Only frequent users of metro-bicycle transfer are likely to be successfully identified. However, this is the best way we can think of to validate the results without offense of individual privacy. More insightful findings about travel behaviors are expected to be found by harmonizing smart card technology across different transport modes. Speeding up the process of data integration and combining different smart cards into one will not only facilitate the traveler, but also provide data support for efficient transportation decision making.

Conflicts of Interest

The authors declare that they have no conflicts of interest.

Acknowledgments

This research is supported by the National Natural Science Foundation of China (71701047 and 51478112) and Projects of International Cooperation and Exchange of the National Natural Science Foundation of China (5151101143).

References

[1] Y. Ji, Y. Fan, A. Ermagun, X. Cao, W. Wang, and K. Das, "Public bicycle as a feeder mode to rail transit in China: The role of gender, age, income, trip purpose, and bicycle theft experience," *International Journal of Sustainable Transportation*, vol. 11, no. 4, pp. 308–317, 2017.

[2] E. W. Martin and S. A. Shaheen, "Evaluating public transit modal shift dynamics in response to bikesharing: a tale of two U.S. cities," *Journal of Transport Geography*, vol. 41, pp. 315–324, 2014.

[3] J. Bachand-Marleau, J. Larsen, and A. M. El-Geneidy, "Much-anticipated marriage of cycling and transit: how will it work?" *Transportation Research Record*, vol. 2247, pp. 109–117, 2011.

[4] R. Nair, E. Miller-Hooks, R. C. Hampshire, and A. Bušić, "Large-scale vehicle sharing systems: analysis of Vélib," *International Journal of Sustainable Transportation*, vol. 7, no. 1, pp. 85–106, 2012.

[5] Q. Zou, X. Yao, P. Zhao, H. Wei, and H. Ren, "Detecting home location and trip purposes for cardholders by mining smart card transaction data in Beijing subway," *Transportation*, vol. 45, no. 3, pp. 919–944, 2018.

[6] M. Trépanier, N. Tranchant, and R. Chapleau, "Individual trip destination estimation in a transit smart card automated fare collection system," *Journal of Intelligent Transportation Systems: Technology, Planning, and Operations*, vol. 11, no. 1, pp. 1–14, 2007.

[7] M.-P. Pelletier, M. Trépanier, and C. Morency, "Smart card data use in public transit: a literature review," *Transportation Research Part C: Emerging Technologies*, vol. 19, no. 4, pp. 557–568, 2011.

[8] C. Morency, M. Trépanier, and B. Agard, "Measuring transit use variability with smart-card data," *Transport Policy*, vol. 14, no. 3, pp. 193–203, 2007.

[9] M. A. Munizaga and C. Palma, "Estimation of a disaggregate multimodal public transport Origin-Destination matrix from passive smartcard data from Santiago, Chile," *Transportation Research Part C: Emerging Technologies*, vol. 24, pp. 9–18, 2012.

[10] X. Ma, Y. J. Wu, Y. Wang, F. Chen, and J. Liu, "Mining smart card data for transit riders' travel patterns," *Transportation Research Part C: Emerging Technologies*, vol. 36, pp. 1–12, 2013.

[11] X. Ma, C. Liu, H. Wen, Y. Wang, and Y. Wu, "Understanding commuting patterns using transit smart card data," *Journal of Transport Geography*, vol. 58, pp. 135–145, 2017.

[12] A.-S. Briand, E. Côme, M. Trépanier, and L. Oukhellou, "Analyzing year-to-year changes in public transport passenger behaviour using smart card data," *Transportation Research Part C: Emerging Technologies*, vol. 79, pp. 274–289, 2017.

[13] R. Agrawal and R. Srikant, "Fast algorithms for mining association rules in large databases," in *Proceedings of the 20th International Conference on Very Large Data Bases (VLDB '94)*, Santiago, Chile, 1994.

[14] J. Zhao, J. Wang, and W. Deng, "Exploring bikesharing travel time and trip chain by gender and day of the week," *Transportation Research Part C: Emerging Technologies*, vol. 58, pp. 251–264, 2015.

[15] S. Tao, D. Rohde, and J. Corcoran, "Examining the spatial-temporal dynamics of bus passenger travel behaviour using smart card data and the flow-comap," *Journal of Transport Geography*, vol. 41, pp. 21–36, 2014.

[16] M. Munizaga, F. Devillaine, C. Navarrete, and D. Silva, "Validating travel behavior estimated from smartcard data," *Transportation Research Part C: Emerging Technologies*, vol. 44, pp. 70–79, 2014.

[17] T. Kusakabe and Y. Asakura, "Behavioural data mining of transit smart card data: a data fusion approach," *Transportation Research Part C: Emerging Technologies*, vol. 46, pp. 179–191, 2014.

[18] X.-L. Ma, Y.-H. Wang, F. Chen, and J.-F. Liu, "Transit smart card data mining for passenger origin information extraction," *Journal of Zhejiang University SCIENCE C*, vol. 13, no. 10, pp. 750–760, 2012.

[19] J. Zhao, A. Rahbee, and N. H. M. Wilson, "Estimating a rail passenger trip origin-destination matrix using automatic data collection systems," *Computer-Aided Civil and Infrastructure Engineering*, vol. 22, no. 5, pp. 376–387, 2007.

[20] E. Fishman, "Bikeshare: a review of recent literature," *Transport Reviews*, vol. 36, no. 1, pp. 92–113, 2016.

[21] M. Ahillen, D. Mateo-Babiano, and J. Corcoran, "Dynamics of bike sharing in Washington, DC and Brisbane, Australia: implications for policy and planning," *International Journal of Sustainable Transportation*, vol. 10, no. 5, pp. 441–454, 2016.

[22] B. Caulfield, M. O'Mahony, W. Brazil, and P. Weldon, "Examining usage patterns of a bike-sharing scheme in a medium sized city," *Transportation Research Part A: Policy and Practice*, vol. 100, pp. 152–161, 2017.

[23] R. Beecham and J. Wood, "Exploring gendered cycling behaviours within a large-scale behavioural data-set," *Transportation Planning and Technology*, vol. 37, no. 1, pp. 83–97, 2014.

[24] M. Bordagaray, L. dell'Olio, A. Fonzone, and Á. Ibeas, "Capturing the conditions that introduce systematic variation in bike-sharing travel behavior using data mining techniques," *Transportation Research Part C: Emerging Technologies*, vol. 71, pp. 231–248, 2016.

[25] X. Ma and Y. Wang, "Development of a data-driven platform for transit performance measures using smart card and GPS data," *Journal of Transportation Engineering*, vol. 140, no. 12, Article ID 04014063, 2014.

[26] R. Agrawal, T. Imieliński, and A. Swami, "Mining association rules between sets of items in large databases," in *Proceedings of the ACM SIGMOD International Conference on Management of Data*, pp. 207–216, Washington, DC, USA, 1993.

[27] B. Keuleers, G. Wets, T. Arentze, and H. Timmermans, "Association rules in identification of spatial-temporal patterns in multiday activity diary data," *Transportation Research Record*, no. 1752, pp. 32–37, 2001.

[28] B. Keuleers, G. Wets, H. Timmermans, T. Arentze, and K. Vanhoof, "Stationary and time-varying patterns in activity diary panel data: explorative analysis with association rules," *Transportation Research Record*, vol. 1807, pp. 9–15, 2002.

[29] D. Kusumastuti, E. Hannes, D. Janssens, G. Wets, and B. G. C. Dellaert, "Scrutinizing individuals' leisure-shopping travel decisions to appraise activity-based models of travel demand," *Transportation*, vol. 37, no. 4, pp. 647–661, 2010.

[30] M. Diana, "Studying patterns of use of transport modes through data mining," *Transportation Research Record*, vol. 2308, pp. 1–9, 2012.

[31] K. K. A. Chu and R. Chapleau, "Augmenting transit trip characterization and travel behavior comprehension: multiday location-stamped smart card transactions," *Transportation Research Record*, vol. 2183, pp. 29–40, 2010.

[32] Y. Zhang, T. Thomas, M. Brussel, and M. van Maarseveen, "Exploring the impact of built environment factors on the use of public bikes at bike stations: Case study in Zhongshan, China," *Journal of Transport Geography*, vol. 58, pp. 59–70, 2017.

[33] T.G.o. Nanjing, *Planning Standards for Public Facilities*, T.G.o. Nanjing, Nanjing, China, 2015.

3

Improving Electric Vehicle Charging Station Efficiency through Pricing

Rick Wolbertus (ID)[1,2] **and Bas Gerzon**[3]

[1] *Transport and Logistics Group, Department of Engineering Systems and Services, Faculty of Technology, Policy and Management, Delft University of Technology, Jaffalaan 5, 2628 BX Delft, Netherlands*
[2] *Department of Urban Technology, Faculty of Technology, Amsterdam University of Applied Sciences, Weesperzijde 190, 1097 DZ Amsterdam, Netherlands*
[3] *PitPoint Clean Fuels, Gelderlandhaven 4, 3433 PG Nieuwegein, Netherlands*

Correspondence should be addressed to Rick Wolbertus; r.wolbertus@tudelft.nl

Academic Editor: Aboelmaged Noureldin

Recent studies show that charging stations are operated in an inefficient way. Due to the fact that electric vehicle (EV) drivers charge while they park, they tend to keep the charging station occupied while not charging. This prevents others from having access. This study is the first to investigate the effect of a pricing strategy to increase the efficient use of electric vehicle charging stations. We used a stated preference survey among EV drivers to investigate the effect of a time-based fee to reduce idle time at a charging station. We tested the effect of such a fee under different scenarios and we modelled the heterogeneity among respondents using a latent class discrete choice model. We find that a fee can be very effective in increasing the efficiency at a charging station but the response to the fee varies among EV drivers depending on their current behaviour and the level of parking pressure they experience near their home. From these findings we draw implications for policy makers and charging point operators who aim to optimize the use of electric vehicle charging stations.

1. Introduction

The transport sector in Europe, which accounts for a quarter of greenhouse gas emissions, is the only main sector that has not been able to reduce emissions over the past 25 years [1]. Electric vehicles (EVs) show great promise to meet CO_2 reduction targets in the transport domain and to reduce local air pollution [2]. Adoption of these vehicles is starting to take off [3] as the main barriers, being the purchase price and the limited range due to high battery costs [4], are overcome by the introduction of more affordable, long range EVs into the market. One of the opportunities EVs offer in comparison to other Alternative Fuel Vehicles (AFVs) [5] is the possibility of charging the car while being parked. This reduces the need for fast refuelling stations. Cars are parked 90-95% of the time [6], which provides the opportunity to overcome problems of limited range and long recharging times even with currently available short range vehicles. This requires instalment of (public) charging infrastructure at places where users park their cars such as at home, at work, or at public facilities such as shopping centres [7].

Investments in the necessary charging infrastructure have been trailing due to chicken-and-egg related problems. In order to solve this, governments stepped in to facilitate basic public charging infrastructure. Efficient use of the limited available charging stations is important in early adoption phases to ensure a positive experience for early adopters and to reduce resistance among nonadopters [8]. Effective usage triggers high throughput which in turn creates a positive business case for charging point operators [9]. Descriptive statistics in the scientific literature [10, 11] and experiences in the field [12], however, show that efficiency at both slow and fast charging stations is not optimal. At slow (level 2) public charging stations (up to 11 kW) only 20 to 40% of the time connected to the charging station is actually used for charging. At fast charging stations these rates are better, but idle times are more costly because charging speeds are higher.

Currently, many charging point operators use a business model that is based upon the sales of the energy transferred, not providing an incentive for the driver to move the vehicle once fully charged. Charging point operators are seeking ways to improve the efficiency of their operations without interfering with the user experience. Learning from parking studies (e.g., [13, 14]), the introduction of time-based fees could help to increase the efficiency of charging station capacities. Although it is known that fees influence the decision to charge [15], there is little knowledge about how fees influence the decision to move the vehicle once fully charged. Straightforward implementation of a time-based fee could prove not to be the optimal solution, because it could interfere with a 'parking is charging' regime; the advantage EVs have over other AFVs. Moreover there are large differences in the way EV drivers use public charging infrastructure. This depends among others on the location (e.g., home or work) and the time of day [16]. Besides such circumstantial differences, there is a diversity among drivers in their parking and charging patterns [17, 18]. Such differences could also influence the way time-based fees are influencing the behaviour of EV drivers. For a successful implementation of a time-based pricing structure, heterogeneity among EV drivers in their parking and charging behaviour is important to understand and take into account.

This paper aims to add to the understanding of the effect of time-based fee structures on charging behaviour and the underlying factors that drive heterogeneity of EV drivers' responses to a new pricing scheme. The effect of a time-based fee during different situations is estimated using a stated choice survey in which respondents are asked whether or not they would move their EV once fully charged. Heterogeneity is addressed using sociodemographic characteristics of respondents. In addition, since all respondents were actual EV drivers, their regular charging behaviour and vehicle characteristics were also used as underlying explanatory variables. By using a latent class discrete choice model, different user types are identified across which the effect of a time-based fee differs.

In Section 2 a literature overview is presented, which is followed by an outline of the structure of this paper. In Section 3 the methodology of the stated preference choice experiment is further explained, followed by the data collection process in Section 4. Results of the model estimations are shown in Section 5, followed by an interpretation of the results and their meaning in the policy context in Section 6.

2. Literature

This literature review addresses two topics, first the heterogeneity in charging behaviour and the factors that drive the decisions to charge and second literature on the influence of pricing on charging behaviour. The relevant knowledge gaps are identified and the last paragraph describes how these gaps are filled with this contribution.

2.1. Heterogeneity in Charging Behaviour. The field of charging behaviour has been found to be under increasing interest of scholars. The number of studies that model charging behaviour based upon assumptions or criteria (e.g., [19–21]) or driving data from conventional cars (e.g., [22–24]) for infrastructure planning is increasing. More recently, attention has shifted towards analysing differences in charging patterns from actual EV drivers. Studies that discuss heterogeneity in charging behaviour fall into two categories, those that discuss heterogeneity in charging patterns (e.g., home, workplace and public charging) and those that study heterogeneity in the factors that drive charging decisions (e.g., pricing and routine behaviour).

The number of studies that investigate heterogeneity in charging patterns using actual driving- or charging data from EVs is small due to the limited number of vehicles on the road. However, with the growing number of EVs on the road, it can be observed that the number of such studies also begins to increase. A number of studies such as by Azadfar, Sreeram, and Harries [25], Robinson, Blythe, Bell, Hübner, and Hill [26], and Morrissey et al. [7] describe charging behaviour and try to derive general conclusions from this. They identify patterns often corresponding to home and workplace charging, the two most dominant modes currently used. Heterogeneity among charging profiles was more systematically addressed by several studies such as Robinson et al. [26] and Desai et al. [10] which both used cluster analysis to identify several charging profiles. Helmus and Van den Hoed [16] identified 6 different user types based on charging data in the city of Amsterdam. Franke and Krems [17, 18] identified two different user battery interaction styles among EV drivers in a trail in Germany; some users preferred to interact with the battery level of the vehicle, while others displayed more opportunity driven recharge styles. Sadeghianpourhamami, Refa, Strobbe, and Develder [27] make use of charging data to determine different user types to assess their flexibility in charging behaviour and therefore their suitability for load shifting purposes. They identify three different user groups using k-means clustering: home, workplace, and park-to-charge charging. The results are largely in line with Robinson et al. [26].

In studies that investigate the factors that drive charging decisions, heterogeneity among EV drivers is often modelled by using random parameter logit models [28–32]. These studies find differences in how EV drivers interpret, e.g., distances to charging stations and different charging speeds. Latent class analysis is used to investigate heterogeneity among the determining factors of charging decisions by Wen, Mackenzie, and Keith [15]. Although they identified three different user groups, these were not linked to actual recharge patterns found in studies based on actual charging behaviour such as in Robinson et al. [26], Van den Hoed and Helmus [16], and Sadeghianpourhamami et al. [27] but on sociodemographic and vehicle characteristics. The only study that does make such a link is by Kim, Yang, Rasouli, and Timmermans [33] who used a latent class hazard duration model to identify differences in user groups in intercharging session duration. The predefined two groups were based upon charging (ir)regularity. Latent class analysis showed that charging behaviour and vehicle characteristics can predict whether users are (ir)regular chargers.

The overview shows that random parameter models are mostly used to capture heterogeneity in decision rules in charging decisions. Descriptive studies, however, more focus on clustering users based on their behaviour. Linkage between these methodologies is mostly missing with the exception of Kim et al. [33].

2.2. Price Incentives for Charging Behaviour. The effect of pricing strategies to steer charging behaviour has mainly been studied in the context of so-called smart charging [34]. Smart charging is the concept in which pricing is used to prevent peaks in grid loads, to let charging coincide with renewable energy production or to feed back into the grid during high energy demand. An overview of the various modes of smart charging is given by García-villalobos et al. [35] and Tamis, van den Hoed, and Thorsdottir [36]. Price setting usually happens in a centralized manner by so-called aggregators as individual users do not have enough volume to trade on energy markets. Setting the price is done dynamically based on current energy prices or using more static time-of-use prices in which differences are made between, e.g., day and night [34]. Generally in studies based on stated choice experiments, a significant positive effect of price on the decision to postpone or to leave control to an aggregator is found [37]. There are, however, studies indicating that too complex pricing strategies have a negative effect on reaching set goals [38].

Besides the influence of price incentives for "smart charging" a few studies have looked into the influence of pricing on more general charging behaviour. Latinopoulos, Sivakumar, and Polak [39] looked into price setting in relation to charging decisions combined with parking reservations. They find that EV drivers are willing to pay more to ensure charging station availability. Wen, MacKenzie, and Keith [15] model the choice to start charging with mixed and latent class models, in which they include the price of the charging session based upon a stated preference survey among EV drivers. In the latent classes they do find differences on price sensitivity between respondents.

In studies that make use of charging data Sun, Yamamoto, and Morikawa [40] find that EV drivers in Japan are willing to make longer detours for free charging stations from their route than for paid chargers. Motoaki and Shirk [41] find that installing a flat fee at fast charging stations resulted in longer charging sessions and less energy transfer per minute connected. Users wanted to get the most out of the money they paid. Consequently, users also fill their car beyond 80% after which charging becomes less efficient. Such inefficient use of the time connected to a charging station with flat fees or other nontime based fees was found to be even worse at slower (level 2) charging stations in Netherlands. Wolbertus and van den Hoed [11] found that only 20% of the time connected to a charging station was actually used for charging. Charging behaviour at "lower" power outlets is more related to parking behaviour in which vehicles stay in the same place for much longer times than is needed to recharge the car. Also on level 2 charging stations in the

United States, Francfort [42] found that installing time-based fees reduced charging times. The report however does not quantify the precise reduction the fee caused after charging was first free.

To summarize, there are various indications that pricing strategies can have an influence on charging behaviour. The studies indicate the location, timing, duration, and the willingness to give up control over the charging process can be influenced. The charging station choice could also be influenced if prices vary enough. However, a quantification of the effect of pricing strategies is missing, especially for time-based strategies.

2.3. Knowledge Gaps and Contributions. In sum, this overview has shown that a growing body of literature is investigating charging behaviour of EV drivers using revealed preference data. Descriptive studies and random parameter models show that heterogeneity is present in charging patterns and in the determining factors which drive the decisions regarding where, how long, and how much to charge. Understanding this heterogeneity is crucial to correctly predict charging demand. Links between descriptive studies which often show clear habitual patterns and studies that model heterogeneity in charging decision rules are sparse. Furthermore, the literature on determining factors focusses on the decision to charge (or not) and not on the duration of the charging session.

The effect of price on the charging sessions is mainly studied in the context of "smart charging" in which the user is asked to hand over a certain amount of control over the charging process for a lower price. Information about price sensitivity mostly comes from stated preference studies or studies that investigate the difference between paid- and free chargers. These studies often find significant effects of such price changes. Literature from other domains, such as parking [43, 44], suggests that behaviour could be well steered by setting the price level and pricing mechanism.

This study contributes by shedding more light on the effect of pricing mechanisms on charging behaviour while taking the heterogeneity of EV drivers in their charging behaviour into account. It does so by looking more at current charging patterns described in the literature. Using a stated preference study on the decision to end a charging session once completely charged, given a certain price per hour, it is investigated how such a pricing strategy can lead to more efficient charging station use. Actual charging patterns are used to simulate scenarios about the timing, location, and parking pressure of charging sessions under which the effect of a time-based fee is tested. Moreover, the participants, all EV owners, are asked about their recharging patterns. This information is used in a discrete choice latent class model to determine if these charging patterns lead to a different evaluation of the proposed pricing mechanism.

3. Methodology

A stated choice study was performed among EV drivers, in which they were asked to imagine that they were charging their electric vehicle at a level 2 public charging station. They

TABLE 1: Overview of variables used in stated choice experiment.

Variable	Levels
Fee (€)	€0,25/hour
	€1/hour
	€1,75/hour
Time to move car	5min
	10min
	15min
Time until next drive	2 hours
	5 hours
	8 hours
Time of day and location	9:00 at work
	14:00 at home
	17:00 at home

TABLE 2: Exemplary choice set.

Situation 2	
Location	Home
Time of arrival	17:00
Time finished charging	19:00
Expected departure	9:00 next morning
Time required to move car into different parking sport	10 minutes
Fee if car Is not moved 1 hour after charging	€1,00/hour

If you do not move your car between 19:00 and 20:00 you will pay an additional fee of €4,00.
2. Would you move your car between 19:00 and 20:00
☐ Yes
☐ No

were presented with the scenario in which the EV was fully charged two hours after having started the charging session. The two hours is the average time needed to recharge [11]. The driver is asked to make the choice to move his vehicle away from the charging station within the next hour. If the driver does not comply, he will be faced with an additional time-based fee. Such a fee was not applicable between 23:00 and 8:00 hours as this would hamper overnight charging sessions and would only create empty charging spots due to the fact that during these hours demand for charging is generally very low.

Different charging scenarios were constructed including the most important factors. These factors were determined by a literature review and interviews with policy makers and EV drivers. Three factors were identified as most relevant in the decision to move the vehicle once the charging session was finished: first, the timing of the charging session in the day, which often coincides with location due to habitual patterns of drivers such as charging at home or work; second, the time until the next drive was relevant; drivers indicated that they would not likely move their car if the parking period after a finished charging session was very short. Last, drivers also indicated that parking pressure or the ability to park somewhere close without too much hassle was relevant. An overview of the variables and their levels is shown in Table 1.

As input to establish the right levels to represent the timing of the charging session, evidence from charging patterns in literature was taken. Jabeen et al. [29] and Hoed, Helmus, Vries, and Bardok [45] showed that significant differences exist between home and workplace charging, the two most dominant modes of charging. These are represented in the survey as 9:00 at work and 17:00 at home. During weekends different patterns arise, in which charging peaks are observed during the afternoon, represented by the 14:00 at home level in the experiment.

The times until the next drive variable levels are based upon typical charging patterns observed in Netherlands [46]. Three levels are chosen based upon a review of the data: removal of the vehicle within 2 hours, 5 hours, and 8 hours after a finished charging session. The two-hour level resembles short sessions mainly observed during the morning and

afternoon, the five-hour level resembles morning sessions ending in the afternoon, and the 8-hour level represents sessions of more than 10 hours, often overnight.

During interviews with policy makers and EV drivers about a potential fee, an often mentioned comment was that EV drivers were willing to move the vehicle once fully charged, but they did not have the opportunity to park elsewhere without cruising for a parking spot for a considerable amount of time. Parking pressure in the surroundings of the charging station is resembled by the *time to move the car* variable. The variable represents the time cruising for a parking spot and the additional walking time to reach the destination. The variable is set with a 5 minute interval with a maximum of 15 minutes as it was expected that drivers would not remove their car if cruising time would be longer.

Finally we resemble an hourly fee for using the charging station without actually charging with a variable that was set on three levels from *low* (€0.25/hour) to *medium* (€1.00/hour; similar as the regular charging costs) and *high* (€1.75/hour). Levels are still below average parking costs. Total fee costs, based upon the fee level multiplied with the remaining number of hours of parking and with exceptions between 23:00 and 8:00, are precalculated. An exemplary choice set (translated from Dutch) is showed in Table 2.

The experimental design was based upon Taguchi's [47] orthogonal arrays. The design uses 3^4 dimensions, resulting into nine different choice sets. Each respondent was faced with each of these nine choices. In the second part of the survey respondents were asked about their social demographic characteristics. Additional information about their electric vehicle (type), reason of purchase, and their recharging behaviour on public charging stations was asked at the end of the survey.

To analyse the data both a binary logit and a latent class discrete choice model were estimated. The time and location, time until next drive, and the time to move the car variables were effect coded. For each of the categorical variables the first value was chosen as a reference point. This reference level is indicated in the results. In effect coding the sum of all the coefficients equals zero. This implies that the coefficient for the reference category can be calculated as the negative sum of the coefficients [48]. Z-values and p-values are not derived for these reference levels. The continuous fee variable was calculated with the shown fee multiplied with the time until the next drive variable in order to capture the total cost of not moving the car. Nonlinear versions of the fee variable were tested but did not provide a better model fit. The logit model was estimated using BIOGEME [49].

To capture the heterogeneity among the EV drivers a latent class discrete choice model was estimated. Latent class choice models are particularly useful in this case, since they divide behaviour into groups of different EV drivers. As seen in the analyses by Jabeen et al. [29] and Helmus and Van den Hoed [16] based upon real charging data, defining different user types is very well possible. Other models, such as mixed logit models, assume a continuous distribution of the taste parameters, making it impossible to link the heterogeneity to the discretely defined user groups. Latent class models are therefore the most suited in this case and can provide the most insight for policy makers as such a discrete distribution into classes provides a richer and often more understandable interpretation of the heterogeneity among EV drivers.

For the latent class model, predictor variables for class membership were entered as covariates in the model. The model is estimated using Latent GOLD 4.0 [50]. The number of classes was determined using ρ^2 and Bayesian Information Criterion (BIC) values.

4. Data Collection

Respondents were recruited via email using the database from the Dutch association for electric drivers (Vereniging Elektrische Rijders). In total 559 people were contacted of whom 128 (23%) responded. Additional EV drivers were recruited via an online EV driver platform and through a message by Dutch charging station organisation "ELaadNL" on social medium platform Twitter. In total 168 respondents completed the online survey. After filtering out incomplete surveys and unrealistic responses, 119 responses were useful. Each respondent was asked to fill in 9 different choice sets, resulting in 1058 choices in total which were used for the model estimation.

The respondents were mainly male (92%) and the income level was distributed upwards in comparison the Dutch average (CBS, 2015). This profile is consistent with the average Dutch EV owner [51]. Table 3 presents the sample distributions of sociodemographic and background characteristics. In contrast to the average Dutch EV owner, the respondents mostly consisted of Full Electric Vehicle (FEV) owners [52]. Nearly 90% of Dutch EV owners have a plug-in hybrid electric vehicle (PHEV), while in the sample this is only 32.2%. Moreover they were more likely to own the car instead of leasing it, which is also inconsistent with the current population of EV owners. The majority of the respondents indicated to have a private charging point at home instead of relying on on-street parking and public charging overnight.

TABLE 3: Sociodemographic figures of respondents to the survey.

Gender	
Male	92.2%
Female	7.8%
Age	
0-30	2.1%
30-60	79.2%
60+	16.6%
Unknown	2.1%
Annual income	
<€50.000	18.7%
€50.000 - €75.000	23.3%
€75.000 - €100.000	14.5%
€100.000 - €125.000	24.9%
>€125.000	18.5%
Type of EV	
FEV	67.8%
PHEV	32.2%
Car ownership	
Privately owned	67.3%
(Company) Leased	32.7%
Private charging point	
Private at home	77.2%
Public at home	22.8%

5. Results

5.1. The Logit Model. First, a standard logit model is estimated to assess the overall effects of the attributes on the choice to move the EV from the charging station to another parking spot (once fully charged). Table 4 shows the results of this analysis and the estimated coefficients for the standard model.

The results show that, as expected, a fee increases respondents' utility and thus increases the probability to move the car. For the *time of day* variable we find that users are more willing to move their vehicle during the evening hours than at the middle of the day. An explanation might be that drivers are not going elsewhere after 19:00 hours and are willing to move their car for neighbours. The interpretation of the *time to move* variable is not straightforward as only the "10 minutes" value has a positive and significant effect. It is unclear why the "15 minute" value is not significantly different from zero. A similar effect can be seen in the time until the next drive variable, where a longer parking time gives a higher utility for the "5 hour" value, but no significant effect is found for the "8 hour" value. A possible explanation is that the fee is relatively high when there are 8 hours until the next drive regardless of the hourly based fee. The effect of the 8 hour variable would then be partially captured by the fee variable.

In general, the model yields plausible results, but nonlinear effects in the time to move the car and time until next drive variables are hard to interpret. The effect of implementing a fee is significant and has the highest relative contribution of the variables in the model. The model provides a reasonable

TABLE 4: Results of binary logit model estimation.

Attribute	Coefficient	z-value
Constant	-0.413∗∗	-3.172
Fee	0.297∗∗	8.521
Time to move car		
5 min (ref. cat.)	-0.208	
10 min	0.299∗∗	2.266
15 min	-0.090	-0.868
Time until next drive		
2 hours (ref. cat.)	-0.521	
5 hours	0.500∗∗	3.950
8 hours	0.021	0.135
Time of day and location		
9:00 at work (ref. cat.)	0.080	
14:00 at home	-0.479∗∗	-3.900
17:00 at home	0.399∗∗	3.054
Model fit		
Null log likelihood	- 699.033	
Final log likelihood	- 547.409	
ρ^2	0.217	

∗∗Significant at the 0.05 level.
∗Significant at the 0.10 level.

fit to the data; the ρ^2 value of 0.217 indicates a substantial reduction of the Final LL compared to the Null LL.

5.2. The Latent Class Discrete Choice Model. To assess heterogeneity in the responses of respondents to the pricing scheme, a latent class choice model was estimated. In this model it is assumed that there exist latent (unobserved) segments in the population, which have different sets of parameters along which the population in these segments asses the choice attributes. For example, there may be a group which is very price-sensitive (high parameter value for the "fee" variable), while another group is very sensitive to parking pressure (high parameter value for the "time to move" variable). The latent classes are inferred from the distributions of the choice parameters emerging from the observed choices using the maximum likelihood principle.

A benefit of using a latent class choice model to reveal heterogeneity in the parameters is that additional explanatory variables can be included in the model to explain latent class membership. For example, it may be plausible to assume that a lease driver who does not have to pay the price of charging (or staying connected) himself is less likely to belong to a "price-sensitive" class/segment. A systematic overview of the model is shown in Figure 1.

In the present application, the following four variables are entered into the model as predictors of class membership: having a full electric (FEV) or plug-in hybrid electric vehicle (PHEV), whether the car was owned or leased, if the participant already moved their car away from the charging station once fully charged, and if the participant experienced high parking pressure in the neighbourhood near their home.

Sociodemographic variables were also included as predictors of class membership, but these turned out to be insignificant. In line with Kim et al. [33] we therefore focused on the vehicle and charging characteristics. Overall, predictors were found to vary across the different classes in a meaningful way.

To estimate the optimal number of classes, consecutive Latent class models (LCMs) were estimated with the number of classes ranging from 1 to 5. Table 5 shows the various model fit indicators for each of the estimated models. The Bayesian Information Criterion (BIC) indicator points to a 3 or 4 class model. To determine the optimal number of classes the predictors in the 3 and 4 class models were assessed. The parameter estimates in the 4 class model could not be meaningfully interpreted, especially as the class sizes became too small. Therefore the 3-class model was chosen as the best fit.

The results of the latent class model estimation are shown in Table 6. In general the LCM provides a substantial improvement in model fit (ρ^2 =0.483 versus 0.217). The classes have clear different meanings when we look at how they interpret the coefficients.

Class 1: members of class 1 do seem sensitive to all four variables. A time-based fee increases the chance of moving the car for respondents in the first class. For the members of the first class the *time to move the car* variable only has a significant negative parameter for the 15-minute level. This shows that severe parking pressure can be of influence on the decision to move the car. This effect was already captured in the membership model for class 3. The *time until the next drive* variable has an expected effect for the 2 and 5 hour levels but surprisingly has no significant effect for the 8 hour level in class 1. As predicted, the longer the duration of the remaining parking time is, the more likely drivers are willing to move their car. The insignificance of the 8 hour parameter could be explained by the effect of the duration and could be partly captured by the fee. The *time of day* and *location* variables are in line with the binary logit model, in which we see that drivers are more likely to move their car in the evening at home than during the afternoon.

Classes 2 and 3: they are relatively insensitive to most of the variables as we see that none of the variables is significant. This is especially relevant for the time-based fee and can be explained by the fact they either nearly always move (class 2) or nearly always stay (class 3). The intercepts (although not significant for classes 2 and 3) play a dominant role in the observed probabilities for members in these two latter classes. Implementing a time-based fee for the latter groups would thus not be as effective. The latter can be related back to the membership model where the same respondents stated that they experienced high parking pressure near their homes and therefore might not see opportunities to park their car elsewhere once fully charged.

The class membership model is displayed in Table 7. For the predictors of class membership the *currently moving* and *parking pressure at home* variables were found to have a significant effect on class membership.

(i) **Class 1**: members did not have a specific profile according to the covariates in the model. Class 1

TABLE 5: Model fit estimators for different number of latent classes.

Number of classes	Number of parameters	Log Likelihood	BIC (LL)	ρ^2
1	11	-547.409	1133.051	0.2169
2	30	-429.102	958.565	0.3861
3	45	-361.750	885.912	0.4825
4	60	-330.085	884.789	0.5277
5	75	-310.849	908.446	0.5553

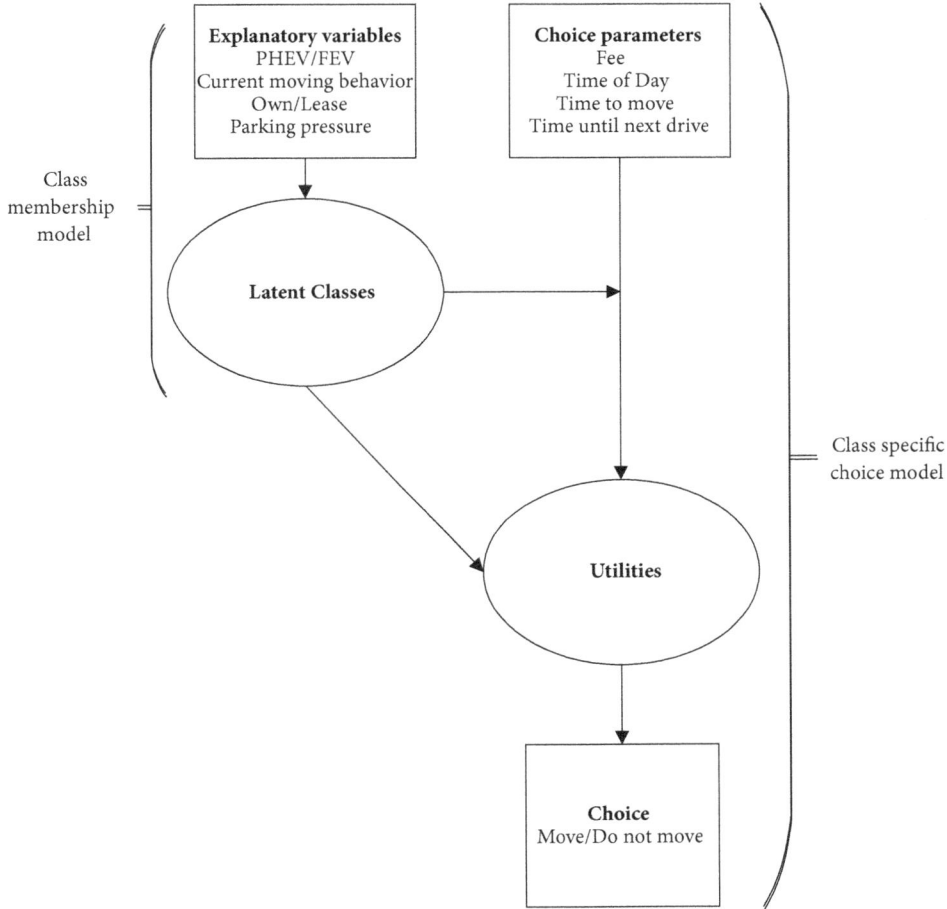

FIGURE 1: Visual representation of the latent class choice model, reproduced from [53].

represents the largest group of respondents (60%) and they are the most responsive to the hourly fee.

(ii) **Class 2:** members nearly always indicated to remove the car from the charging station during the experiment also indicated that this was their current behaviour. They also did not perceive parking pressure at home in comparison to members of the other classes.

(iii) **Class 3:** members experience more parking pressure near their homes. This could be one of the main drivers why they almost never choose to move the EV from the charging point.

6. Conclusion

This paper has examined the influence of a time-based fee on the decision to remove an EV from a charging station once fully charged. Results from a stated choice survey that have been analysed in a binary logit model show that such a fee can be effective and can result in more efficient use of charging stations. Other factors influencing the choice, such as parking pressure, time until next drive, and the time of day were also found to be relevant, although straightforward interpretation was not always possible.

To assess the heterogeneity among EV drivers regarding the time-based fee, a discrete choice latent class model was estimated. Additional variables about the type of EV and charging behaviour of the respondents were added to

TABLE 6: Results of latent class model estimation.

	Class 1		Class 2		Class 3	
	Coefficient	z-value	Coefficient	z-value	Coefficient	z-value
Intercept	-2.333**	-5.836	1.503	1.683	-2.998	-1.294
Predictors						
Fee (€)	0.813**	5.268	0.854	1.128	0.003	0.006
Time to move car						
5min (ref. cat.)	1.175		1.774		1.223	
10min	0.215	0.796	-0.598	-0.460	0.182	0.161
15min	-1.390*	-1.947	-1.177	-0.152	-1.406	-0.931
Time until next drive						
2 hours (ref. cat.)	-1.244		0.590		1.767	
5 hours	1.860**	3.280	0.383	0.050	-1.704	-0.880
8 hours	-0.617	-0.898	-0.973	-0.124	-0.063	-0.110
Time of day and location						
9:00 at work (ref. cat.)	0.456		-1.688		-0.398	
14:00 at home	-1.779**	-3.400	0.756	0.098	-1.320	-0.947
17:00 at home	1.323**	1.989	0.932	0.120	1.718	1.261
Model fit						
Log Likelihood	-361.751					
ρ^2	0.483					

*Significant at the 0.10 level.
**Significant at the 0.05 level.

TABLE 7: Class membership model for 3-class model.

Class membership model	Class 1		Class 2		Class 3	
Class size	60.0%		30.9%		9.1%	
% Choice to move (observed)	53.8%		93.6%		13.2%	
	Coefficient	z-value	Coefficient	z-value	Coefficient	z-value
Intercept	0.981	2.430	0.303	0.683	-1.284	-2.066
Attributes						
Full Electric	0.165	0.437	-0.186	-0.431	0.021	0.038
Lease	0.395	0.950	-0.425	-0.886	0.031	0.050
Currently moving	-0.397	-1.041	1.232**	2.937	-0.835	-1.391
Parking pressure at home	-0.690	-1.544	-0.978*	-1.777	1.668**	2.728

**Significant at the 0.05 level.
*Significant at the 0.10 level.

the model as predictor of class membership. Results show that three types of users could be distinguished: those that responded to the fee, users that always moved their car once fully charged, and those that refused to move, regardless of the set fee level. Membership variables showed that members of the second class indicated that indeed this behaviour belonged to their normal charging behaviour. Members of the third class were more likely to experience parking pressure when parking at home. Users in the third class might not see the opportunity to park their car elsewhere once fully charged. Such distinctions are important for policy makers because those that experience parking pressure are mostly drivers who rely on curb side charging and parking because they make use of public charging infrastructure on a daily basis. Although in some countries the majority of EV drivers

have charging facilities at home; the needs of future drivers, which might be more dependent on on-street parking and charging facilities, have to be taken into account by policy makers. This is especially relevant in more dense urban areas. Municipal policy makers can make distinctions between inhabitants and visitors, possibly relieving the impact of a time-based fee for those that experience parking pressure in the city they live in.

The results show that taking into account the heterogeneity among respondents can be very relevant. Using a discrete choice latent class approach has the benefit that results are easier to interpret for policy makers, as users are divided into clear groups. This allows for assessing biases among respondents groups. In this case early adopters can display distinct different charging behaviour regarding on- and off-street charging at home, which resulted in a different acceptance of the proposed pricing scheme.

7. Discussion

This research is limited by the fact that the respondents are not completely representative for the population of Dutch EV drivers. The research was aimed solely at EV drivers as it was believed that non-EV drivers did not have the experience to correctly predict what their response would be to the scenarios in the choice experiment. This limited our search to members of the Dutch association for EV drivers. The respondents drove more full instead of plug-in hybrid electric vehicles and were less likely to be company lease drivers compared to the population of Dutch EV drivers. From practical experience it is known that company lease drivers very often do not have to pay for charging costs themselves. They are therefore less aware of the costs and they could therefore be more reluctant moving the vehicle once fully charged even when presented with a time-based fee. Although no effect was found for having company lease car in the latent class membership model, future research could look more into differences between private owners and company lease drivers.

For charging point operators the results of this study show that implementing a time-based fee could result in a higher efficiency in charging station usage. The results show that even with a modest fee, to not frustrate EV drivers, a substantial improvement could be reached. In the final design of the fee, the charging point operators would have to take into account the segment of drivers that experience severe parking pressure and are therefore not willing or able to move their vehicle away from the charging station once fully charged. The design of the fee could focus on only preventing very long charging sessions (e.g., >24 hours) as suggested by Wolbertus and Van den Hoed [46]. This would also prevent misuse by EV drivers, who could set the charging speed at a very low rate to prevent them from completing the charging session. Another important factor that has to be taken into account when considering an implementation of a time-based fee is the precondition that the policy is only effective when the fee is communicated clearly. This requires all costs related to the time-based fee to be at least specified in the transaction data and the bill and preferably beforehand at the charging location.

This study builds on various studies that investigate the effects of pricing strategies to influence charging behaviour. The results are in line with previous studies [39–41] which also find that pricing strategy can be an effective strategy to steer charging behaviour. This study has been the first to quantify this effect for a time-based fee. Moreover, in addition to previous studies, this study added the influence of charging behaviour (as a variable in the model). Finally, it has provided a segmentation of EV drivers using characteristics of their car, their current behaviour, and the effect of parking pressure. This segmentation has proved to be useful, as the time-based fee was assessed differently by the three different segments found in this study. Doing so this paper has given additional insight into the motivations of charging behaviour in an urban context.

As the literature review showed, many applications can benefit from dynamic price signals in the context of smart charging, charging station efficiency, or station reservation. Such price signals make sense from the perspective of the problem owner, the grid operator, the charging point operator, or the parking manager, respectively. However as electric vehicle charging is a combination of these different areas, it is evident that implementation of each of these pricing strategies is not in the interest of the EV driver. Dynamic price setting should be considered carefully for each application separately.

Future research could also look at heterogeneity among more charging decisions such as charging station choice. Understanding differences in user groups can be important for policy makers for the spatial planning of a charging infrastructure. Further understanding of pricing effects can also be important in being able to steer charging behaviour to goals of stakeholders. This research and others have shown that clustering users based upon their charging behaviour and vehicle characteristics is useful to capture heterogeneity in charging decision rules.

Conflicts of Interest

The authors declare that they have no conflicts of interest.

Acknowledgments

The authors are grateful for the funding provided by the Sia Raak for the IDOLaad project of which this research is part of. They also like to thank Maarten Kroesen, Caspar Chorus, and Kees Maat for their guidance and critical remarks during the research and writing the paper.

References

[1] European Environment Agency, *Annual European Union Greenhouse Gas Inventory 1990–2015 and Inventory Report*, European Environment Agency, Copenhagen, Denmark, 2017.

[2] G. Razeghi, M. Carreras-sospedra, T. Brown, J. Brouwer, D. Dabdub, and S. Samuelsen, "Episodic air quality impacts of plug-in electric vehicles," *Atmospheric Environment*, vol. 137, pp. 90–100, 2016.

[3] International Energy Agency, "Global EV Outlook 2016: Beyond one million electric cars," Tech. Rep., 2016.

[4] B. Nykvist and M. Nilsson, "Rapidly falling costs of battery packs for electric vehicles," *Nature Climate Change*, vol. 5, no. 4, pp. 329–332, 2015.

[5] P. C. Flynn, "Commercializing an alternate vehicle fuel: Lessons learned from natural gas for vehicles," *Energy Policy*, vol. 30, no. 7, pp. 613–619, 2002.

[6] E. Paffumi, M. De Gennaro, and G. Martini, "Assessment of the potential of electric vehicles and charging strategies to meet urban mobility requirements," *Transportmetrica A: Transport Science*, vol. 11, no. 1, pp. 22–60, 2015.

[7] P. Morrissey, P. Weldon, and M. O. Mahony, "Future standard and fast charging infrastructure planning: An analysis of electric vehicle charging behaviour," *Energy Policy*, vol. 89, pp. 257–270, 2016.

[8] S. Bakker, K. Maat, and B. van Wee, "Stakeholders interests, expectations, and strategies regarding the development and implementation of electric vehicles: The case of the Netherlands," *Transportation Research Part A: Policy and Practice*, vol. 66, no. 1, pp. 52–64, 2014.

[9] C. Madina, I. Zamora, and E. Zabala, "Methodology for assessing electric vehicle charging infrastructure business models," *Energy Policy*, vol. 89, pp. 284–293, 2016.

[10] R. R. Desai, R. B. Chen, and W. Armington, "A pattern analysis of daily electric vehicle charging profiles: operational efficiency and environmental impacts," *Journal of Advanced Transportation*, vol. 2018, Article ID 6930932, p. 15, 2018.

[11] R. Wolbertus and R. van den Hoed, "Benchmarking charging infrastructure utilization," in *In Proceedings of the EVS29 Symposium*, Montreal, Quebec, Canada, 2016.

[12] D. Z. Morris, "Tesla Will Charge Users for Hogging the Supercharger," 2016, http://fortune.com/2016/12/17/tesla-supercharger-fees/.

[13] D. Shoup, *The High Cost of Free Parking*, American Planning Association, 2005.

[14] G. Pierce and D. Shoup, "Getting the prices right," *Journal of the American Planning Association*, vol. 79, no. 1, Article ID 787307, pp. 67–81, 2013.

[15] Y. Wen, D. Mackenzie, and D. R. Keith, "Modeling the charging choices of battery electric vehicle drivers by using stated preference data," *Transportation Research Record: Journal of the Transportation Research Board*, vol. 2572, pp. 47–55, 2016.

[16] J. Helmus and R. Van Den Hoed, "Unraveling user type characteristics: Towards a taxonomy for charging infrastructure," in *Proceedings of the 28th International Electric Vehicle Exhibition, EVS 2015*, KINTEX, Republic of Korea, May 2015.

[17] T. Franke and J. F. Krems, "Understanding charging behaviour of electric vehicle users," *Transportation Research Part F: Traffic Psychology and Behaviour*, vol. 21, pp. 75–89, 2013.

[18] T. Franke and J. F. Krems, "Understanding charging behaviour of electric vehicle users," *Transportation Research Part F: Psychology and Behaviour*, vol. 21, pp. 75–89, 2013.

[19] I. Frade, A. Ribeiro, G. Gonçalves, and A. Antunes, "Optimal location of charging stations for electric vehicles in a neighborhood in Lisbon, Portugal," *Transportation Research Record*, no. 2252, pp. 91–98, 2011.

[20] S. Guo and H. Zhao, "Optimal site selection of electric vehicle charging station by using fuzzy TOPSIS based on sustainability perspective," *Applied Energy*, vol. 158, pp. 390–402, 2015.

[21] F. He, D. Wu, Y. Yin, and Y. Guan, "Optimal deployment of public charging stations for plug-in hybrid electric vehicles," *Transportation Research Part B: Methodological*, vol. 47, pp. 87–101, 2013.

[22] R. P. Brooker and N. Qin, "Identification of potential locations of electric vehicle supply equipment," *Journal of Power Sources*, vol. 299, pp. 76–84, 2015.

[23] N. Shahraki, H. Cai, M. Turkay, and M. Xu, "Optimal locations of electric public charging stations using real world vehicle travel patterns," *Transportation Research Part D: Transport and Environment*, vol. 41, no. Part D, pp. 165–176, 2015.

[24] L. Zhang, B. Shaffer, T. Brown, and G. S. Samuelsen, "The optimization of DC fast charging deployment in California," *Applied Energy Journal*, vol. 157, pp. 111–122, 2015.

[25] E. Azadfar, V. Sreeram, and D. Harries, "The investigation of the major factors influencing plug-in electric vehicle driving patterns and charging behaviour," *Renewable and Sustainable Energy Reviews*, vol. 42, pp. 1065–1076, 2015.

[26] A. P. Robinson, P. T. Blythe, M. C. Bell, Y. Hübner, and G. A. Hill, "Analysis of electric vehicle driver recharging demand profiles and subsequent impacts on the carbon content of electric vehicle trips," *Energy Policy*, vol. 61, pp. 337–348, 2013.

[27] N. Sadeghianpourhamami, N. Refa, M. Strobbe, and C. Develder, "Quantitive analysis of electric vehicle flexibility: A data-driven approach," *International Journal of Electrical Power & Energy Systems*, vol. 95, pp. 451–462, 2018.

[28] C. Hou, M. Ouyang, H. Wang, and L. Xu, "An assessment of PHEV energy management strategies using driving range data collected in Beijing," in *In Proceedings of the World Electric Vehicle Symposium and Exhibition (EVS27)*, Barcelona, Spain, 2013.

[29] F. Jabeen, D. Olaru, B. Smith, T. Braunl, and S. Speidel, "Electric vehicle battery charging behaviour: Findings from a driver survey," in *Proceedings of the 36th Australasian Transport Research Forum, ATRF 2013*, Queensland, Australia, October 2013.

[30] M. Xu, Q. Meng, K. Liu, and T. Yamamoto, "Joint charging mode and location choice model for battery electric vehicle users," *Transportation Research Part B: Methodological*, vol. 103, pp. 68–86, 2017.

[31] H. Yu and D. Mackenzie, "Modeling charging choices of small-battery plug-in hybrid electric vehicle drivers by using instrumented vehicle data," *Transportation Research Record*, vol. 2572, pp. 56–65, 2016.

[32] S. Zoepf, D. Mackenzie, D. Keith, and W. Chernicoff, "Charging choices and fuel displacement in a large-scale demonstration of plug-in hybrid electric vehicles," *Transportation Research Record*, no. 2385, pp. 1–10, 2013.

[33] S. Kim, D. Yang, S. Rasouli, and H. Timmermans, "Heterogeneous hazard model of PEV users charging intervals: Analysis of four year charging transactions data," *Transportation Research Part C: Emerging Technologies*, vol. 82, pp. 248–260, 2017.

[34] M. D. Galus, M. G. Vaya, T. Krause, and G. Andersson, "The role of electric vehicles in smart grids," *Wiley Interdisciplinary Reviews: Energy and Environment*, vol. 2, no. 4, pp. 384–400, 2013.

[35] J. García-villalobos, I. Zamora, J. I. S. Martín, F. J. Asensio, and V. Aperribay, "Plug-in electric vehicles in electric distribution networks: A review of smart charging approaches," *Renewable and Sustainable Energy Reviews*, vol. 38, pp. 717–731, 2014.

[36] M. Tamis, R. van den Hoed, and R. H. Thorsdottir, "Smart Charging in the Netherlands," in *In Proceedings of the European Battery , Hybrid & Electric Fuel Cell Electric Vehicle Congress*, Geneva, Switzerland, 2017.

[37] N. Daina, A. Sivakumar, and J. W. Polak, "Electric vehicle charging choices: Modelling and implications for smart charging services," *Transportation Research Part C: Emerging Technologies*, vol. 81, pp. 36–56, 2017.

[38] P. Layer, S. Feurer, and P. Jochem, "Perceived price complexity of dynamic energy tariffs: An investigation of antecedents and consequences," *Energy Policy*, vol. 106, pp. 244–254, 2017.

[39] C. Latinopoulos, A. Sivakumar, and J. W. Polak, "Response of electric vehicle drivers to dynamic pricing of parking and charging services: risky choice in early reservations," *Transportation Research Part C: Emerging Technologies*, vol. 80, pp. 175–189, 2017.

[40] X. H. Sun, T. Yamamoto, and T. Morikawa, "Fast-charging station choice behavior among battery electric vehicle users," *Transportation Research Part D: Transport and Environment*, vol. 46, pp. 26–39, 2016.

[41] Y. Motoaki and M. G. Shirk, "Consumer behavioral adaption in EV fast charging through pricing," *Energy Policy*, vol. 108, pp. 178–183, 2017.

[42] J. E. Francfort, "The EV project price/fee models for publicly accessible charging," Tech. Rep. INL/EXT–15-36314, Idaho Falls, Idaho, USA, 2015.

[43] G. Pierce and D. Shoup, "Response to Millard-Ball et al.: parking prices and parking occupancy in San Francisco," *Journal of the American Planning Association*, vol. 79, no. 4, pp. 336–339, 2014.

[44] Z. Pu, Z. Li, J. Ash, W. Zhu, and Y. Wang, "Evaluation of spatial heterogeneity in the sensitivity of on-street parking occupancy to price change," *Transportation Research Part C: Emerging Technologies*, vol. 77, pp. 67–79, 2017.

[45] R. Hoed, Van. Den, J. R. Helmus, R. Vries, and D. Bardok, "Data analysis on the public charge infrastructure in the city of Amsterdam," in *In Proceedings of the 2013 World Electric Vehicle Symposium and Exhibition (EVS27)*, Barcelona, Spain, 2013.

[46] R. Wolbertus and R. van den Hoed, "Charging station hogging?: A data-driven analysis," in *in Proceedings of the Electric Vehicle Symposium 30*, Stuttgart, Germany, 2017.

[47] G. Taguchi, *System of Experimental Design: Engineering Methods to Optimize Quality and Minimize Costs*, Unipub/Kraus International Publications, 1987.

[48] M. Bech and D. Gyrd-Hansen, "Effects coding in discrete choice experiments," *Health Economics*, vol. 14, no. 10, pp. 1079–1083, 2005.

[49] M. Bierlaire, "BIOGEME, A free package for the estimation of discrete choice models," in *In Proceedings of the 3rd Swiss Transportation Research Conference*, Ascona, Switzerland, 2003.

[50] J. K. Vermunt and J. Magidson, *ATENT GOLD ® 4 . 0 User's guide*, Statistical Innovations, Belmont, Mass, USA, 2006.

[51] A. Hoekstra and N. Refa, "Characteristics of Dutch EV drivers," in *In Proceedings of the 30th International Electric Vehicle Symposium & Exhibition*, Stuttgart, Germany, 2017.

[52] RVO.nl., "Special: Analyse over 2015. The Hague," 2016, https://www.rvo.nl/sites/default/files/2016/01/Special%20Analyse%20over%202015.pdf.

[53] J. L. Walker and J. Li, "Latent lifestyle preferences and household location decisions," *Journal of Geographical Systems*, vol. 9, no. 1, pp. 77–101, 2006.

4

Decade-Long Changes in Disparity and Distribution of Transit Opportunity in Shenzhen China: A Transportation Equity Perspective

Qingfeng Zhou ⓘ,[1] Donghui Dai ⓘ,[1] Yaowu Wang ⓘ,[1] and Jianshuang Fan ⓘ[2]

[1]*Harbin Institute of Technology Shenzhen Graduate School, Shenzhen, Guangdong 518055, China*
[2]*Zhejiang University of Technology, Hangzhou, Zhejiang 310014, China*

Correspondence should be addressed to Donghui Dai; dai_donghui@hotmail.com

Academic Editor: Paola Pellegrini

Efficiency and equity have always been the two points of focus of transport projects. Compared with efficiency, equity is easily overlooked in the evaluation of transport projects. Many studies emphasize that defining and operationalizing costs and benefits and the distributive principle are critical parts in the assessment of transportation equity. However, the scope and time frame of the assessment target are also critical. In this paper, we took China's fastest urbanizing city, Shenzhen, as a case study to assess transport equity by comparing accessibility among groups. First, the public transport system was divided into bus and subway, and the residents were divided into two groups: urban village and nonurban village. Second, we adopted an enhanced potential opportunity model to measure residents' bus and subway accessibility and summarized them as transit opportunity. Third, we used the Dagum Gini coefficient decomposition and kernel density estimation method to explore the fair distribution of transit opportunity among groups and districts from 2011 to 2020. Decade-long changes in disparity and distribution of transit opportunity gave us a clear picture. On the one hand, the development of Shenzhen public transport system had a positive effect. All populations are benefiting, and their accessibility is increasing. On the other hand, it also had a negative effect to exacerbate inequality between populations. For the absolute value of the opportunity, Shenzhen's urban village populations do have fewer transportation opportunities than nonurban villages, and this gap between them will be wider more and more. The public transport system is more inclined to improve the population with high initial opportunity and make them higher. The results illustrated the importance of examining transportation equity over an extended period and could provide information on urban development strategies.

1. Introduction

Public transport is an effective way to solve the problem of traffic congestion and environmental pollution in high population density metropolitan. More importantly, it provides the necessary motorized transport to access jobs and social activity needed especially for low-income people without cars [1]. Many cities are aware of the importance of urban public transport and are planning to enhance public transport services, but the improvement of transit may not help low-income people. Decision-makers need to know the scope and scale of the benefit of people from the public transportation system to make more sensible investment decisions for an equitable transportation system. In recent years, the public is increasingly concerned about the impact of traffic policy and investment on equity.

Transport-related equity involves a wide range of topics and previous studies can be divided into four areas: (1) research on the consequences of transport inequity and people who are vulnerable to suffer transport inequity, exploring the relationship between transit supply and time-poverty, social exclusion, and well-being [2–5]; (2) study on the conceptual frameworks to integrate equity assessment in transport project appraisals, such as discussing cost-benefit analysis (CBA) and multicriteria analysis (MCA) which is more suitable for transport equity assessment [6–8]; (3) focus on the match between transit supply and transit demand based on spatial mismatch theory concerning access to different opportunities among socioeconomic population [9–11]; (4) research on equity aspects of distribution effects of public transport policies and infrastructure projects through accessibility [12–14], guiding assessing the distribution of benefits generated by transport investment projects for transport agencies. In this field many studies recognized that defining and operationalizing costs and benefits and the distributive principle are critical parts in the assessment of transport equity [15]. However, the scope and time frame of the assessment target are also critical. At present, the topic of incorporating equity consideration is involved very little in transport projects evaluation and decision-making in China. This paper focuses on the fourth area taking the China fastest urbanizing city, Shenzhen, as a case study. The purpose of this paper is to assess transport equity from changes of disparity and distribution of transit opportunity over the period from 2011 to 2020. Does the development of transit have a different impact on different group of people and to what extent? What is the trend of change in transit equity effect?

This paper is organized as follows: Section 2 presents a literature review including equity types, variables, and measures involved in transport equity analysis. Section 3 briefly describes the variables and measures used in this paper, an enhanced transit opportunity measure, the decomposition of the Gini coefficient, and the kernel density estimation method. Section 4 introduces the basic situation of public transportation in Shenzhen, the data used in the research, and the concept of "urban village" to classify people. In Section 5, we assess transport equity from changes of disparity and distribution of transit opportunity in Shenzhen. Finally, Section 6 summarizes the results of this study and provides direction for improving the equity of transit opportunity distribution in future studies.

2. Previous Work

2.1. Type of Transportation Equity. Conducting transport equity analysis first involves conceptual issues of equity. The definition of equity has extensive discussions in all fields from philosophy to economics. It differs in different historical periods and different perspectives of research. The definition of equity used in this study is "the distribution of benefits and costs over members of society" [16]. There are two major types of transport equity: horizontal equity and vertical equity [17]. Horizontal equity advocates that individuals or groups are considered with the same weight considered equal in ability and need; transport should provide service equally regardless of need or ability and avoid favoring specific individual or group over another. Vertical equity recognizes that the ability and needs of individuals or groups are not the same; transport should favor spatially, economically, and socially disadvantaged individual or groups to compensate for overall inequities. Vertical equity comprises three components: spatial, social, and economic. Spatial equity refers to providing the equitable transport services and

improvements in transport infrastructure, especially in peripheral or rural areas. Economic equity is related to transport services designed for users with lower incomes or without transport affordability. Social equity is related to the availability of special transport services adapted for persons with mobility impairments. Vertical equity can be divided into equity of opportunity and equity of outcome from another dimension. Equity of opportunity means that disadvantaged people have adequate access to social activity opportunities. Equity of outcome means that society must ensure disadvantaged people succeed in social activities. Many works of literature agree that people should have equity of opportunity; scholars focus on equal opportunity rather than equitable outcome; this study also holds this view.

2.2. Analysis of Transportation Equity. Three issues need to be clear when conducting an equity assessment of transport policy or infrastructure projects [18]. Frist, it is about inequality of what variable. We need to define costs and benefits of transport system; second, it is about what unit of inequality measured. The definition of equity introduced before involves the categorizing people; we need to choose target population groups over which costs and benefits are distributed; third, it is about the distributive principle used to judge what distribution of costs and benefits is "morally proper" and "socially acceptable." The distribution principle is related to the type of equity and equity measures.

2.2.1. The Variable of Costs and Benefits in Transportation Equity. Many variables can be used to represent the costs and benefits in transportation equity, and the most often used are transport affordability, access to transport, and accessibility to opportunities. Transport affordability measures individuals or household's actual expenditure on public transport usually as the percent of household disposable income [19]. Public transport affordability equity addresses that a city should enable their poorest citizens to afford at least the motorized trip rates reached by average income stratum. Equity strongly relates to distribution effects, so the main limitation of affordability is that it does not reflect the distribution effect well. Access to transport is the ability of a person to reach transit facilities [20]. It measures transport service characteristics and physical proximity to transport service. Many papers use it to assess the equity of public transit service provision [21–23]. Its disadvantage is that it does not consider whether the transport system can provide the individual with the desired destination. Having transport services does not mean that the transport system can enable people to go to the destination where they want to go to. Accessibility is the ability and ease of achieving activities, opportunities, and goods which is frequently used to evaluate transit provision concerning equity [14, 24, 25]. It can reflect the distribution effect; Martens K. [19] claims that accessibility is the most appropriate measure of benefits from transportation plans and investments.

Accessibility to opportunities is related to cumulative-opportunity and potential/gravity measures which sum the number of destinations/jobs reachable within certain times by transport mode; substantial literature discusses measure [26]. This measure is particularly useful in describing how well transportation networks perform about the distribution of destinations and the needs for subgroups. Measure accessibility to opportunities usually requires an attractiveness indicator, a transport network, and an impedance function of travel cost. The attractiveness indicator is expressed regarding the number of jobs or variables that represents opportunity size. The transport network is used to obtain travel costs, and travel cost can be travel time, distance, fares, or a combination of the three [27, 28]. There are many ways to get travel costs, such as simulation of the transport network in traffic software, calculation from simplified transport network in Geographic Information System (GIS), or calculation from a realistic transport network information database. The impedance function is an inverse function which indicates the increase in travel costs will reduce the opportunities of attraction.

2.2.2. The Categorizing People in Transportation Equity. Equity analysis needs to define a unit that can be distinguished, and units usually are groups of people/households or regions. Many studies use demographic and geographic factors categorizing people to identify transport disadvantaged people in equity evaluation [17], and these factors include income, car ownership, age, gender, career, household composition, and location of residence. Most papers distinguish between high income and low income, as well as car owners and car-less individuals or households. Indeed, disadvantaged status is multidimensional; some studies combine these factors to determine whether a person or area is a transport disadvantaged.

2.2.3. The Equity Measures in Transportation Equity. Well-known horizontal equity measures are the Gini coefficient, Theil index, and coefficient of variation; they are expressed as ratios which are compared among groups to measure equity performance. Gini coefficient initially indicates level of equality of income distributions in economic studies; it ranges from 0 to 1; 0 means absolute equality, and 1 indicates absolute inequality; in transportation equity it is used to evaluate the degree of accessibility concentration level of different regions or groups of people and compare the level of equity before and after implementing a policy or transport infrastructure. Some argue that the Gini coefficient fails to indicate the structure of inequality; the same Gini coefficients can have different income distributions by the group. Theil index is using the information entropy concept to measure individual or interregional income inequality named. The Theil index has good decomposability as a measure of inequality when the sample is divided into multiple groups. The Theil index can measure the contribution of the intragroup gap and the intergroup gap to the total gap, so it provides more interpretation of the inequality among different groups. The primary approach of vertical equity is to evaluate transport

policy or infrastructure projects according to how they affect accessibility between disadvantaged people/households or regions. It is fairer if transportation disadvantaged group benefits, like transport service improvements, favor lower-income areas, and groups, or transportation services provide more access to job opportunities and other "basic" activities.

The keys to equity analysis of public transportation are the measure of accessibility and the equity measure of distribution. Our work complements previous research from four aspects. First, when evaluating the equity impact of public transport, only one mode of public transport is concerned in previous studies, so the result of equity evaluation could be bias. We combined subways and buses to consider equity issues in this paper and proposed an enhanced potential opportunity model to measure residents' bus and subway accessibility considering public service reliability, attractiveness, and frequency. Second, due to the limitations of data acquisition, scholars discuss the impact on equity of transport policy or infrastructure projects during a relatively short period (before and after the implementation of the target). Our research used a long-term data to examine equity situation dynamic change, and it was helpful to capture the trends of equity influence on different groups in the development of public transport. Third, indicators of transport distribution effects were further explored and applied. We used the Dagum Gini coefficient decomposition which is more convenient than Theil measure and kernel density estimation method to investigate the fair distribution of potential opportunities. At last, existing research focused on Europe and the United States, and we took the China fastest urbanizing city, Shenzhen, as a case study to assess transport equity by comparing accessibility among social groups. The results can provide a reference for the study of the impact of transportation equity in the world.

3. Methodology

In this study, the public transport system was including bus and subway, and we divided the residents into two groups: urban village population and nonurban village population which will be explained in Section 4. We used an enhanced potential opportunity model to measure residents' bus and subway accessibility and summarized them as transit opportunity. Then, we used the Dagum Gini coefficient decomposition and kernel density estimation method to explore the fair distribution of transit opportunity among groups and districts from 2011 to 2020.

3.1. Measure of Accessibility. This research adopted cumulative-opportunity and potential/gravity measures models to measure transit-based job accessibility and made some enhancements. Given that accessibility measurement is especially important for the analysis of traffic equity, this section will detail how this research calculates accessibility.

Step 1. Calculate the service range of each transit stop. The service radius of the bus stop is 500 meters, and the service radius of the metro station is 700 meters.

Step 2. Calculate the population in each transit stop service area and calculate the job opportunity in each transit stop service area. In our study, job opportunity is represented by the floor area of factory, company, and government office.

Step 3. For transit line *l*, consider both *i* and *j* are transit stops of line *l* and calculate A_{ijl}; the job opportunity of *i* can be assessed at *j* as follows.

Calculate per capita service frequency:

$$S_{ijl} = \frac{V_{ijl}U}{P_i} \qquad (1)$$

V_{ijl} is average of vehicles (one day or a week) of transit line *l* from *i* to *j*, *U* is transit vehicle capacity, and P_i is the population of transit stop *i*. S_{ijl} reflects different transit service capabilities of bus and subway.

Calculate T_{ijl}, travel time from *i* to *j*.

$$T_{ijl} = T_{access} + T_{wait} + T_{in-vehicle} + T_{egress} \qquad (2)$$

The access and egress times are assumed to be 5 min of walking time, which transit users are generally willing to undertake, waiting time at transit stops is assumed to be one-half of the scheduled headway when the average headway of transit service is around 10 min, and the in-vehicle travel time is calculated using the scheduled arrival and departure time that is obtained from transit service schedules.

Calculate the distance decay factor:

$$f_{ijl} = \frac{1}{1 + \alpha e^{-\beta T_{ijl}}} \qquad (3)$$

α and β are two coefficients which need to be calibrated, and we used an average travel time survey. The survey data show that, in Shenzhen, 96.3% of the people make their work trips in less than 60 min. Therefore, we assumed that the connectivity from an origin to a destination that takes 60 min travel time would be 0.037. The estimates of the parameters are $\alpha = 0.0024321$ and $\beta = -0.143161$.

Calculate the job opportunity of *i* which can be assessed at *j*:

$$A_{ijl} = S_{ijl}O_jf_{ijl} \qquad (4)$$

where O_j is the opportunity at *j*.

Step 4. Consider the job opportunity of *i* can be assessed at *e* that requires a transfer between transit lines *l* and *m*:

$$A_{ie}^k = \frac{1}{2}\left(\frac{A_{ikl}}{f_{ikl}} + \frac{A_{kem}}{f_{kem}}\right)f_{ie}^k \qquad (5)$$

$$f_{ie}^k = \frac{1}{1 + \alpha e^{-\beta(T_{ikl}+20+T_{kem})}} \qquad (6)$$

where *i* is the transit stop of line *l*, *e* is the transit stop of line *m*, and *k* is the transfer center between *l* and *m*.

Step 5. Calculate the cumulative opportunities of i:

$$A_i = \sum_j A_{ijl} + \sum_k \sum_e A_{ie}^k \qquad (7)$$

Step 6. Convert residential area to the centroid and calculate the sum of job opportunities of bus stops in the 500-meter buffer of residential centroid and the number of metro job opportunities of metro stations in the 700-meter buffer of residential centroid (if there are more than 1 transit stops belonging to the same line in the buffer, they will be averaged as 1 transit stop). The sum of bus and metro opportunities is the transit opportunity for residential areas.

3.2. Decomposition of the Gini Coefficient. The method of decomposition of the Gini coefficient in a discrete space is proposed by Dagum [29]. The Dagum Gini coefficient is calculated using the following.

$$G = \frac{\sum_{j=1}^k \sum_{h=2}^k \sum_{i=1}^{nj} \sum_{r=1}^{nh} |y_{ji} - y_{hr}|}{2n^2\overline{y}} \qquad (8)$$

G is the Gini coefficient, $y_{ji}(y_{hr})$ is individual income of *i(r)* belong to subgroup *j(h)*, $\overline{y}$ is the average income of all population, *n* is the number of all population, *k* is the number of subgroups, and *nj(nh)* is the number of people in the *j*(h) subgroup. Dagum decomposes the Gini coefficient into three components [26]: (1) G_w, contribution of within groups income inequalities to *G*; (2) G_b, the net contribution of the between-group inequalities to *G* measured on all population; (3) G_t, the contribution of the transvariation between the subpopulations to *G*. Detailed derivation and calculation process for each component are listed in [29]; the equation of Gini coefficient decomposition in three components is shown as follows:

$$G = G_w + G_b + G_t \qquad (9)$$

Decomposition of the Gini coefficient not only effectively solves the source of group disparities but also describes the distribution of subgroups and solves the problem of overlap between groups (shows the structure of inequality). In this paper, the Gini coefficient of transit opportunity is calculated and decomposed; the population in Shenzhen is divided into different groups; it helps us to know if the transit opportunity gaps within groups generate the inequalities or if the transit opportunity gaps between groups engender the inequalities.

3.3. Kernel Density Estimation. Kernel density estimation (KDE) is a nonparametric way to estimate the probability density function of a random variable in statistics [30]. Let $(x_1, x_2, ..., x_n)$ be a univariate independent and identically distributed sample drawn from some distribution with an unknown density *f*. Its kernel density estimator is shown as follows:

$$\widehat{f_h}(x) = \frac{1}{nh}\sum_{i=1}^n K\left(\frac{x - x_i}{h}\right) \qquad (10)$$

where K is the kernel, a nonnegative function that integrates to one; the normal kernel is often used. $h > 0$ is a smoothing parameter called the bandwidth.

As mentioned in Section 2.2, Gini coefficient or Theil measure are ratios which are compared among groups to measure equity. KDE enables us to shape the distribution of a variable and analyze differences from the perspective of visual graphics comparison. So, we use KDE as a complementary way to draw probability distribution curves of transit opportunity of different population groups and grasp disparity and distribution of transit opportunity changes in decade-long urban developments in Shenzhen.

4. Study Area

4.1. The Social Group. Shenzhen is located in the Pearl River Delta region with a land area of 1996.8 km^2 and an urban population of over 14 million in 2016. It is the first Special Economic Zone (SEZ) city after the institution of reform and the Open-Door Policy in China in 1979. In the past 30 years, the operation of a market economy has made Shenzhen's economy develop rapidly, bringing with it a dramatic population increase and spatial expansion. In the study of transport equity, an important part is to group residents according to their socioeconomic level. In Shenzhen, detailed data on residents' occupations and income is not easily accessible, so we use three characteristics of a resident's residence to reflect his/her socioeconomic status. These three characteristics are average house price of residence, the average rental price of residence, and whether the residence is in the urban village; the third feature especially is an essential basis for judging a resident as disadvantaged people.

Urban village (Cheng Zhong Cun in Chinese), some scholars preferring the term "urbanized villages" or "villages in the city" to avoid the confusion with the Western planning concept "urban village", is an outcome of China's rapid urbanization and its associated rural-urban migration. When urban expansion encroaches into rural land; the city government needs to acquire land rights from the rural collective to convert the rural land into urban land. The city government only expropriates the farmland of the village to avoid the costly compensation to relocate villagers, and the housing land remains in the hands of the collective. Over time, the village settlement is surrounded by urban built-up area, creating the so-called urban village. **The bond between the urban village and the disadvantaged people is mainly caused by the residence registration (hukou) system and rural-urban migration tide in China**. The hukou system divides the population into the rural population and the urban population. Change from the rural to urban population needs to be approved by the authorities. Rural migrants can stay in large cities as temporary residents (do not have a local urban hukou), and they are excluded from the formal urban housing market. Commercial housing is generally expensive and thus not affordable to migrants who are employed in low-paid jobs. More affordable units provided by urban housing provision system generally require a local urban hukou and are thus not available to rural migrants. China's land policies have enabled the native farmers in the urban villages to construct inexpensive housing units and rent out these units to the rural migrants. In many cities, the urban village is a significant type of settlement for both local landless peasants and migrants, which are two groups with high urban poverty incidence.

The data supplied by the Shenzhen Urban Planning Bureau (SUPB) and the Urban Planning and Design Institute of Shenzhen (UPDIS) shows that there are 2,942 urban village residential lands and 4,683 nonurban village residential lands in Shenzhen 2009. The Municipal Building Survey 2009 provides information for all buildings in Shenzhen, including the urban villages. There are 615,702 buildings in 2009 and 333,576 (54%) in urban villages. Urban villages in Shenzhen, which are thought to accommodate approximately seven million, meet the basic needs of people, particularly poor and low-income residents [31, 32]. There are also some studies proving that immigrants in urban villages are vulnerable. Concerning economic sectors, the largest proportion of migrants is employed in retail, hotel, catering, and other services (50.8%). The second most common category is manufacturing (19.3%) and construction (9.2%). The proportion of people employed by highly paid public and finance sectors is tiny [32]. With the relatively poor employment profile, income among migrants is low in comparison with the city average. In 2004 the Municipal Government found that the average monthly income among migrants was only 1149 yuan, far below the average personal income in the city (2195 yuan) (Shenzhen Municipal Government Housing System Reform Office, 2004).

Our study first divided the population into two groups: residents living in urban villages called the urban village group (UVG) and residents not living in urban villages called the Not urban village group (NUV). The UVG is mainly composed of low-income migrants and contains some high-income local residents. A study shows that the ratio of local residents to migrants in the urban village is 1:88 [32]. Therefore, most of the residents in the urban village group are low-income people, and they are more dependent on transit to access jobs and social activity. The residents in NUV live in formal urban houses meaning that they have owned a house or can afford the higher rental price (the rent of a formal house is 2.5-5 times the rent in the urban village in Shenzhen), so they have higher disposable income than UVG. Second, according to housing prices and rents in different districts of Shenzhen, each group is further divided into subgroups. We collected more than 4,000 rental information and more than 3,000 residential house price information through the website (https://sz.fang.lianjia.com), which cover all areas in Shenzhen. Table 1 lists the detailed information; note that the price and rent in the table refer to the formal house, not the urban village. In each area, the rent price of the urban village is the lowest. Figure 1 is a map of Shenzhen and distributions of urban village lands and nonurban village lands in 2011.

Futian, Luohu, and Nanshan are the core areas of Shenzhen. Investments have been made in the service industry and high-technology companies in the three areas, so the

TABLE 1: Average house price and average rental price of ten districts in Shenzhen in 2017.

Spatial Location	District	Average house price in 2017 (yuan/m2)	Average rent in 2017 (yuan/m2)
Core area	Futian	52968	119.5
	Luohu	38143	94.7
	Nanshan	56597	121.7
Subcenter	Baoan	22580	76.9
	Longgang	27567	50.3
	Longhua	36432	61.2
	Yantian	33970	59.4
Suburb	Dapeng	13377	27.3
	Pingshan	12601	39.3
	Guangming	10278	24.6

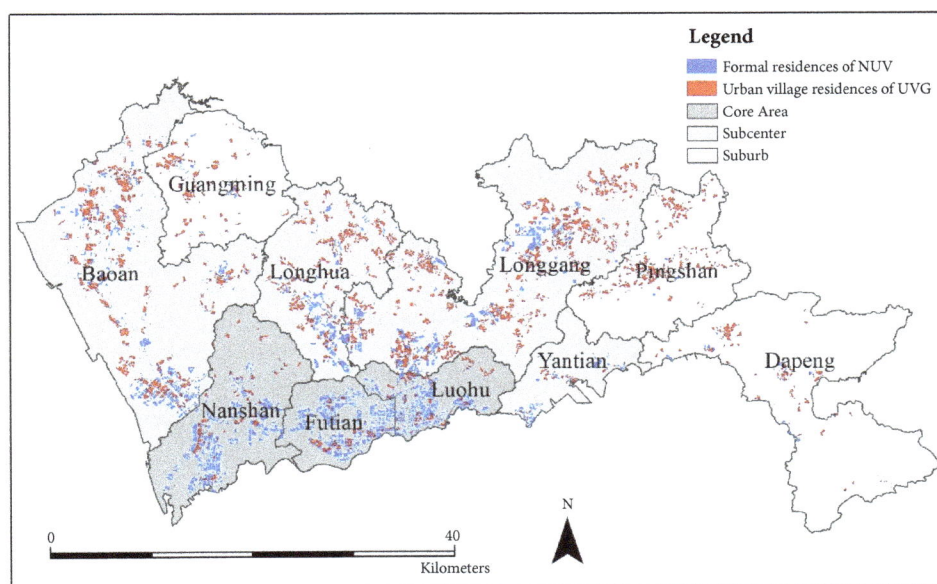

FIGURE 1: Distributions of urban village lands and nonurban village lands in 2011.

house price and rental price are the highest. Baoan, Longhua, Longgang, and Yantian are the subcenter of Shenzhen, and manufacturing industry provides many jobs in these three regions, so the house price and rental price are lower than in core areas. The remaining three regions are relatively far from the city center, and they have the lowest house price and rental price. For core areas, subcenters, and suburbs, there is a clear difference between house prices and rents, so NUV and UVG are each divided into three subgroups. The descriptions are shown in Table 2.

4.2. The Transit of Different Periods.
This study mainly analyzes the changes in public transport opportunities in the three periods from 2011 to 2020. 2011 is a base scenario, and 2020 is analyzed as a planning scenario. Calculating transportation opportunity requires transportation network and active location information. Public transport data includes

subway and bus network as shown in Figure 2. The data and operation information of subway in 2011 and 2016 come from the website (http://www.szmc.net) of Shenzhen Metro Group Co., Ltd., and the data and operation information of subway in 2020 is provided by SUPB. Due to the limitations of data acquisition, we use bus network and operation information of 2016 provided by Shenzhen Urban Transport Planning Center (SUTPC) in three measured years. We used the same level of bus service for the 10-year analysis period, and this operation would cause deviations in the calculation results of transit opportunities. This is one of the limitations of our study. The bus data contains 1,700 routes and more than 8,000 bus stops. Compared with the subway system, the changes in bus services are relatively small, and we assumed that using the same bus data had a smaller impact on the trends of the transit equity. As seen in Figure 2, subway lines are mainly concentrated in the core areas. The coverage of subway services in suburb areas will be improved

TABLE 2: Subgroups of NUV and UVG in Shenzhen.

Group	Subgroup	Description	Population of 2011	Percent
NUV	1	Residents who live in formal urban housing at core area (Futian, Luohu, and Nanshan)	4007619	27.8%
	2	Residents who live in formal urban housing at subcenter (Longhua, Yantian, Longgang, and Baoan)	3505463	23.69%
	3	Residents who live in formal urban housing at suburb (Dapeng, Pingshan, and Guangming)	162948	0.37%
		Total	7676030	51.86%
UVG	4	Residents who live in the urban village at core area (Futian, Luohu, and Nanshan)	1402041	9.48%
	5	Residents who live in the urban village of subcenter (Longhua, Yantian, Longgang, and Baoan)	4724083	31.93%
	6	Residents who live in the urban village at suburb (Dapeng, Pingshan, and Guangming)	995467	6.73%
		Total	7121591	48.14%

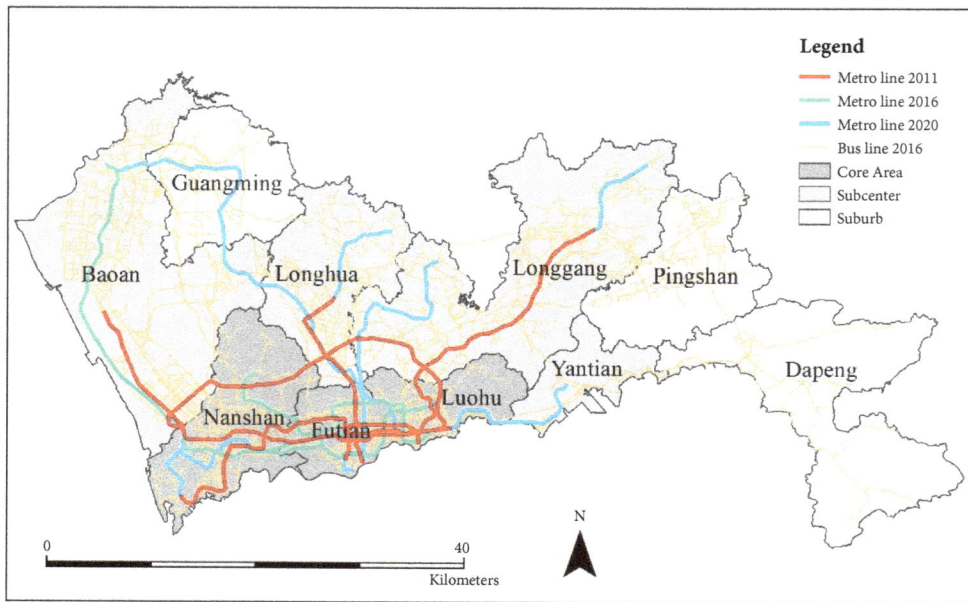

FIGURE 2: Maps of the transit network in three periods.

until 2020, so it is essential to analyze the impact of public transport networks on fairness. When calculating potential opportunity of residents, our definition of opportunity refers to jobs. UPDIS provides a detailed building data of Shenzhen, so the floor area of building in employment site is calculated to represent the opportunity.

5. Results and Discussions

5.1. The Average Transit Opportunity of Two Groups in Each District. The minimum unit for calculating transit opportunity is the residential land unit, and transit opportunity at the different aggregate levels are calculated by population weights. Figure 3 shows transit opportunity for the different years in Shenzhen at the community level. These maps

illustrate how varied are the distributions of transit opportunity among the region by public bus and metro.

We examined the average transit opportunity of two groups in each district. Table 3 shows the average transit opportunity of two groups at the region level. Shenzhen's core area has the most significant opportunity. Futian which is the city center has the highest transit opportunities. Luohu is the former city center in the 1990s and has the second highest accessibility; Nanshan is the critical development area in the future with the third highest transit opportunities. These three regions are spatially adjacent and possess the most public transport resources which have many subway lines and bus routes. Baoan, Longhua, and Longgang are located outside the core area, which is the subcenter of Shenzhen. The transit opportunity in these areas is about one-third

TABLE 3: Transit opportunity of two groups in each region.

Year	2011		2016		2020	
	NUV	UVG	NUV	UVG	NUV	UVG
Futian	12381	13156	19592	21091	23095	23848
Luohu	12477	13041	16143	15106	18823	17035
Nanshan	5713	5886	8056	7372	10231	9035
Baoan	3367	2012	3821	2185	4197	2323
Longgang	4602	3337	4919	3543	6340	4820
Longhua	4499	2736	4805	2848	6170	4572
Yantian	1630	1285	2030	1411	4931	2858
Dapeng	274	373	431	420	431	420
Pingshan	1163	1107	1231	1134	1231	1134
Guangming	1862	909	1881	927	2835	1297

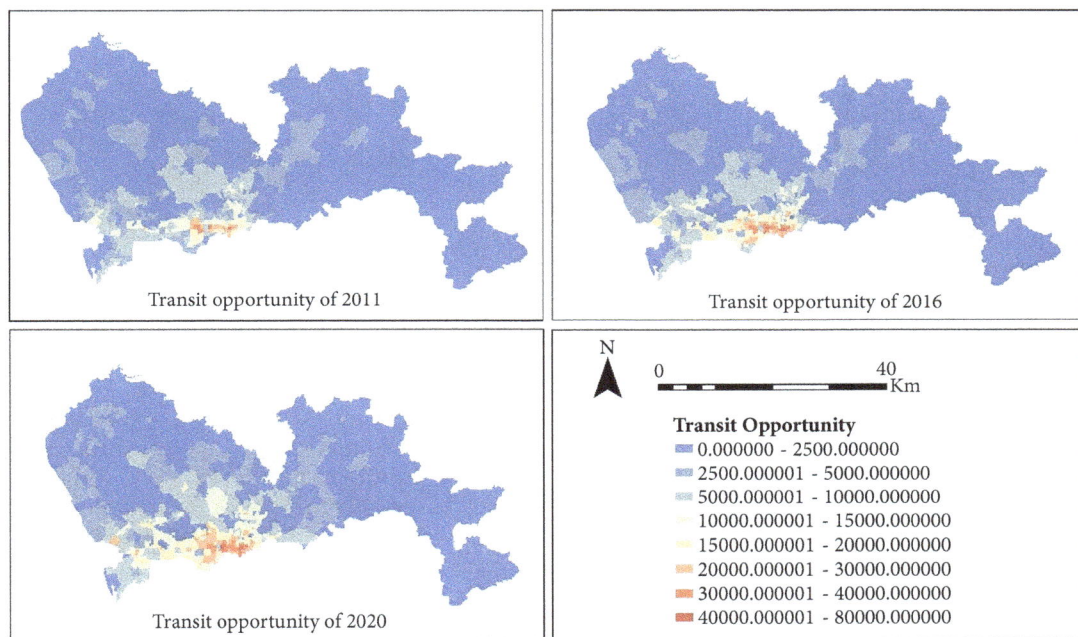

FIGURE 3: Maps of transit opportunity in three periods.

of the core area. Yantian District is a natural scenic tourist area and port area. Although it is located in the subcentral area, its public transportation system is not well-developed, so the transit opportunities are lower than other subcenters. The rest regions are in the outer suburbs of Shenzhen, which are far from the core area with a small population and minimal opportunity. The opportunity of all regions is growing over time for both groups. Transit opportunity of core area (especially in Futian) proliferated more than 25% from 2011 to 2016. Subcenters and suburbs had a smaller growth lower than 10%. Meanwhile, the differences in the absolute value of transit opportunity between the core area, subcenter area, and the suburbs were unusually large and will be more significant in 2020. It indicates that the residents in the city center have the most significant benefit from public transport system.

Table 4 shows the average transit opportunity in the whole city. In each measure year, transit opportunity of NUV is greater than UVG. The transit opportunity of UVG will increase 45% from 2011 to 2020 and NUV will increase 63% at the same time. The difference in growth rate is 22%. When comparing the absolute value of transit opportunity, the difference between the two groups has almost doubled from 2011 to 2020; in 2011 NUV's transit opportunity was 3279 more than UVG's, and the gap will be 6102 in 2020. From the perspective of the city, NUV has more public transport advantages than UVG. For subgroups, it is clear that the core area has the highest transit opportunities, whether it is UVG or NUV. Subgroup 4 (residents who live in the urban village at core area) even has higher opportunities than subgroup 2 (residents who live in formal urban housing at subcenter).

TABLE 4: The average transit opportunity of two groups in the whole city.

Year		2011	2016	2020
Group	NUV	7307	9959	11972
Subgroup	1	10597	15367	18249
	2	3819	4172	5255
	3	1457	1530	2090
Group	UVG	4028	4846	5870
Subgroup	4	10825	14397	16480
	5	2662	2835	3726
	6	943	987	1102

TABLE 5: Decomposition of the Gini coefficient between UVG and NUV.

Year		2011	2016	2020
Total Gini	G	0.5725	0.5916	0.5736
Gini within	G_{nuv}	0.5049	0.5160	0.4920
	G_{uvg}	0.6267	0.6480	0.6385
Gini between	G_{uvg} VS G_{nuv}	0.6026	0.6303	0.6140
Contribution	G_w	48.14%	47.63%	47.38%
	G_b	25.04%	28.84%	29.45%
	G_t	26.81%	23.53%	23.17%

G_w: contribution of the Gini inequality indexes within subpopulations to the total Gini ratio.
G_b: contribution of the Gini inequality ratios between subpopulations to the total Gini ratio.
G_t: contribution of the transit opportunity intensity of transvariation between subpopulations to the total Gini ratio.

5.2. Changes of Transit Opportunity Distribution

5.2.1. Decomposition of Transit Opportunity Gini Coefficient between NUV and UVG. To analyze the horizontal equity of public transport, we calculated and decomposed the Gini coefficient using the transit opportunity of 7625 residential units; the total populations of Shenzhen was divided into the following subgroups: UVG and NUV. Table 5 presents the decomposition of the Gini coefficients estimated for the two groups. The total Gini G reflects the equity situation of transit opportunity distribution in all populations of Shenzhen. From 2011 to 2020 its value changed from 0.5725 to 0.5736. Although the changes are not significant, we still can see the trends of transit opportunity equity. Compared with the public transport network in 2011, there were three new subway lines added to the public transport network in 2016 (in Figure 2, green lines). These lines are mainly located in the core area, and it caused an increase of inequity for all populations. In 2020, the subway network will expand from the core area to the subcenter area (in Figure 2, blue lines). The total Gini coefficient will be reduced; that is, the distribution of transit opportunity in 2020 will be more even than in 2016. For horizontal equity of the total population, the gap in transit opportunity between the overall population increases first and then decreases.

As for within group inequity, G_{uvg} is the equity index of transit opportunity distribution for UVG population and G_{nuv} is the equity index for NUV population. The result shows that in each measure year G_{nuv} is less than G_{uvg}, so the transit opportunity distribution of NUV is more equitable than UVG, and the gap of Gini coefficients between UVG

and NUV is significant from 0.1218 in 2011 to 0.1465 in 2020. The impact of changes in the public transport system on transportation equity is consistent for both NUV and UVG. In 2016 the Gini coefficient increased in both groups, and the Gini coefficient of both groups will decrease by 2020. However, in 2002 the NUV's distribution of opportunity is more equitable than the distribution in 2011, UVGs will not return to the level in 2011, which means that G_{nvg} (2020) > G_{nvg} (2011).

The between-group inequity (G_{uvg} VS G_{nuv}) results show that the public transport network in 2016 resulted in the most considerable inequality between the two groups. With the expansion of the subway network from the core area to other areas in 2020, the inequality between the two groups will decrease. The analysis of inequality contribution shows that the within groups inequality (G_w) has contributed the most to the total inequity in three periods. The contribution of inequality between groups (G_b) is growing, so the development of public transport has led to growing inequality between the two groups.

5.2.2. Decomposition of Transit Opportunity Gini Coefficient between Subgroups. We know that within groups inequality has the most significant impact on overall inequality in Table 5, so we decomposed the Gini coefficient for each group and explored about the influence of spatial location of residence on equity. Table 6 presents the decomposition of transit opportunity Gini coefficient G_{nuv} between subgroups of NUV. G_i means the equity index of transit opportunity distribution of subgroup i. The result shows that most equitable

TABLE 6: Decomposition of the Gini coefficient of NUV between subgroups.

Year		2011	2016	2020
G_{nuv}		0.5049	0.5159	0.4920
Gini within NUV	G_1	0.4060	0.3685	0.3398
	G_2	0.5045	0.5192	0.5131
	G_3	0.5097	0.5101	0.5458
Contribution	$G_{w(nuv)}$	42.6%	38.75%	38.19%
	$G_{b(nuv)}$	47.32%	55.68%	56.36%
	$G_{t(nuv)}$	10.8%	5.57%	5.45%

TABLE 7: Decomposition of the Gini coefficient of UVG between subgroups.

Year		2011	2016	2020
G_{uvg}		0.6266	0.6480	0.6385
Gini within UVG	G_4	0.4333	0.3748	0.3567
	G_5	0.5441	0.5525	0.5760
	G_6	0.5424	0.5463	0.5572
Contribution	$G_{w(uvg)}$	32.76%	28.85%	31.53%
	$G_{b(uvg)}$	59.51%	65.66%	62.46&
	$G_{t(uvg)}$	7.73%	5.49%	6.01%

TABLE 8: The Gini coefficients and transit opportunities of all subgroups.

| | | | | Year | | | | |
| | 2011 | | | 2016 | | | 2020 | |
subgroup	G_i	ATO	subgroup	G_i	ATO	subgroup	G_i	ATO
1	0.406	10597	1	0.3685	15367	1	0.3398	18249
4	0.4333	10825	4	0.3748	14397	4	0.3567	16480
2	0.5045	3819	3	0.5101	1530	2	0.5131	5255
3	0.5097	1457	2	0.5192	4172	3	0.5458	2090
6	0.5424	943	6	0.5463	987	6	0.5572	1102
5	0.5441	2662	5	0.5525	2835	5	0.576	3726

G_i: transit opportunity Gini coefficient of subgroup i.
ATO: average transit opportunity.

distribution of transit opportunity in NUV is subgroup 1 in all three years. Compared with G_2 and G_3, G_1 is the smallest and is consistently decreasing in three periods, and its decrease is also the largest. G_2 and G_3 are almost equal in 2011 and 2016, but the value of G_3 will be larger in 2020. The analysis of inequality contribution of G_{nuv} shows that the between groups inequality ($G_{b(nuv)}$) has contributed the most to the total inequity in three periods of NUV.

Table 7 presents the decomposition of transit opportunity Gini coefficient G_{uvg} between subgroups of UVG. For UVG, Table 7 shows that G_4 is significantly different from G_5 and G_6, and the most equitable distribution of transit opportunity is subgroup 4. From the changes in the Gini coefficient, the inequality of all UVG population has increased, and the opportunity distribution gap of UVG population in core area tends to be smaller. Subgroup 5 is the most unfair distribution of opportunities in UVG. G_6 is slightly smaller than G_5. The construction of subway infrastructure from 2011 to 2020 will have a negative impact on the distribution of transit opportunity in subcenters and suburbs. From the

analysis of the contribution to overall inequality of UVG, the inequality among the groups $G_{b(uvg)}$ contributes the most, accounting for about 60%. This shows that the unfairness between regions mainly causes the overall inequality of UVG.

Table 8 summarizes the Gini coefficients and transit opportunities of all subgroups. The order of the subgroups in the table is arranged according to the value of the Gini coefficient. For each region (core area, subcenters, and suburb), the Gini coefficient of NUV is always smaller than UVG in all three years, and transit opportunity of NUV is larger than of UVG. Residents of NUV in the core areas have the highest opportunities and the most equitable distribution. Residents of UVG in the core areas have the second highest opportunities and equitable distribution. For subcenters and suburbs, the opportunity gap between the two regions is significant, but the difference in equity distribution is not apparent. With the development of urban public transportation systems, the absolute value of residents' transit opportunities is improved, but it may lead to unfair

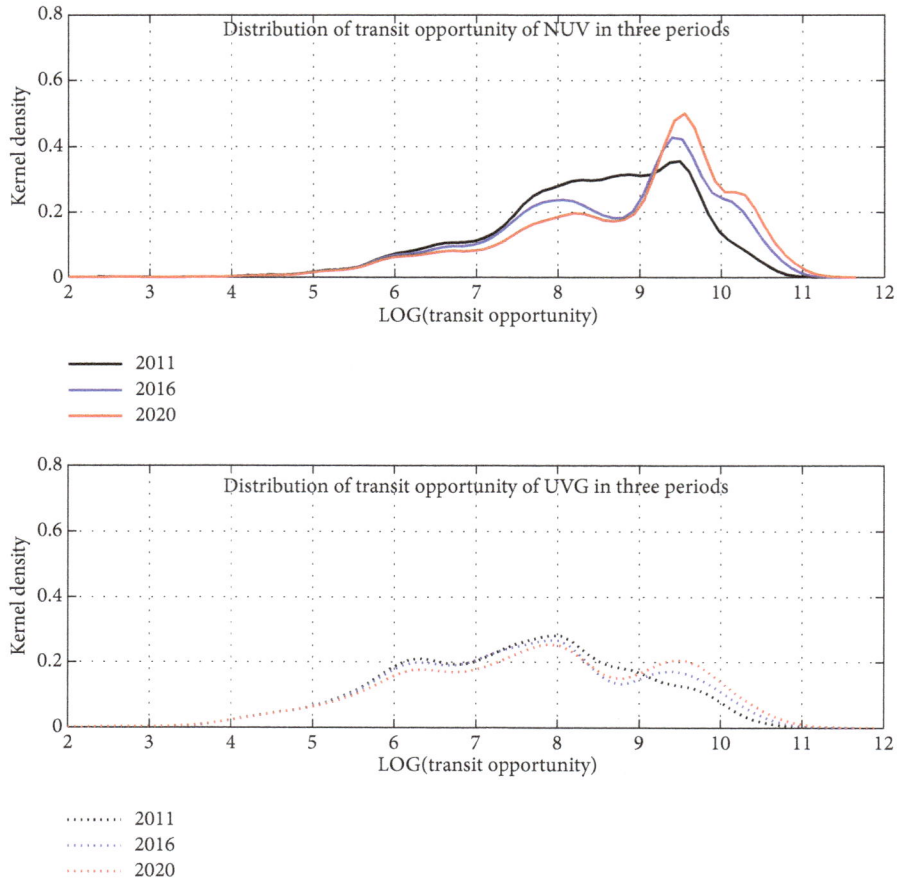

FIGURE 4: Distribution of transit opportunity probability of NUV and UVG population.

distribution. From 2011 to 2020, transit opportunities of subgroup 5 and subgroup 6 will rise, but equity performance will fall. Spatial location and low-income both have impact on equity, but spatial location plays a significant role in the difference and opportunities distributions between groups especially for the UVG.

5.3. Changes of Transit Opportunity Probability Distribution.
Based on the transit opportunity of 7825 settlements, we used kernel density estimation to get the probability of transit opportunity and plotted the probability density distribution map.

5.3.1. The Comparisons of Probability Density Maps between UVG and NUV.
Figure 4 shows the distribution of transit opportunity probability of NUV and UVG. The horizontal axis is the value of transit opportunity. Since the range of opportunity in different residential areas is too large, a LOG conversion is performed. The vertical axis is the probability of transit opportunity.

The top of Figure 4 is transit opportunity probability of NUV. From 2011 to 2020, there is a slight right movement of the curve which means the value of the opportunity will increase. The density curve in 2016 is quite different

from in 2011, and it shows that the subway line in 2016 has a significant influence on the opportunity distribution of NUV population. In 2011, the probability of opportunity value "8" and opportunity value "9" was not much different. Considering vertical equity, an ideal situation is that the public transport system prioritizes raising the odds of lower opportunity population and reduces the probability of lower opportunity. The person with the opportunity "8" should obtain the priority of improvement. In fact, the probability of "9" was reduced more than the probability of "8" in 2016 which means that the public transport system is more inclined to improve the population with high initial opportunity and make them higher. Therefore, the distribution of unfairness increased. The probability density curve shape in 2016 is very similar to that in 2020, and it indicates that the subway line in 2020 will have a little influence on the opportunity distribution of NUV population. The tendency of concentration will be more obvious and further reduce the probability of around "8" in 2020, and distribution of unfairness will decrease. The bottom of Figure 4 shows the distribution of transit opportunity of UVG population. There is no significant change in the density curve shape compared with NUV, and it means that from 2011 to 2020 the improvement of the public transportation system has a smaller impact than NUV. It has the same phenomenon with

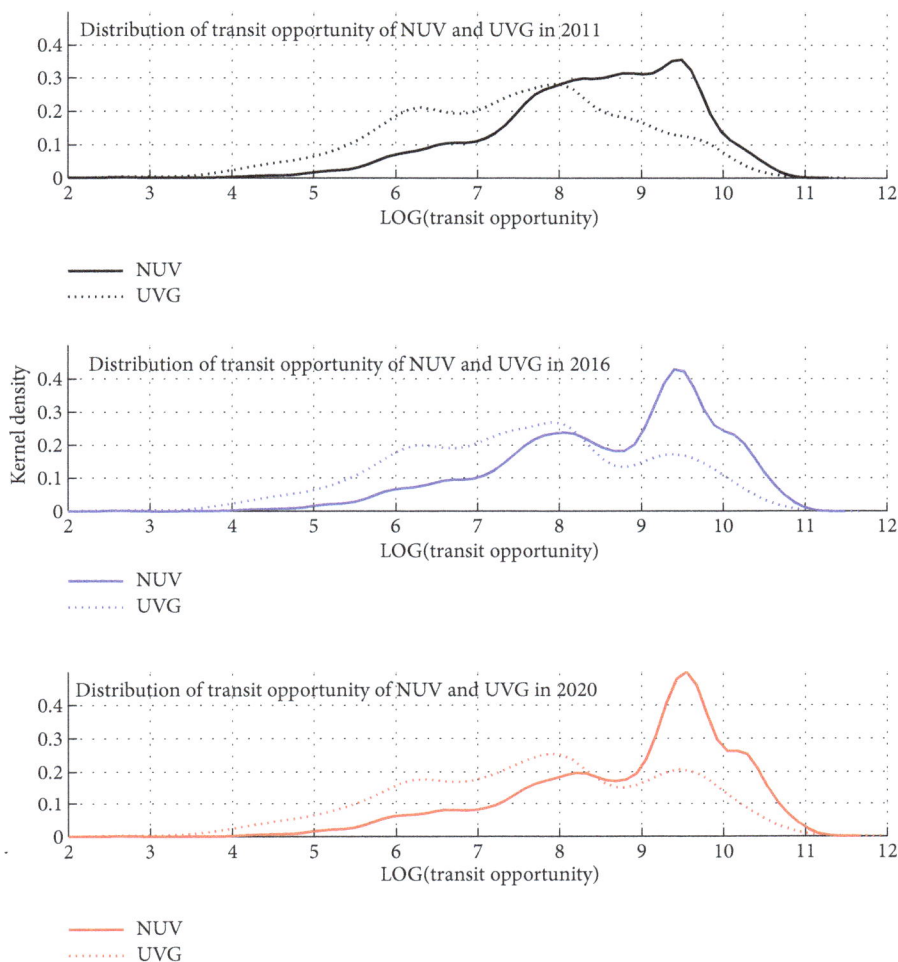

Figure 5: Comparisons of the probability distribution of the two groups in the whole city.

NUV which has a priority to improve the high opportunity population. Besides, the curves have three peaks in 2016 and in 2020, which shows that multipolarity is more severe and noticeable.

Figure 5 shows the comparisons of the probability distribution of the two groups at the whole city level. We can observe that the opportunity value "8" is a dividing point in 2011 and 2016. For transit opportunity which value is less than 8, the probability of UVG is higher than NUV. This means that the UVG population is more likely to have low public transport opportunities. For transit opportunity which value is larger than 8, the probability of UVG is less than NUV, so the NUV population is more likely to have high public transport opportunities. In 2020, the dividing point of the two group is 8.5, and this means that the opportunities of both groups will be improved. The comparison of the two groups in the three periods illustrates Shenzhen public transportation system is very favorable to NUV in all three periods. The probability gap of high transit opportunity between the two groups is increasing in different periods; this also shows that NUV benefits more than UVG from the improvement of the urban transport system.

5.3.2. The Comparisons of Probability Density Maps between Subgroups. Figure 6 is the comparisons of the probability distribution of the two groups in core areas. The distributions of the two groups are very similar. The width of the wave in the distribution curve of the two groups becomes narrower over time indicating that the difference of transit opportunity between population is smaller. Figure 7 is comparisons of the probability distribution of the two groups in subcenter. Compared with Figure 6, the difference in probability distribution curves between the two groups is noticeable, and the width of the wave in the distribution curve is wider than the core areas. From 2011 to 2020, the probability gap of high transit opportunity between UVG and NUV is getting larger. The distributions of transit opportunities between the two groups of suburbs have the same characteristics as that of the subcenter in Figure 8, including broader wave and significant distribution differences between UVG and NUV. The three figures show that there is a difference in the distribution of opportunities between different regions, and there are also differences between the two groups in the same region. Spatial location and type of group have an impact on access to transit opportunities.

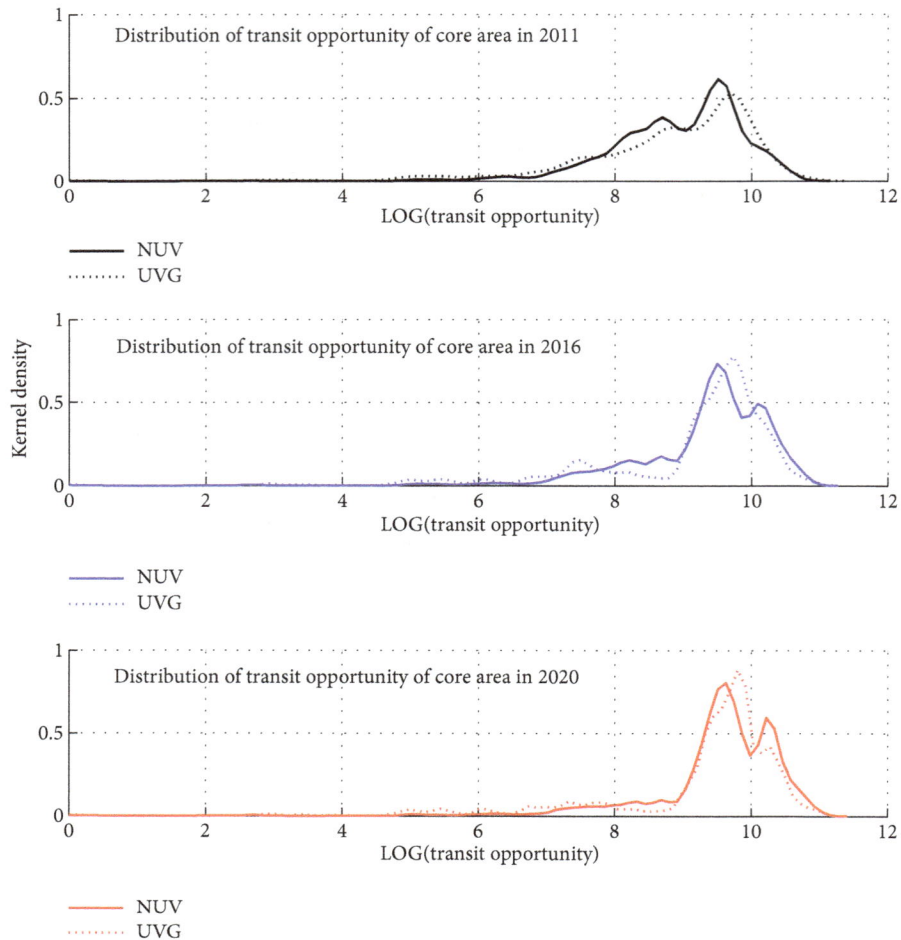

FIGURE 6: Comparisons of the probability distribution of the two groups in core areas.

6. Conclusions

Decade-long changes in disparity and distribution of transit opportunity gave us a clear picture, and the results illustrated the importance of examining transportation equity over a long period. (1) For the absolute value of the opportunity, Shenzhen's core area has the most significant opportunity, and urban village populations do have fewer transportation opportunities than nonurban villages. People in the different regions and groups inevitably do not have equal transit opportunity, and this is not necessarily problematic. Transport policy is fair if it distributes transport investments and services in ways that reduce inequality of opportunity. With the development of Shenzhen public transport infrastructures, although all populations are benefiting from increasing transit opportunity, transit opportunity of NUV is greater than UVG. Their gap is widening in each measure year, and it is necessary to limit the highest levels of accessibility of social groups when a marginal improvement of accessibility at the upper levels would harm those groups at the bottom. (2) For the vertical equity, the public transport system is more inclined to improve the population with high initial opportunity and make them higher. The

improvement of the public transportation system has a smaller impact than NUV in all three periods, and policies should prioritize disadvantaged groups. (3) From the horizontal equity of transit opportunity distribution, for all population, the development of public transport in different periods in Shenzhen first exacerbated unfair distribution of transit opportunity and then increased fairness. The development of transit also has a different impact on the different group of people. In each measure year, the distribution of NUV is more equitable than UVG, and the gap of Gini coefficient between the two groups is significant and is widening. (4) People's social status and spatial location are both factors that contribute to the total inequity. The spatial location has more impact on equity, and low-income people do not necessarily have traffic disadvantages. Low-income people living in the core area of Shenzhen such as Futian and Luohu have high transit opportunity and a more equitable distribution.

Our research is beneficial for providing information to adjust the planning of future Metro routes and urban development strategies in Shenzhen. Since social status and spatial location are factors of impact equity, Shenzhen should

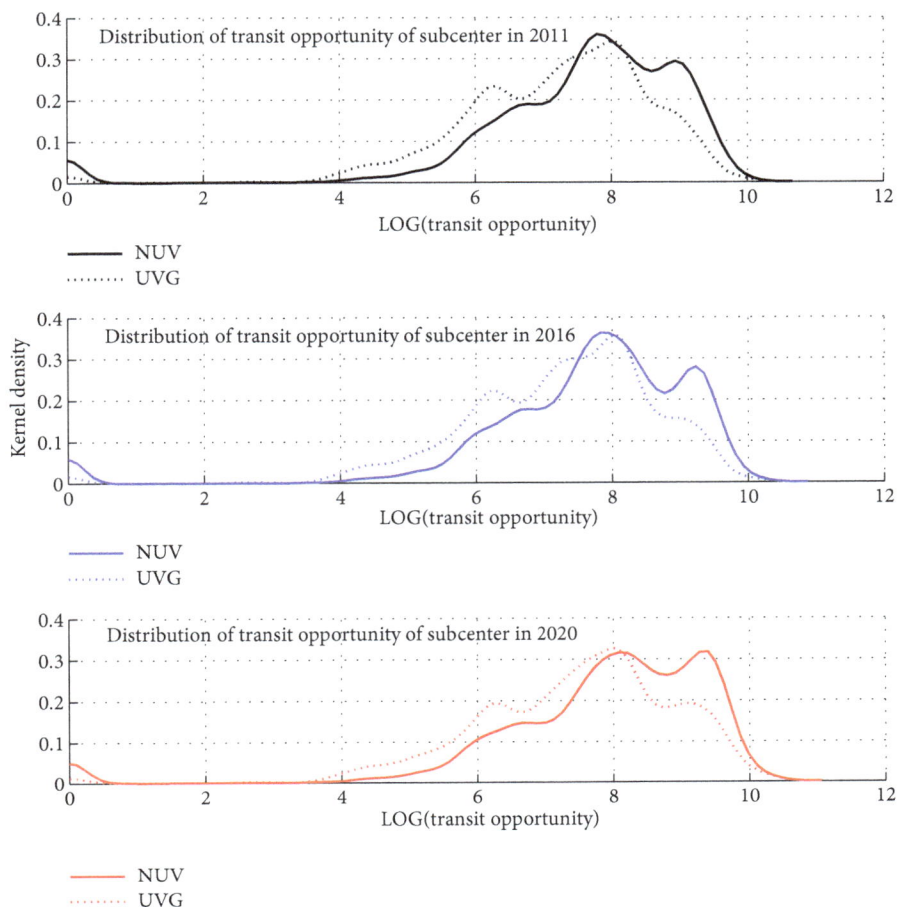

FIGURE 7: Comparisons of the probability distribution of the two groups in subcenter.

strengthen public transport services in subcentral areas, especially Baoan and Longgang. The two regions have a large population for both NUV and UVG, and improvement of public transport in these areas will be the most effective way for improving public transport accessibility and fair distribution. Our study also has some limitations. First, we used the same bus network for the 10-year analysis period, and this operation would cause deviations in the calculation results of transit opportunities. So, the analysis of equity ignored the changes in fairness brought about by the improvement of bus service level. Second, transportation equity not only is an infrastructure issue but also involves land use planning as well. The transit opportunity of a resident is related to the public transport system and to the distribution and size of the opportunities. The distribution of opportunities is related to urban planning, especially land use. This study does not explore the impact of land use on transportation equity. Third, it is not only the distribution of access to destinations that matter, but also in some cases the absolute level of access for those who are worse. We only discussed the changes of the residents' opportunities and the changes of disparity and distribution of transit opportunity, and we do not discuss whether transit opportunities meet the needs of different groups of people and their satisfaction of transportation opportunities. In the future research, those are our next research direction.

Conflicts of Interest

The authors declarethat they have no conflicts of interest.

Acknowledgments

The authors gratefully acknowledge financial support from the China Scholarship Council. The authors gratefully acknowledge support from Program on Chinese Cities of Center for Urban and regional Studies, University of North Carolina at Chapel Hill.

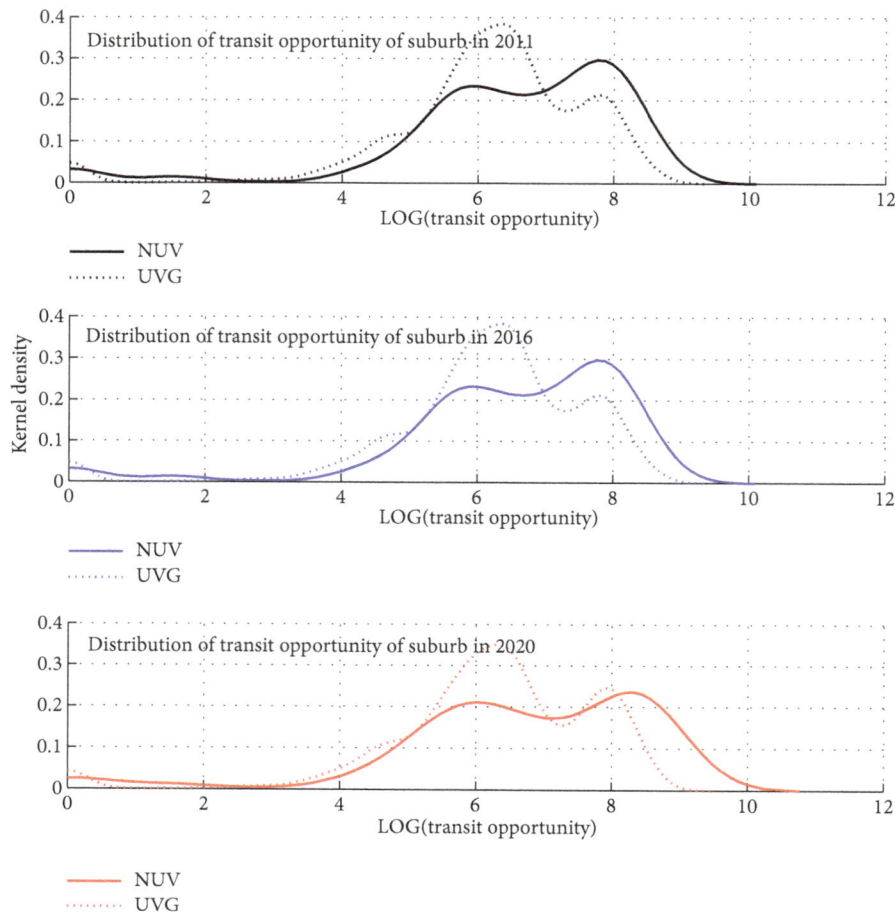

FIGURE 8: Comparisons of the probability distribution of the two groups in the suburb.

References

[1] J. Jain, T. Line, and G. Lyons, "A troublesome transport challenge? Working round the school run," *Journal of Transport Geography*, vol. 19, no. 6, pp. 1608–1615, 2011.

[2] G. Currie and A. Delbosc, "Modelling the social and psychological impacts of transport disadvantage," *Transportation*, vol. 37, no. 6, pp. 953–966, 2010.

[3] A. Lubin and D. Deka, "Role of public transportation as job access mode: lessons fromfbfbsurvey of people with disabilities in New Jersey," *Transportation Research Record*, vol. 2277, pp. 90–97, 2012.

[4] A. Matas, J.-L. Raymond, and J.-L. Roig, "Job accessibility and female employment probability: The cases of Barcelona and Madrid," *Urban Studies*, vol. 47, no. 4, pp. 769–787, 2010.

[5] À. Cebollada, "Mobility and labour market exclusion in the Barcelona Metropolitan Region," *Journal of Transport Geography*, vol. 17, no. 3, pp. 226–233, 2009.

[6] K. Martens, "Substance precedes methodology: On cost-benefit analysis and equity," *Transportation*, vol. 38, no. 6, pp. 959–974, 2011.

[7] K. Martens, A. Golub, and G. Robinson, "A justice-theoretic approach to the distribution of transportation benefits: Implications for transportation planning practice in the United States," *Transportation Research Part A: Policy and Practice*, vol. 46, no.

4, pp. 684–695, 2012.

[8] N. Thomopoulos and S. Grant-Muller, "Incorporating equity as part of the wider impacts in transport infrastructure assessment: An application of the SUMINI approach," *Transportation*, vol. 40, no. 2, pp. 315–345, 2013.

[9] G. Currie, "Quantifying spatial gaps in public transport supply based on social needs," *Journal of Transport Geography*, vol. 18, no. 1, pp. 31–41, 2010.

[10] B. Epstein and M. Givoni, "Analyzing the gap between the QOS demanded by PT users and QOS supplied by service operators," *Transportation Research Part A: Policy and Practice*, vol. 94, pp. 622–637, 2016.

[11] C. Jaramillo, C. Lizárraga, and A. L. Grindlay, "Spatial disparity in transport social needs and public transport provision in Santiago de Cali (Colombia)," *Journal of Transport Geography*, vol. 24, pp. 340–357, 2012.

[12] E. M. Ferguson, J. Duthie, A. Unnikrishnan, and S. T. Waller, "Incorporating equity into the transit frequency-setting problem," *Transportation Research Part A: Policy and Practice*, vol. 46, no. 1, pp. 190–199, 2012.

[13] A. Monzón, E. Ortega, and E. López, "Efficiency and spatial equity impacts of high-speed rail extensions in urban areas," *Cities*, vol. 30, no. 1, pp. 18–30, 2013.

[14] J. P. Bocarejo S. and D. R. Oviedo H., "Transport accessibility

and social inequities: a tool for identification of mobility needs and evaluation of transport investments," *Journal of Transport Geography*, vol. 24, pp. 142–154, 2012.

[15] B. van Wee and K. Geurs, "Discussing equity and social exclusion in accessibility evaluations," *European Journal of Transport and Infrastructure Research*, vol. 11, no. 4, pp. 350–367, 2011.

[16] D. Boucher and P. J. Kelly, *Social justice: from Hume to Walzer*, Psychology Press, 1998.

[17] T. Litman, "Evaluating transportation equity," *World Transport Policy & Practice*, vol. 8, no. 2, pp. 50–65, 2002.

[18] F. Di Ciommo and Y. Shiftan, "Transport equity analysis," *Transport Reviews*, vol. 37, no. 2, pp. 139–151, 2017.

[19] C. Falavigna and D. Hernandez, "Assessing inequalities on public transport affordability in two latin American cities: Montevideo (Uruguay) and Córdoba (Argentina)," *Transport Policy*, vol. 45, pp. 145–155, 2016.

[20] A. T. Murray, "Strategic analysis of public transport coverage," *Socio-Economic Planning Sciences*, vol. 35, no. 3, pp. 175–188, 2001.

[21] A. Delbosc and G. Currie, "Using Lorenz curves to assess public transport equity," *Journal of Transport Geography*, vol. 19, no. 6, pp. 1252–1259, 2011.

[22] K. Fransen, T. Neutens, S. Farber, P. De Maeyer, G. Deruyter, and F. Witlox, "Identifying public transport gaps using time-dependent accessibility levels," *Journal of Transport Geography*, vol. 48, pp. 176–187, 2015.

[23] T. Saghapour, S. Moridpour, and R. G. Thompson, "Public transport accessibility in metropolitan areas: A new approach incorporating population density," *Journal of Transport Geography*, vol. 54, pp. 273–285, 2016.

[24] E. C. Delmelle and I. Casas, "Evaluating the spatial equity of bus rapid transit-based accessibility patterns in a developing country: The case of Cali, Colombia," *Transport Policy*, vol. 20, pp. 36–46, 2012.

[25] J. Grengs, "Nonwork Accessibility as a Social Equity Indicator," *International Journal of Sustainable Transportation*, vol. 9, no. 1, pp. 1–14, 2015.

[26] A. Páez, D. M. Scott, and C. Morency, "Measuring accessibility: Positive and normative implementations of various accessibility indicators," *Journal of Transport Geography*, vol. 25, pp. 141–153, 2012.

[27] A. El-Geneidy, D. Levinson, E. Diab, G. Boisjoly, D. Verbich, and C. Loong, "The cost of equity: Assessing transit accessibility and social disparity using total travel cost," *Transportation Research Part A: Policy and Practice*, vol. 91, pp. 302–316, 2016.

[28] M. Niehaus, P. Galilea, and R. Hurtubia, "Accessibility and equity: An approach for wider transport project assessment in Chile," *Research in Transportation Economics*, vol. 59, pp. 412–422, 2016.

[29] C. Dagum, "A new approach to the decomposition of the Gini income inequality ratio," in *Income Inequality, Poverty, and Economic Welfare*, pp. 47–63, Physica-Verlag HD, 1998.

[30] M. Rosenblatt, "Remarks on some nonparametric estimates of a density function," *Annals of Mathematical Statistics*, vol. 27, pp. 832–837, 1956.

[31] J. Zacharias and Y. Tang, "Restructuring and repositioning Shenzhen, China's new mega city," *Progress in Planning*, vol. 73, no. 4, pp. 209–249, 2010.

[32] Y. P. Wang, Y. Wang, and J. Wu, "Housing migrant workers in rapidly urbanizing regions: A study of the Chinese model in Shenzhen," *Housing Studies*, vol. 25, no. 1, pp. 83–100, 2010.

5

Individual Travel Behavior Modeling of Public Transport Passenger Based on Graph Construction

Quan Liang,[1] **Jiancheng Weng**(ID)**,**[1] **Wei Zhou,**[2] **Selene Baez Santamaria,**[3] **Jianming Ma,**[4] **and Jian Rong**[1]

[1]*Beijing Key Laboratory of Traffic Engineering, Beijing University of Technology, Beijing 100124, China*
[2]*Ministry of Transport of the People's Republic of China, Beijing 100736, China*
[3]*Computer Science Department, Vrije Universiteit Amsterdam, 1081 HV Amsterdam, Netherlands*
[4]*Texas Department of Transportation, Austin, TX 78717, USA*

Correspondence should be addressed to Jiancheng Weng; youthweng@bjut.edu.cn

Academic Editor: Luca D'Acierno

This paper presents a novel method for mining the individual travel behavior regularity of different public transport passengers through constructing travel behavior graph based model. The individual travel behavior graph is developed to represent spatial positions, time distributions, and travel routes and further forecasts the public transport passenger's behavior choice. The proposed travel behavior graph is composed of macronodes, arcs, and transfer probability. Each macronode corresponds to a travel association map and represents a travel behavior. A travel association map also contains its own nodes. The nodes of a travel association map are created when the processed travel chain data shows significant change. Thus, each node of three layers represents a significant change of spatial travel positions, travel time, and routes, respectively. Since a travel association map represents a travel behavior, the graph can be considered a sequence of travel behaviors. Through integrating travel association map and calculating the probabilities of the arcs, it is possible to construct a unique travel behavior graph for each passenger. The data used in this study are multimode data matched by certain rules based on the data of public transport smart card transactions and network features. The case study results show that graph based method to model the individual travel behavior of public transport passengers is effective and feasible. Travel behavior graphs support customized public transport travel characteristics analysis and demand prediction.

1. Introduction

Public transport has been one of the main travel modes in urban areas due to its comprehensive service for travelers and big influence on urban traffic systems. Taking Beijing as example, there are about 11.29 million public transport commute trips per day [1]. The average trip time and distance are 54 minutes and 19.4 kilometers, respectively [2]. There exist also many different kinds of public transport modes, such as metro, traditional bus, customized bus, and microcirculation shuttle bus. In order to better meet travel demands and further improve public transport service level, it is essential to obtain the demand of transport scheduling accurately and hierarchically. To achieve this, it is first necessary to extract the personal travel characteristics with respect to temporal and spatial distribution and their travel behavior choice.

In previous studies, many researchers have attempted to analyze public transport travel characteristics and to explore extraction methods for travel behavior regularity. Generally, these previous studies mainly focused on statistical analysis based on conventional trip survey of relatively small sample sizes or depended on limited public transport data [3–7]. On one hand, common feature parameters of public transport travel were extracted. Examples include the following: Walle and Steenberghen [4] analyzed the influence of temporal and spatial factors on public transport travel mode selection. Zhou et al. [5] presented comparative analysis of resident travel characteristics between typical cities in developed countries and in China, according to individual trip amount, purpose, and mode. On the other hand, travel behavior models of public transport were established to further analyze passengers' travel behavior choice. Structural equation model was usually used to reflect public transport selection behavior based on trip chain extraction [6]. A travel mode selection model of trip chain was also developed and verified with high accuracy by taking travel time and cost into account [7].

As limited to data collection approaches, travel behavior characteristic parameters and travel behavior choice of public transport travelers were partially analyzed in previous studies. However, limited sample data does not sufficiently reflect the accurate travel behavior features in a public transport system. It is necessary to study the travel behavior features of public transport in larger scales. Fortunately, emerging technologies such as cloud computing and Internet+ have promoted the development of advanced public transport system (APTS). Combined with network communication, geographic information systems (GIS), global position systems (GPS), and electronic controls, APTS lays a good foundation for multimode data collection.

In fact, many researchers have attempted to apply public transport smart card data and GPS data for travel behavior characteristic analysis. Most of these relative studies focused on estimating public transport travelers' origin, transfer, and destination. These researches tried to explore travel features in local areas or for given travel modes. Zhang [8] analyzed the distribution and transfer of passenger flow of platform metro passenger at operational stages. Cao et al. [9] proposed models that demonstrated the characteristics of entering station, exiting station, and transferring and waiting of metro passengers. Meanwhile, an index algorithm was developed to optimize public transport vehicles operation [10]. In summary, compared to manual surveys, these studies have lower cost and higher accuracy. However, these statistical results were applied to obtain the average features and total attributes of public transport travelers; the characteristics of individual passenger were largely ignored.

As with different personal characteristics, such as travel habits, job type, and income, not all public transport travelers have the same travel regularity. Personal characteristics and habits would affect the regularity of passenger travel behavior apparently [11]. Recently, a few research studies have taken individuals travel characteristics into account, but the individual features were not studied adequately. For example, Yang et al. [12] constructed a binomial model to analyze the microcosmic factors that influence individual choice of bike sharing services. But influence factors were not comprehensive and accurate enough because of limited sample data. Besides, Xu [13] studied the individual passenger behavior in urban subway hub, but only walkways and downward and upward stairways were built to illustrate the relationships between individual passengers' walking speed and space. Many other types of infrastructure, such as the platform, were not fully considered.

Summarizing above, there is no doubt that grasping travel characteristics of individual passengers based on personal travel data could potentially help to predict personal trip behavior more accurately. However, as being subjected to the limitation of data sample, conventional statistical analysis method, and mathematic model, there is still lack of approaches to exhibit the space-time changing process of travel behavior intuitively, and thus it is difficult for visual grasping or quantitative comparing of the whole and partial travel behaviors of a traveler. Therefore, it is challenging to obtain the travel demand of public transport passenger precisely.

In recent years, the knowledge graph, consisting of nodes, arcs, and/or transfer possibility, has been widely used in traffic, medicine, library, information, and many other areas recently [14–16]. For Chen et al.' study [14], driver behavior graph was constructed to identify the driving habit of different drivers (e.g., whether it is safe or not) and to express the process of his/her whole performance during driving intuitively. Integrating medical data from different languages, medical graph could seek the best knowledge presentation scheme [15]. Medical knowledge graph helped doctors make a rational diagnosis by using dynamic reasoning of graphs and patients understood the diagnosis results because of the intuitive information from graphs [15]. Similarly, in library and information studies, reference citation graph could be automatically generated to display the relationship between literatures and authors and further to analyze the structure, relationship, and evolutionary process of scientific knowledge [16]. Overall, three advantages of this graphical method were apparent. Firstly, knowledge graph can organize and express the features of mass and heterogeneous data effectively. Secondly, knowledge graph has excellent ability in deep reasoning depending on powerful knowledge base. Thirdly, knowledge graph could achieve cognitive ability when further combined with other methods (e.g., deep learning and dynamic fuzzy logic).

Based on these above analyses, the main innovation of our research is to model travel behavior from the aspect of individual passengers by adopting knowledge graph. On one hand, the whole travel characteristics of a given passenger during any time period could be completely and intuitively expressed in one graph. On the other hand, quantitative comparing, deep mining, forecasting, and reasoning of travel regularity would be achieved when establishing the database of travel behavior graphs from various passengers under the assist of computer programming. After this, the accurate travel regularity of one or more passengers could be easily got even if the initial travel data missed partially.

In this study, we mainly discussed how to establish individual travel behavior graph referencing the elements, rules, and other requirements of knowledge graph, especially the reference citation graph and measurement visualization analysis in Chinese National Knowledge Infrastructure (CNKI, http://www.cnki.net). Firstly, the travel chain extraction method of individual public transport traveler was proposed based on multimode public transport data. Then, the travel association map, including individual space position, travel time, and route, was constructed and integrated to form the structure of travel behavior graph. Probabilities of the graph arcs were considered to estimate the transfer probability of passenger's next trip. The proposed travel behavior graph of public transport intends to provide new insight into individual travel behavior modeling, which supports the intuitive individual public transport travel presentation and implicit regularity mining. The study results lay a foundation to more accurate travel demand prediction considering

TABLE 1: Effective fields of public transport smart card transaction data.

	Bus	Metro	Bike sharing services
Smart card transaction	(i) Line code (ii) Card code (iii) Boarding station number (iv) Boarding time (v) Alighting station number (vi) Alighting time	(i) Card code (ii) Entry line number (iii) Entry station number (iv) Entry station time (v) Exit line number (vi) Exit station number (vii) Exit station time	(i) Card code (ii) Rent station number (iii) Rent time (iv) Return station number (v) Return time

TABLE 2: Effective fields of public transport network features data.

	Bus	Metro	Bike sharing services
Network features	(i) Arc length (ii) Arc code (iii) Station longitude and latitude (iv) Station spaces	(i) Line code (ii) Station code (iii) Station name (iv) Station longitude and latitude (v) Station spaces	(i) District name (ii) Station code (iii) Station name (iv) Station latitude and longitude

the personal characteristics of public transport individuals.

2. Multimode Data Collection and Matching

In order to construct individual travel behavior graph and further extract travel characteristics of public transport travelers in depth, two types of public transport base data were collected and matched in this study. Namely, multimode data of public transport smart card transaction and network features were used.

2.1. Smart Card Transaction Data of Public Transport. The smart card transaction data contains bus Integrated Circuit (IC) card data, metro automated fare collection (AFC) data, and bike sharing services card data. The data used in the current study were obtained from the government agency of the Beijing Transportation Operations Coordination Center (TOCC). Each day, there are about eight million bus IC card data, five million metro AFC data, and sixty thousand bike sharing services card data, respectively. Public transport data from April 2017 was used in this study.

Six effective fields of bus IC card data, seven effective fields of AFC data, and five effective fields of bike sharing services card data were extracted from smart card terminal. The detailed effective fields of smart card transaction data were shown in Table 1.

2.2. Data of Multimode Public Transport Network Features. The data of network features used in this study contains bus line and station data, metro line and station data, and the station data of bike sharing services. The detailed fields of network features data are shown in Table 2.

Bus line and station data were collected by geographic information system (GIS) map analysis. The arc length and code were obtained directly by GIS map. The longitude

and latitude of stations were acquired through station aggregation. The space between each station and the other was calculated by searching bus route.

For the metro line and station data, AFC station code was specified by uniform rules in Beijing and the station names were confirmed by combining the line code with the station code. Station latitude and longitude were obtained by coordinate aggregation of each exit and entry of the station. The travel distance of any metro station pairs is acquired through the shortest route searching method.

For the station data of bike sharing services, district name, station code, and station name were obtained directly from base data. Station latitude and longitude were calculated by coordinate aggregation.

2.3. Data Matching. As multimode public transport data is obtained from different approaches, it is necessary to reconstruct the travel process of a given passenger by using corresponding rules. Thus, the main objective of data matching is to get integration data of public transport smart card from different travel modes and further to obtain the travel chain of individual passenger.

Firstly, public transport smart card data are integrated from different travel modes together by using the same card code of an individual public transport passenger.

Secondly, public transport smart card integration data are ordered by departure time. Departure time represents bus boarding time, metro entry station time, and bike rent time.

Thirdly, travel chain data of an individual passenger are obtained to record the whole travel process per travel through travelers' transfer judgment. The travel chain consists of one or more travel stages. The method used for travel chain extraction from smart card integration data was mainly based on the approach proposed by Wang [17]. Table 3 is an example of the travel chain data for a passenger in April 2017 in Beijing.

TABLE 3: Example of travel chain data of one passenger.

Card code	24050273	24050273	. . .	24050273
Travel mode	Metro	Metro	. . .	Bus
Departure time	2017/4/1 8:28	2017/4/1 17:53	. . .	2017/4/30 17:04
Arrival time	2017/4/1 8:55	2017/4/1 16:29	. . .	2017/4/30 17:39
Departure line	4	1	. . .	114
Arrival line	1	4	. . .	114
Path distance (km)	8.115	8.115	. . .	8.620
Departure station name	BeiJingNan Zhan	Mu Xi Di	. . .	Bai Yun Qiao Xi
Arrival station name	Mu Xi Di	BeiJingNan Zhan	. . .	Kai Yang Qiao Nan
Longitude of departure station (°)	116.3779	116.3369	. . .	116.3395
Latitude of departure station (°)	39.8641	39.9075	. . .	39.8973
Longitude of arrival station (°)	116.3369	116.3779	. . .	116.3471
Latitude of arrival station (°)	39.9075	39.8641	. . .	39.8666

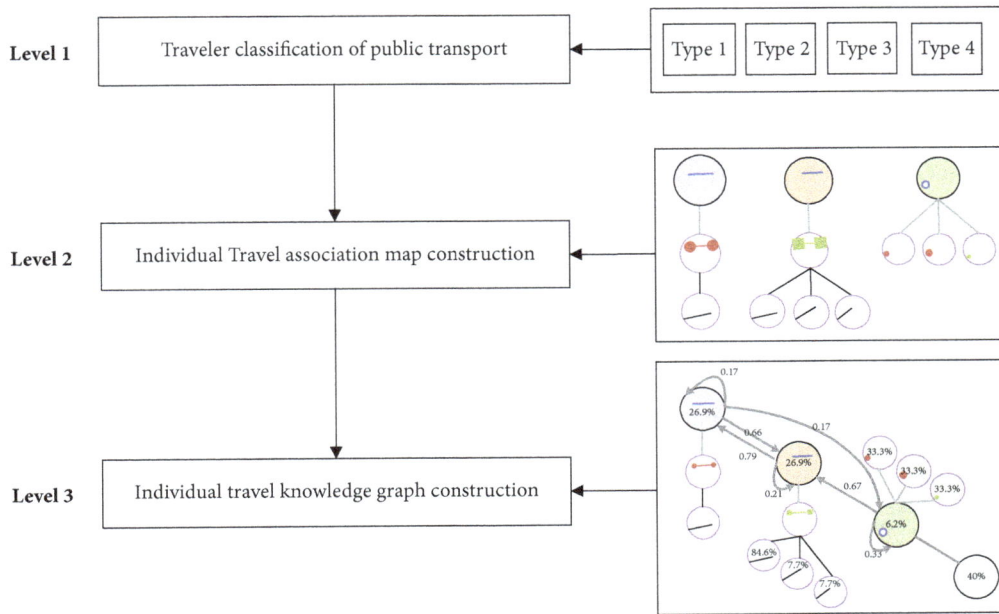

FIGURE 1: The flowchart of individual travel behavior graph construction.

3. Individual Travel Behavior Graph Construction

The main function of a travel behavior graph is to transform the low-level numerical data into a high-level abstracted form and thus to display the individual travel characteristics intuitively. The flowchart for constructing a travel behavior graph is shown in Figure 1, which includes three levels from top to bottom. After the individual travel chain data is preprocessed, the typical type of passengers can be classified (for level 1). Next, for level 2, the individual travel association maps are constructed mainly based on the travel space position, travel time, and routes. Each travel association map can be regarded as a macronode of the travel behavior graph for level 3. The arcs connecting each macronode and its corresponding possibility representing travel behavior transfer are created in level 3. Thus, the travel behavior graph can be used to analyze the temporal and spatial characteristics and travel behavior choice of an individual passenger.

3.1. Traveler Classification of Public Transport. The dependency degree of an individual passenger on public transport is an important indicator to reflect his or her travel characteristics [18], which are typically represented by travel days and travel times. Thus, passengers were classified into different types based on their dependency on public transport in this study. On one hand, applying passenger type classification provides a uniform rule for establishing travel graphs (i.e., the layout of travel behavior graph). On the other hand, it becomes easy to grasp and compare the travel features of various public transport passengers.

Passengers' travel chain data in April 2017 in Beijing was analyzed to set the indicators' thresholds for passenger classification. For travel days, the threshold used for

TABLE 4: Symbols used for travel time classification.

Symbol	Morning			Noon	Afternoon				
	●	●	⬤	▲	▪	▪	■	■	■
Size (pound)	9	12	15	9	7.5	10.5	14	16	22
Travel time	5–7:00	7–9:00	9–11:00	11–13:00	13–15:00	15–17:00	17–19:00	19–21:00	21–23:00

identifying higher or lower public transport travel frequency for a passenger was defined as average 4 days in a week and thus 16 days in one month, while for travel times, the threshold for higher frequency of public transport travel times was supposed to be at least twice in one day and thus 32 times in a month.

As passengers are divided into two groups by each indicator, a total of four types of passengers could be classified, such as passengers with high frequency of both travel days and times or passengers with high frequency of travel days and low frequency of travel times.

3.2. Travel Association Map Construction. The travel association map, which consists of passengers' travel spatial positions, time distributions, and routes, represents macronodes of the individual travel behavior graphs. This part will discuss the method for building travel association maps from travel chain data.

3.2.1. Individual Travel Space Position Clustering. Based on the travel chain data ordered by time of occurrence, the longitude and latitude of the travel space position could be obtained through matching the departure and arrival stations with the geolocation data. Then, the method of hierarchical system clustering is applied to merge spatial position clusters according to the longitude and latitude data. The cluster method is based on connections between groups and the measurement standard was Euclidean distance. Thus, the positions are merged into the same set if the ODs of travel chains fall within a certain range.

In the first and second layers, the lines in the node were represented by the origin and destination points that were confirmed through travel data of longitudes and latitudes. According to data of public transport network features in Beijing, the longitude within the Five Rings is from 116.22 to 116.54, and the latitude within the Five Rings is from 39.76 to 40.02. So the horizontal and vertical coordinates were set as the range of longitude and latitude, respectively. After setting the node size for these two layers and removing the horizontal and vertical axes, the point positions were uniquely confirmed.

Figure 2 shows an example of cluster distribution results for a given passenger (P1). A total of three OD groups are created. The horizontal and vertical axes represent that the longitudes and latitudes changed from origin to destination. From group 1 (13 travel chains) and group 2 (13 travel chains), it is intuitive that two major travels existed for this passenger and the public transport travel was a round-trip. For group 3 (3 travel chains), as the distance of this paired OD is

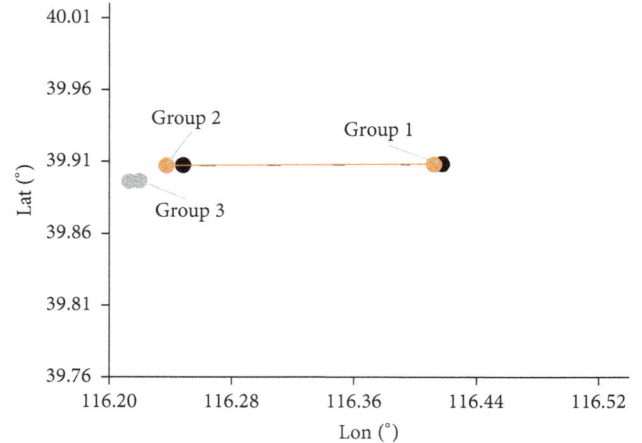

FIGURE 2: Example of cluster results of departure and arrival stations position.

relatively closer, the average geolocation is nearly overlapping.

3.2.2. Individual Travel Time Classification. Individual travel time is further classified based on the results of travel space position clustering. In each OD pair, travel time is classified by the departure time because of its typical representation of travel behavior. The departure time of each trip falls into one of nine different intervals, which are divided into two-hour periods from 5:00 a.m. to 23:00 p.m. The time periods cover the main operating time of public transport in Beijing. To illustrate the complete travel time, the classification result refers to passenger's both departure and arrival times in the OD groups.

Table 4 shows the symbols used to mark the individual travel time classification results. Different shapes (circle, triangle, and square), sizes (7.5 to 22 pounds), and colors (red, orange, and green) are defined to represent the different travel time for each OD pair.

Figure 3 shows the results of individual travel time classification of three ODs for P1 in Figure 2. The travel time classifications, respectively, correspond to groups (1), (2), and (3) shown in Figure 2. Figure 3(a) illustrates that the departure and arrival times are both from 7:00 to 9:00 when the passenger's travel belongs to the first OD pairs (group 1). Figure 3(b) displays that the departure and arrival times are both from 17:00 to 19:00 when the passenger's travel is in group 2. For the third group, the travel time is irregular, which is distributed among three intervals as 5:00 to 7:00, 9:00 to 11:00, and 13:00 to 15:00 (see Figure 3(c1–c3)).

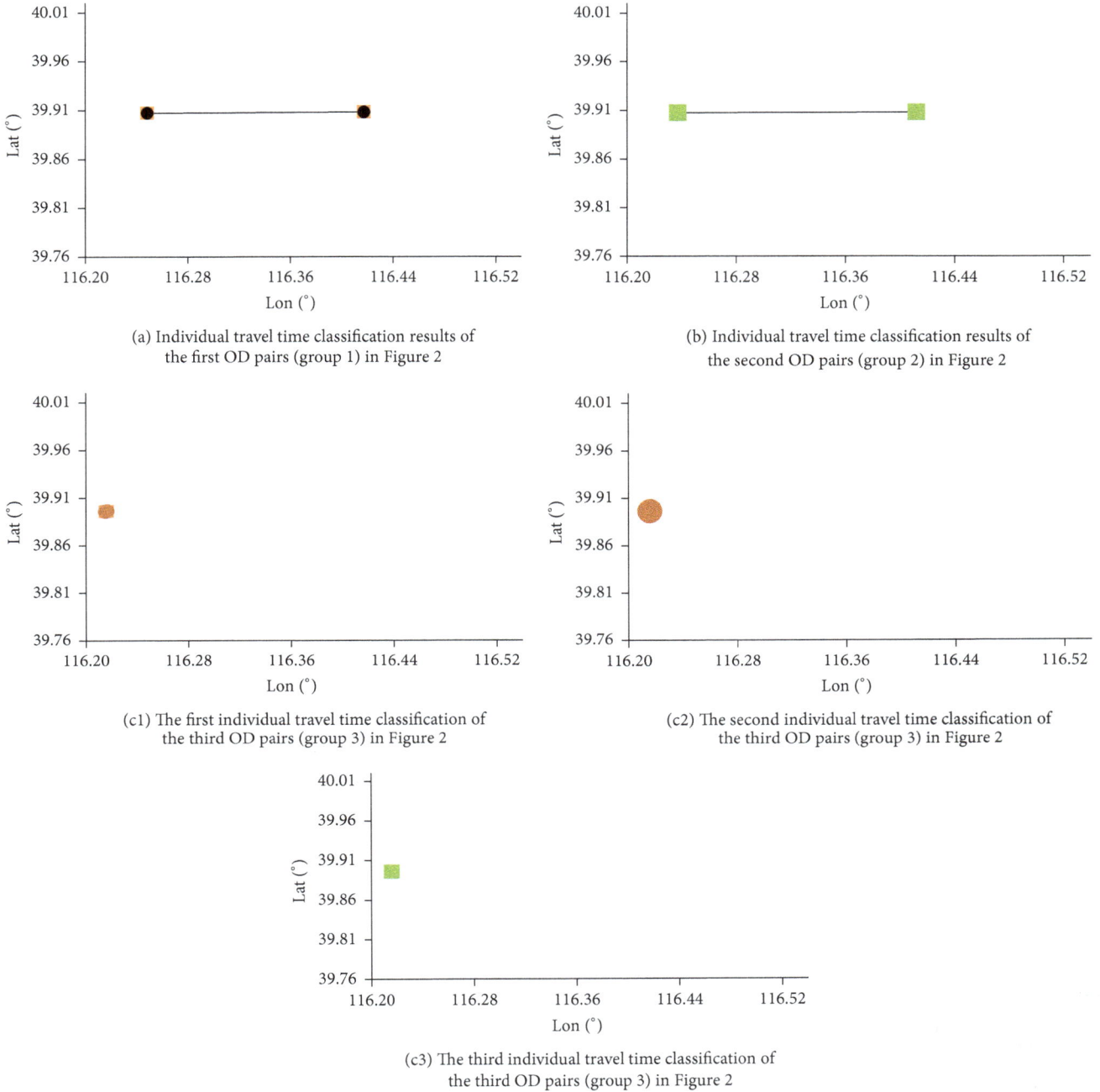

(a) Individual travel time classification results of the first OD pairs (group 1) in Figure 2

(b) Individual travel time classification results of the second OD pairs (group 2) in Figure 2

(c1) The first individual travel time classification of the third OD pairs (group 3) in Figure 2

(c2) The second individual travel time classification of the third OD pairs (group 3) in Figure 2

(c3) The third individual travel time classification of the third OD pairs (group 3) in Figure 2

FIGURE 3: Examples of passenger travel time classification results of P1 in Figure 2.

Also, as the distance of this paired OD is relatively closer, the time distribution in Figure 3(c1–c3) is almost overlapping.

3.2.3. *Travel Route Clustering.* The result of travel route clustering represents travel modes of individual passengers. It is a further step based on space position clustering and travel time classification. Travel distance and route direction are selected to reflect travel modes. If both the route distance and directions are close in a certain range, the travel chains will be merged into the same cluster. This step is repeated until the difference between the two closest travel chains is not significant. Different travel route clusters would be created for each travel time classification subset.

The travel distance is the actual path distance of each travel stage, such as the distance from departure to transfer stations and transfer to arrival stations. Travel distance is obtained by matching station information to network features data. Similarly, the travel direction is also calculated for each travel stage. Given a travel chain n, let lo_{no} and lo_{nd} be the

(a) Travel route clustering results of the time classification group (1) in Figure 3.

(b1) The first travel route clustering results of the time classification group (2) in Figure 3.

(b2) The second travel route clustering results of the time classification group (2) in Figure 3.

(b3) The third travel route clustering results of the time classification group (2) in Figure 3.

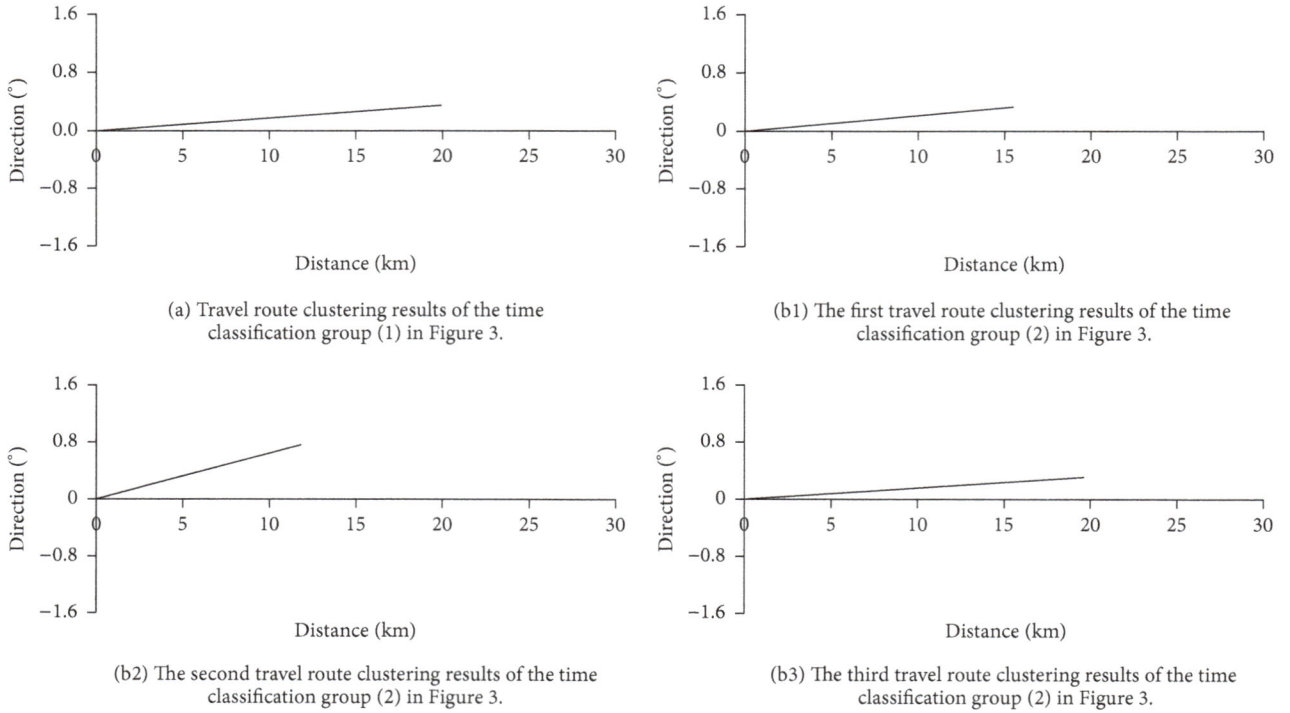

FIGURE 4: Travel route clustering results of P1 in Figure 3.

longitude of origin and destination and let la_{no} and la_{nd} be the latitude of origin and destination; the route direction can be obtained by the following equation:

$$Ang = \frac{\text{la}_{nd} - \text{la}_{no}}{7.5 * 10^{-6}} \times \frac{10^{-5}}{\text{lo}_{nd} - \text{lo}_{no}} \quad (1)$$

$$D_{\text{chain}}(n) = \text{degree}(\text{ATAN}(Ang)),$$

where Ang is the tangent value of the distance between the origin and destination of travel stage. D_{chain} is the direction of each travel stage.

For the third layer, the diagram was demonstrated by travel distance (x-axis) and travel direction (y-axis). According to the statistical data in Beijing, the average travel distance is about 16 kilometers, and the 85th percentiles value is about 26 kilometers. Thus, the range of x-axis was defined as 0 to 30 kilometers in this study. The direction of each travel stage was calculated by (1) in the paper, the range of which would be 0 to 360 degrees. Correspondingly, the antitangent value of the travel direction would be about (−1.6, 1.6). In order to display travel distance and direction in the same figure with more clear visibility, the antitangent value of real angle was set as y-axis. Similarly, after setting the node size for the third layer and removing the horizontal and vertical axes, the point positions were uniquely confirmed.

In addition, the travel modes are represented by different line attributes. The single and black solid lines represent bus travel, the dotted and red lines represent metro travel, and the double and blue solid lines indicate bike sharing service. Figure 4 shows an example of travel route clustering result. These route clustering results in Figure 4 correspond

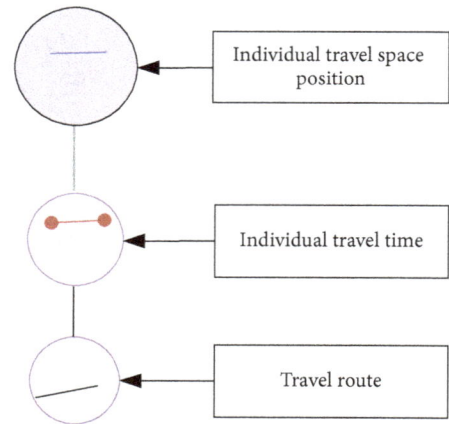

FIGURE 5: Example of one travel association map of P1.

to groups (1) and (2) shown in Figure 3. In Figure 4, group (1) has only one path and group (2) has three different paths (see Figure 4(b1–b3)). In addition, the distance of group (3) is too short to display and thus its route clustering result was not considered.

3.2.4. Individual Travel Association Map Construction. The travel association map is regarded as a macronode of the whole travel behavior graph. Travel spatial positions, time distributions, and the routes are zoomed out and, respectively, merged into three layers of a travel association map (see Figure 5). Referring to the multilayer programming theory,

TABLE 5: Four layouts of travel behavior graph.

Travel times	Travel days	
	High	Low
High	↘	→
Low	↙	↓

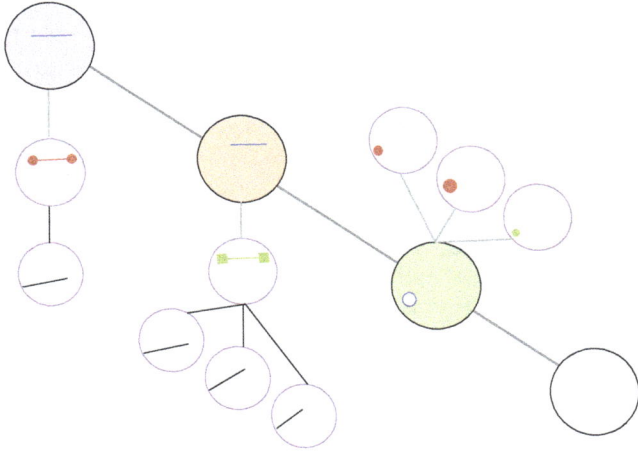

FIGURE 6: Travel association map integration of P1.

travel space position (big circle) is set as the first layer of the travel behavior graph, and travel time (median circle) and route (small circle) are treated as the second and third layers. Every layer is considered as a node of travel association map. In each layer, the horizontal and vertical coordinates are uniformed as well as the symbols and lines. After further abstraction processing, one of the travel association maps of passenger P1 can be drawn as Figure 5. According to the information displayed in Figure 2, there should be a total of three travel association maps of P1. Each travel association map represents a specific travel behavior.

3.3. Travel Behavior Graph Construction.
After each travel association map constructed, the travel behavior graph is developed by using arcs to connect and arrange every association map. Meanwhile, the statistical probability within and among each travel association map was also calculated to represent travel behavior choice.

3.3.1. Travel Association Map Integration.
Firstly, according to these four types of passengers defined by travel days and times in this study, the travel behavior graph is arranged as four different layouts as below, shown in Table 5.

For P1 mentioned earlier, the travel days and times are 18 and 29 in one month, which means that the traveler has high frequency of travel days and times. The arrangement of integrated travel association map is shown in Figure 6 (i.e., diagonally downwards from left to right). As the space positions of P1 consist of four categories, three public transport (solid circle) and one nonpublic transport (hollow circle) travel, the first layer is presented by four uniformed big

circles connected and abstracted space position coordinate axis within each circle. The information of the first layer exhibited is consistent with Figure 2. Then, the second layer is set as middle circles and abstracted time classification coordinate axes are within it. The information of circles in the second layer is corresponding to the findings in Figure 3. For the third layer, small circles are selected and abstracted route clustering coordinate axes are within these circles. It is worth noting that as travel distance for the third space position (shown as green circle) is too short to display, the travel route clustering is not considered in its third layer. The information in Figure 4 is displayed in the third layer.

3.3.2. Probabilities Calculation within and among Travel Association Map.
In the travel behavior graph, probabilities of nodes (within circles) illustrate the occurrence proportion, and the weight of arcs among association maps indicates the probability of transformation from one association map to another.

For the transfer probability, it is calculated through three steps. Firstly, the travel space position of individual passengers is clustered to get his/her origin-destination categories (ODs). Secondly, the individual travel chain data of one month are ordered by departure time. Thirdly, the transfer probabilities are calculated by statistically analyzing the occupancy percentages from one OD to any others and itself, for example, if there are 11 times when the second OD occurred after the first OD and 3 times when the first OD followed itself; but there was inexistence when the third OD occurred after the first OD. Thus, the transfer probability is 0.79 (11/14) from the first to the second OD, is 0 (0/14) from the first to the third OD, and is 0.21 (3/14) from the first OD to itself.

The total occurrence proportion of each node per layer is assumed as 100 percent. In the first layer, the proportion of nonpublic transport and total public transport proportion are calculated by travel days during a month, respectively. Then proportions of each public transport OD cluster group are obtained through travel times of different ODs. In the second layer, the proportion is the occurrence frequency of different travel time. Similarly, the proportion in the third layer reflects travel routes selection.

As shown in Figure 7, probabilities within and between two travel association maps were displayed. For the first travel association map, the space position accounted for 26.9% in the first layer for all ODs. When it occurred, the departure time was between 7:00 and 9:00 a.m. in the second layer. The travel route took bus all the way in the third layer. It means that the travel time and route are stable for the first OD, while for the second travel association map, the occurrence probabilities were also 26.9% in its first layer. As with the high similarity to the first OD, it implied that these two travels were round-trips. For the second OD, the travel time was concentrated on 17:00–19:00 p.m. The route was focused on the first node in the third layer. In addition, the probability between these two association maps is 0.79, which indicates that one macronode would occur after the other in a large degree.

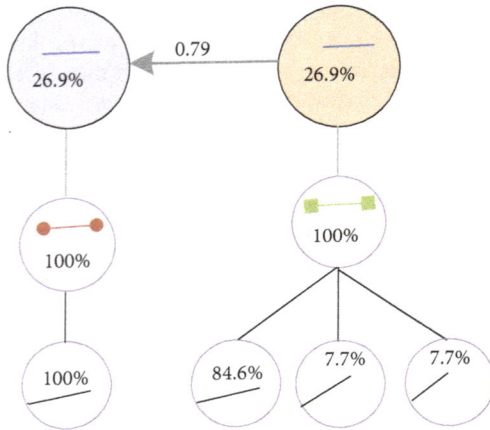

FIGURE 7: A simple example of the probabilities within and among travel association map.

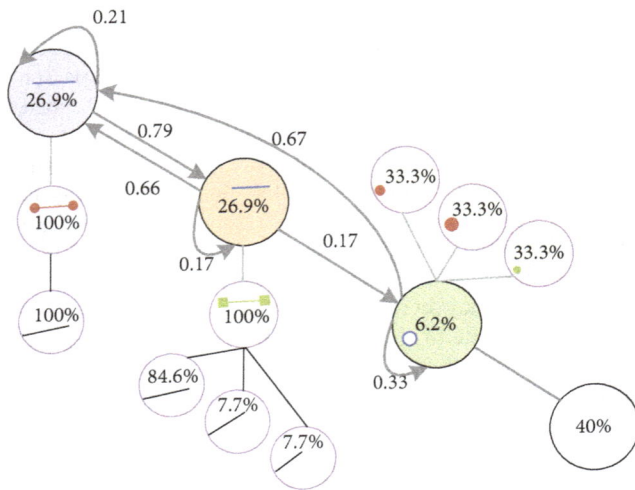

FIGURE 8: Individual travel behavior graph of passenger P1.

4. Case Studies

To illustrate and verify the effectiveness of travel behavior graph based method in modeling individual travel behavior of public transport travelers, four passengers of each type were randomly selected to construct their individual travel behavior graphs (shown in Figures 8–11).

Figure 8 is the travel behavior graph of P1. It shows that the public transport travel accounts for 60% of this individual traveler's entire trips. For this passenger's public transport travel, there are mainly three travel behaviors (i.e., three travel association maps). The first two travel behaviors displayed as purple and pink circles are the main travels and they accounted for equal percentages. For the first travel behavior, both the travel time (morning peak) and path are unique (bus). For the second travel behavior, the travel time (evening peak) is also fixed, while there are three travel different routes and the first route is most likely to be adopted by this traveler (i.e., 84.6%). For the third travel behavior, the travel time is averagely distributed to three periods, and the

travel distance is very short. In addition, according to the transfer probabilities among each travel behavior, it could be drawn that the first two behaviors occurred subsequently between each other in high degree. It is implied that these two travel behaviors might be round-trips. Besides, the second behavior might be generated with possibility of about 70% if the third behavior existed. Summarizing the regularities analysis above, it is implied that this passenger would be a commuter.

Figure 9 demonstrates the travel behavior graph of P2, which is more complicated and irregular than P1. This travel (16 travel days and 25 travel times in one month) belongs to high frequency of travel days and low frequency of travel times. The public transport travel accounts for 53.3% of the entire trips. This passenger's public transport travel could be classified as four travel behaviors. The first two travel behaviors displayed as purple and pink circles are the main travels (i.e., 25.6% and 17.1%). For the first travel behavior, the travel time is concentrated on morning peak and the route is unique (metro). For the second travel behavior, the travel time is mainly in the afternoon especially during evening peak, and metro is the main travel mode. The travel time of his third travel behavior is focused on evening and travel path is also fixed by metro. For the fourth travel behavior, the travel time is equally distributed to morning and evening, and the route includes single metro or the combination of metro and bus. In addition, based on the transfer probabilities among each travel behavior, it could be drawn that the first behavior could follow the second behavior with half probability. Additionally, the first behavior would be generated if the third behavior existed, and the third behavior would occur after the fourth behavior.

The travel behavior graph of P3 is displayed in Figure 10. The travel days are low frequency (13 travel days in one month) and travel times are high frequency (39 travel times in one month). Public transport is not the dominant mode (only 43.3%). There are mainly three public transport travel behaviors for this passenger. The first two travel behaviors displayed as purple and pink circles are the main public transport travels. For the first travel behavior, the travel time is averagely divided into morning and afternoon periods, and bus is the only travel mode. The travel time is also mainly focused on morning and noon (47.6% for each) for the second travel behavior, and travel path with fixed travel mode (i.e., bus) is unique for certain time periods. For the third travel behavior, the travel time is averagely distributed to two periods, and the travel route is not the same between these two periods. In addition, depending on the transfer probabilities among each travel behavior, it is indicated that the second behavior would occur subsequently after the first behavior, while the first behavior might also follow the second one with more than half possibility. Besides, if the third behavior existed, the first and second behaviors might happen with equal probabilities of 50% for each.

Figure 11 shows the travel behavior graph of P4. This travel belongs to low frequency of both travel days and times (15 travel days and 25 travel times in one month). Public transport and nonpublic transport, respectively, account for half of this passenger's travels. For this passenger's public

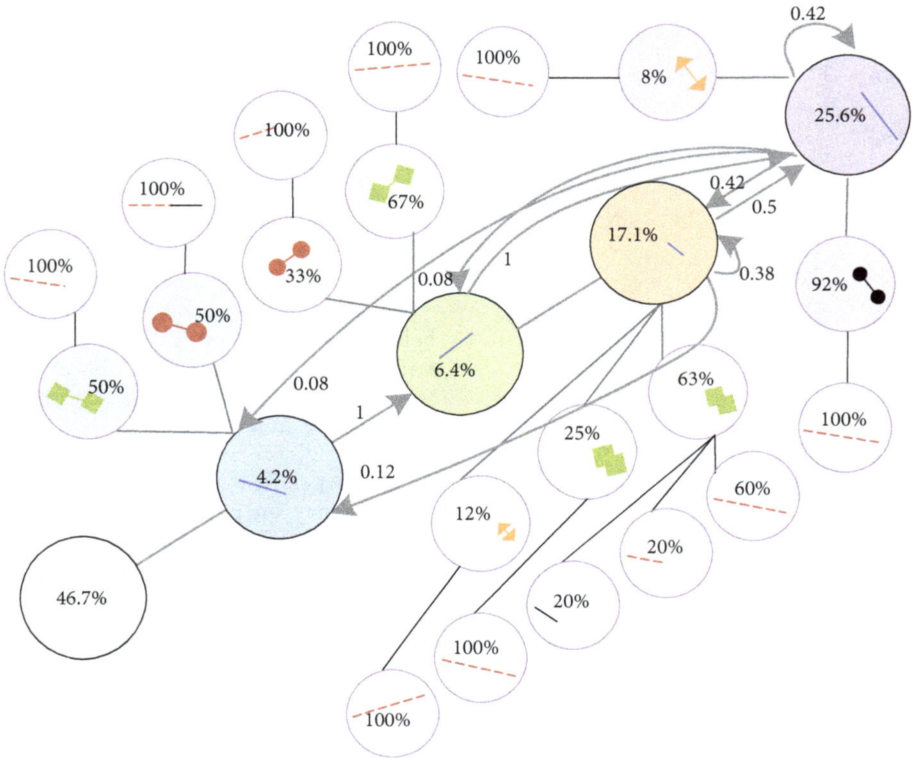

FIGURE 9: Individual travel behavior graph of passenger P2.

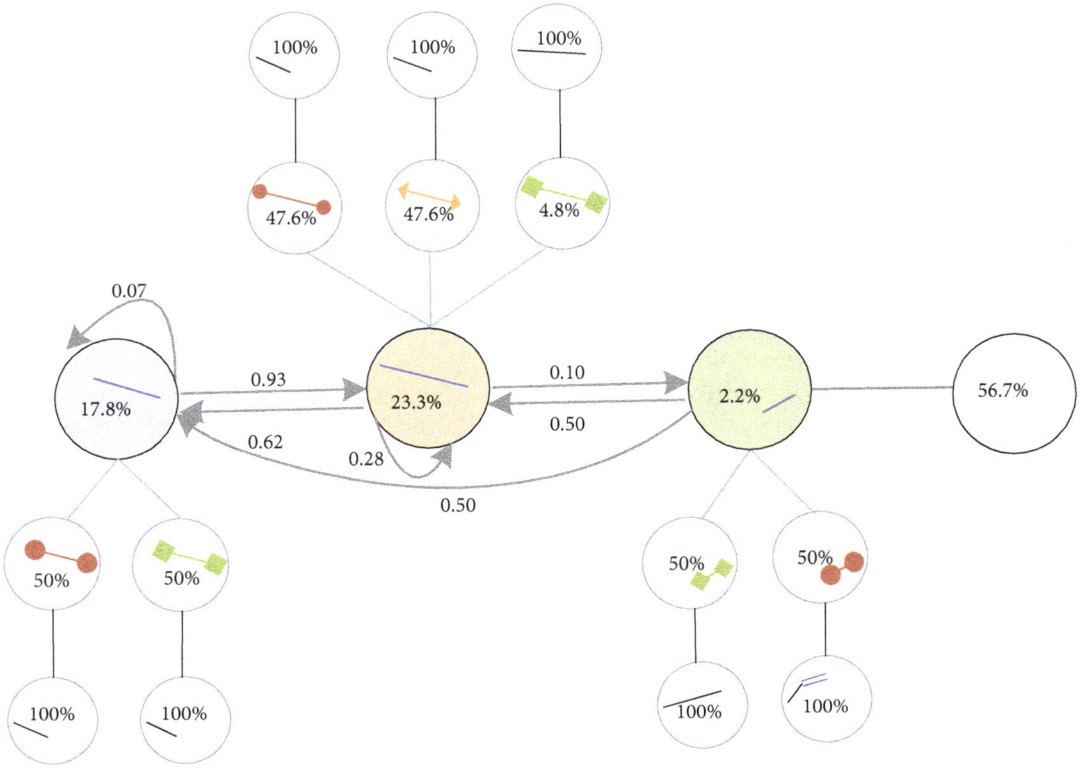

FIGURE 10: Individual travel behavior graph of passenger P3.

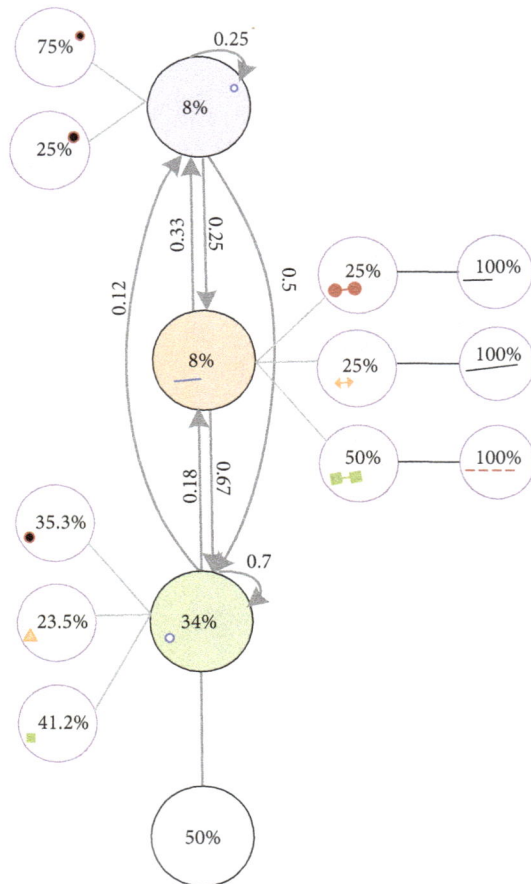

FIGURE 11: Individual travel behavior graph of passenger P4.

transport travel, there are three types of travel behaviors. For the first travel behavior, the travel time is centralized on the early morning, and the travel distance is very short. For the second travel behavior, the travel times are scattered, and half travels are during early afternoon with metro. For the third travel behavior with highest proportion of public transport travel, the travel times are scattered in the morning, at noon, and in the afternoon. The route distance is short as well. In addition, the transfer probabilities among each travel behaviors illustrate that the third travel behavior might be highly possible to happen if the first behavior, second behavior, or itself existed.

5. Discussion

In this study, an individual travel behavior graph constructing method was proposed and then followed by case studies of four passengers. The travel behaviors of these four passengers are not the same and thus the individual travel behavior graphs have different shapes. Travel behavior graph could intuitively display the difference of individual travel pattern. Examples of different travel information drawn from travel behavior graphs include the following: (1) P1 would be a commuter with fixed round-trip characteristics; (2) the travel behavior is relatively complicated for P2 because of

more dispersive travel time and routes; (3) bus is the main travel mode for P3; and (4) the travel destination by public transport is single and the travel distance is short for P4. In addition, although the overall travel features are dissimilar for different passengers, some specific travel behaviors (i.e., travel association map) are similar. For example, the fourth travel association map of P2 is similar to the third travel association map of P3. Both travel times are equally divided into two different periods, and the travel modes are mixed. The third travel association maps of P1 and P4 are also similar, both of which belong to short distance travel.

Besides its advantage in intuitive exhibition of travel behavior information, six feature indexes for travel characteristics description could be also extracted from individual travel behavior graphs. These quantitative indexes could be concluded as follows:

(1) Dependency on public transport: passengers' dependency on public transport is represented by travel days and travel times, which is displayed as different layouts of travel behavior graphs. For example, the dependency on public transport of P1 is higher than that of P4

(2) OD classifications: the first layer of graph includes OD clusters. Taking P1 and P4 as examples, both of them have three main public transport destinations, and two destinations for each occupy the same proportions

(3) Round-trip or not: if public transport round-trip exists, two OD clusters would be significantly similar. Round-trips only existed for P1 for these four travelers

(4) Peak periods: the travel time concentration degree is reflected by peak periods. The travel time of P1 is focused on morning peak and evening peak, but the travel time of P4 is flexible

(5) Route selection: for individual travel with similar travel destination and departure time, the travel route might be unique or multiple. For instance, the travel path is fixed for the first travel behavior of both P1 and P2, while there exist different routes for the second travel association map of P2

(6) Transfer probability of the next trip: this indicator represents the occurrence probability of the subsequent trip. For P2, if the fourth travel association map exists, the next trip would be the third and then follows by the first travel behavior.

Indeed, the advantage of travel behavior graph was not limited to extracting feature indexes of different passengers. Moreover, the graph can exhibit a complete individual travel behavior intuitively, and it contributes to judging the travel behavior's similarity and difference among different passengers. In addition, after generating travel behavior automatically depending on computer assistance, these graphs are the foundation for further intelligent applications (e.g., identifying, forecasting, and reasoning) when combined with other methods (e.g., deep learning).

As stated above, in recent years, knowledge graph has been widely applied in driving behavior identification in traffic areas [14], state of illness analysis in modern medicine [15], and relationship mining between literature and authors in information science [16]. In different research areas, the implication and contents of knowledge information might vary significantly, but the elements and structure were almost consistent. In this research, the shape of travel behavior graph was mainly referenced to the reference citation graph and measurement visualization analysis in CNKI. Thus, travel behavior graph also constituted of nodes and arcs. Besides, transfer possibility was added to travel behavior graph to forecast passengers' travel behavior choice.

In fact, the reference citation graph and measurement visualization analysis could be generated automatically in CNKI searching website within a few seconds. As we defined the uniformed standard of the elements of travel behavior graph (i.e., the method for obtaining node, arc, and transfer possibility of graph), it is believable to generate travel behavior graph automatically in our future work. In addition, we also discussed this issue of automatic generation of travel behavior graph with professional computer programmers. In the future and next study, we think that achieving the automatic generating of travel behavior graph supported by computer programming is practicable. The computing time of an individual travel behavior graph could be controlled into several seconds.

6. Conclusions

This paper develops a novel method for modeling an individual travel behavior based on knowledge graph to mine the travel regularity of different public transport passengers. The study results indicated that travel behavior graph is effective for intuitive presenting and further forecasting individual passengers' travel behavior characteristics, which support customized public transport travel characteristics analysis and demand prediction.

Based on multimode data of public transport smart card transaction and network features of one month in Beijing, travel chain data was obtained to reflect the entire travel process of different trips for an individual passenger by data matching. Depending on travel chain data, individual travel behavior graphs composed of macronodes, arcs, and transfer probability were constructed to represent travel spatial positions, time distributions, routes, and their possible behavior choice of next trip. In addition to intuitive illustration of travel information, the whole and specific travel behavior similarities or differences among different passengers were compared. Besides, six feature indexes were extracted from graphs to analyze the hidden characteristics of individual travel behavior.

In this study, we focused on finding a way to establish individual travel behavior graph referencing the elements, rules, and other requirements of knowledge graph, especially the reference citation graph and measurement visualization analysis in Chinese National Knowledge Infrastructure (CNKI, http://www.cnki.net). In addition, the reasonability and practicability were tested and verified by case studies of

several passengers. In the next studies, we will make efforts to generate travel behavior graph automatically depending on computer programming. The database of individual travel graph from various and large amount of passengers would be established in future researches. Finally, we hope to provide a convenient and precise way to grasp travel regularity through combining travel behavior graph and artificial intelligence. For example, combining with semantic analysis technology, individual travel behavior graphs can be used to further classify the types of public transport passengers in the future, such as high, moderate, and low stability of taking public transport of commuters and noncommuters. Thus, different types of passengers can be estimated to provide better public transport service with more suitable modes, such as rapid bus, customized shuttle bus, and mini bus. These findings could provide a foundation for transportation agencies to predict the public transport scheduling demand more accurately and optimize the transport operating network more reasonably.

Conflicts of Interest

The authors declare that there are no conflicts of interest regarding the publication of this paper.

Acknowledgments

The authors would like to show great appreciation for support from the National Natural Science Foundation of China (project: *Multimode Travel Demand Identification Methods Based on Individual Travel Feature Atlas of Public Transport Commuters (no. 51578028)*) and "Beijing Nova" Program by the Beijing Municipal Science and Technology Commission: *Study on the Feature Extraction Method and Demand Mechanism of Public Transport Travel with Multimodes (no. Z171100001117100)*.

References

[1] J. F. Guo, X. Li, G. C. Wang, H. M. Wen, and Y. Liu, "Beijing transport annual report," Research Report, 2017.

[2] J. Rong and J. C. Weng, *Features Extraction of Public Transport Commute Travel Based on Multi-Mode Data*, Beijing University of Technology, 2014.

[3] M. V. Sezhian, C. Muralidharan, T. Nambirajan, and S. G. Deshmukh, "Attribute-based perceptual mapping using discriminant analysis in a public sector passenger bus transport company: a case study," *Journal of Advanced Transportation*, vol. 48, no. 1, pp. 32–47, 2014.

[4] S. V. Walle and T. Steenberghen, "Space and time related determinants of public transport use in trip chains," *Transportation Research Part A: Policy and Practice*, vol. 40, no. 2, pp. 151–162, 2006.

[5] Z. Zhou, Y. Li, C. Meng, and H. P. Lu, "Analysis of travel demand based on a structural equation model," *Journal of Tsinghua University (Science and Technology)*, vol. 48, no. 5, pp. 879–882, 2008.

[6] H. R. Zhu, "Research on sensitivity of public transport priority in central districts of Beijing," in *Proceedings of Urban and Rural Planning in the Perspective of Conservation Culture, Chinese Urban Planning Conference*, Dalian, China, 2008.

[7] R. Zhang, *Mode Split Model and Its Application on Urban Passenger Transportation System [Master, thesis]*, Beijing Jiaotong University, Beijing, China, 2011.

[8] C. Zhang, *Analysis on the Characteristics of Urban Rail Transit Passenger Flow [Ph.D. thesis]*, Southwest Jiaotong University, Chengdu, China, 2006.

[9] S.-H. Cao, Z.-Z. Yuan, C.-Q. Zhang, and L. Zhao, "LOS classification for urban rail transit passages based on passenger perceptions," *Jiaotong Yunshu Xitong Gongcheng Yu Xinxi/ Journal of Transportation Systems Engineering and Information Technology*, vol. 9, no. 2, pp. 99–104, 2009.

[10] S. H. Chen, Y. Y. Chen, and C. Y. Yin, "Calculation method for bus vehicle performance indexes based on IC data," *Traffic standardization*, no. 21, pp. 52–55, 2011.

[11] J. Wa, *The Travel Characteristics Analysis of Passengers Based on the Bus Data [Master, thesis]*, South China University of Technology, Guangzhou, China, 2016.

[12] C. Yang, J. Lu, W. Wang, and Q. Wang, "A study on the influencing factors of bicycle transportation based on individual mode choice," in *Transportation Systems Engineering and Information Technology*, vol. 7, pp. 131–136, 4 edition, 2007.

[13] Q. W. Xu, *Analysis and Modeling of Individual Passenger Behavior in Urban Subway Hub [Master, thesis]*, Beijing Jiaotong University, Beijing, China, 2008.

[14] S. Chen, C. Fang, and C. Tien, "Driving behaviour modelling system based on graph construction," *Transportation Research Part C: Emerging Technologies*, vol. 26, pp. 314–330, 2013.

[15] K. Q. Yuan, Y. Deng, D. Y. Chen, B. Zhang, K. Lei, and Y. Shen, "Construction techniques and research development of medical knowledge graph," *Application Research of Computers*, vol. 35, no. 7, pp. 1–12, 2017.

[16] X. Jiao, L. Wang, and F. Han, "Application model construction of mapping knowledge domains in the science and technology intelligence research," in *Knowledge, Learning & Management*, vol. 3, pp. 118–129, 2017.

[17] C. Wang, "Extraction method of public transportation trip chain based on the individual travel data," *Transportation Systems Engineering and Information Technology*, vol. 17, no. 3, pp. 68–73, 2017.

[18] R. L. Wang, *Public Transit Passenger Division and Analysis of Trip Characteristics Based on Smart Card Data [Master, thesis]*, Tongji University, Shanghai, China, 2014.

6

Mining Connected Vehicle Data for Beneficial Patterns in Dubai Taxi Operations

Raj Bridgelall ⓘ,[1] Pan Lu ⓘ,[1] Denver D. Tolliver,[2] and Tai Xu ⓘ[3]

[1]College of Business, North Dakota State University, Fargo, North Dakota 58108, USA
[2]UGPTI, North Dakota State University, Fargo, North Dakota 58108, USA
[3]University of Modern Sciences, Dubai, UAE

Correspondence should be addressed to Pan Lu; pan.lu@ndsu.edu

Academic Editor: Eleonora Papadimitriou

On-demand shared mobility services such as Uber and microtransit are steadily penetrating the worldwide market for traditional dispatched taxi services. Hence, taxi companies are seeking ways to compete. This study mined large-scale mobility data from connected taxis to discover beneficial patterns that may inform strategies to improve dispatch taxi business. It is not practical to manually clean and filter large-scale mobility data that contains GPS information. Therefore, this research contributes and demonstrates an automated method of data cleaning and filtering that is suitable for such types of datasets. The cleaning method defines three filter variables and applies a layered statistical filtering technique to eliminate outlier records that do not contribute to distributions that match expected theoretical distributions of the variables. Chi-squared statistical tests evaluate the quality of the cleaned data by comparing the distribution of the three variables with their expected distributions. The overall cleaning method removed approximately 5% of the data, which consisted of errors that were obvious and others that were poor quality outliers. Subsequently, mining the cleaned data revealed that trip production in Dubai peaks for the case when only the same two drivers operate the same taxi. This finding would not have been possible without access to proprietary data that contains unique identifiers for both drivers and taxis. Datasets that identify individual drivers are not publicly available.

1. Introduction

In many major cities of the world, on-demand shared mobility services are disrupting the business model of traditional street hailing and dispatched taxi services. On-demand shared mobility services involve popular transportation network companies (TNCs) such as Uber and Lyft and microtransit services such as Ford-owned Chariot [1]. This escalating competition for passengers has been motivating taxi companies to mine dynamic mobility data to reveal insights that could benefit operations [2], locate more customers [3], and forecast demand [4].

The *goal* of this study is to mine large-scale dynamic mobility data from connected vehicles to reveal potentially beneficial patterns that can help taxi services improve their business in the midst of growing competition from nontraditional shared mobility services. Privacy policies in many parts of the world require that, before making such data

publicly available, the data owner must remove information that could identify persons. By contrast, the authors of this paper obtained a proprietary and unique dataset of Dubai taxi operations with the names of drivers replaced with a unique identification number. This enabled data mining to reveal driver-vehicle sharing patterns, which, to the best of the authors' knowledge, is a gap in the available literature.

This paper *contributes* details of the Dubai case study, the proposed automated data cleaning method, and the *main finding* that a beneficial driver assignment pattern exists. This finding could inform tactics that encourage more of the beneficial assignments to help improve the efficiency and effectiveness of Dubai taxis and perhaps shared mobility services in general.

Typical connected vehicle data from taxis include messages and variables such as geospatial position, meter-engaged, meter-vacant, door-open, idling, speed, timestamps, and dozens of other status indicators. The data size grows rapidly as tens of thousands of vehicles attempt to upload data packets every second to every few minutes. Aside from being so-called big data, dynamic mobility data can also be rather messy [5]. Data cleaning to enhance data quality is critically important in data mining, but the literature on data cleaning methods is sparse [6]. Manually cleaning large-scale mobility data is impractical. Therefore, a *primary objective* of this research is to develop and apply an automated method of data cleaning and filtering that is suitable for large-scale dynamic mobility datasets. A *secondary objective* is to develop a method for validating the quality of the cleaned dataset.

Dirty data from connected vehicles that operate in the vehicle-to-infrastructure (V2I) mode arises from many factors. They include unexpected malfunctions and various errors in the output of on-board sensors, trip meters, and the V2I communications system itself. In particular, standard global positioning system (GPS) receivers produce inaccurate or missing location coordinates because tall buildings and other occlusions distort or block the direct path of the satellite radio frequency signals [7]. Trip meters often encounter radio frequency interference as they attempt to upload data and receive acknowledgements. Hence, they tend to retransmit and create duplicate records. Meter malfunctions or resets due to spurious electrical faults also produce inaccurate timestamps and odometer readings. To minimize the cost of data storage and communications, on-board systems seek to minimize the frequency and regularity of the geospatial position sampling [8]. Therefore, using filtering and interpolation techniques to reconstruct vehicle paths and speed profiles becomes ineffective [9].

Obvious errors such as missing GPS data and incorrect timestamps are easy to detect and remove. However, errors in trip length and trip durations are not as obvious. The literature lacks studies of automated methods to clean such nonobvious errors from large-scale dynamic mobility data. Work by others affirmed that the sampling variability of vehicle position data reduces the accuracy of link travel time and route choice estimates [5]. In general, researchers found that the inaccuracy of GPS data requires some form of data cleaning for route estimation [10]. Other studies confirmed that the nonuniform sampling of GPS data results in large gaps that reduce the accuracy of recovery methods such as linear interpolation and historical averaging [11].

The organization of this paper is as follows: the next section (Methods and Data) describes the dataset in terms of its variables, its original structure, and a distillation process to restructure the data into trip records. The methods section also describes the three key filter variables and the *layered statistical filtering* technique that automatically eliminates nonobvious errors. Section 3 (Results) validates the quality of the cleaned data by comparing the overall distribution of the key variables with their expected theoretical distributions. The results section also describes the data mining results and reveals a potentially beneficial driver assignment pattern that maximizes trip productivity while minimizing overhead. Section 4 (Discussion and Conclusions) discusses the results, concludes the study, and describes future work to leverage the uniqueness of the dataset.

2. Methods and Data

The Road and Transport Authority (RTA) of Dubai, United Arab Emirates, provided the authors with an exclusive dataset of their taxi activity. The data records combine information from both "dispatch-only" and "street-hailed" taxis. Emirates refer to the dispatch-only taxis as Hala taxis. Dispatchers often call on non-Hala taxis when Hala taxis are unavailable. Unlike publicly available dynamic mobility datasets such as those from New York City [12], the Dubai Taxi dataset contains the unique license plate number of each vehicle. The RTA anonymized the driver information by replacing their names with unique identifiers. A literature search indicates that this is the first study to report the results of mining dispatch-taxi mobility data that contain unique identifiers for both drivers and taxis.

2.1. Data Reduction and Restructuring. The dynamic mobility data obtained from Dubai taxis covered a 185-day period from March 15, 2016 to September 15, 2016. Analysis of the data revealed that Dubai taxi companies employed nearly 21,000 drivers who operated nearly 10,000 vehicles during that period. Dubai taxis provide service any time of day, every day. Each taxi has an on-board unit that contains a trip meter, a GPS receiver, and a wireless system that enables V2I communications. On average, the on-board unit transmits the status and position of the taxi approximately every minute. Subsequently, the dynamic mobility database annually accumulates more than five billion records, each with numerous variables.

In addition to the unique taxi and driver identifiers, each data record contains the fleet identifier, the vehicle status, its speed, its position in latitude and longitude, and a timestamp. The vehicle status indicates 45 different events, one of which is when a driver accepts a dispatch request. Therefore, the first step in data distillation was to extract records of dispatched versus street-hailed trips. The status also indicates instants of trip meter engagement and vacancy. When paired, this information forms a trip record containing the times and positions of pick-up and drop-off events. Hence, the second data reduction step was to extract only those records that indicate when the meter engaged and then became vacant. Subsequently, the data distillation process substantially reduced the data size by building approximately 3.4 million trip records from the much larger dynamic mobility dataset.

2.2. Layered Statistical Filtering. The proposed layered statistical filtering technique incorporates three layers of filtering that use *known* likelihood distribution functions for trip duration, trip length, and average trip speed. The main concept of the proposed technique is that likely errors would be outlier records that also collectively do not contribute to the formation of an overall distribution that matches the expected distribution of a key variable. The *first* layer of filtering was the *trip duration*, which is the timestamp difference between paired drop-off and pick-up events. The *second* layer of filtering was the *trip length* in terms of total path distance. However, it was not directly available because

TABLE 1: Summary of data reduction and cleansing.

Records	Description	Low Tail		High Tail		Reduction	
		Count	%	Count	%	Count	%
3,444,310	185-day dispatched trip records						
3,444,304	Remove duplicate records					6	0.0002%
3,444,304	Remove invalid latitude or longitude					0	0.0%
3,391,795	Remove (pickup time) ≥ (drop-off time)					52,509	1.5%
3,391,770	Remove records missing a Driver ID					25	0.001%
3,331,032	Filter by trip time distribution	20,836	0.6%	39,902	1.2%	60,738	1.8%
3,259,001	Filter by trip distance distribution	52,400	1.6%	19,631	0.6%	72,031	2.2%
3,258,218	Filter by mobility index distribution	0	0.0%	783	0.02%	785	0.02%

of the nonuniform sampling of the geospatial positions. That is, the high variability of position update rates resulted in distance gaps that span several kilometres. Such large gaps made it impossible to determine the actual path taken between two points that frame the street grids. However, the geodesic distance between pick-up and drop-off positions provided a suitable proxy. The algorithm used the recursive Vincenty method to derive the geodesic distance [13]. The *third* layer of filtering was the *average trip speed*. However, it was also not directly available. Therefore, the authors developed a proxy dubbed the mobility index (MI), which is the ratio of the geodesic distance to the trip duration. Intuitively, the MI represents the average rate that taxis typically cover the geodesic distance between two geospatial coordinates.

2.3. *Data Reduction and Cleaning Approach.* The first step of the data cleaning method removed obvious errors. More than 1.5% of the records contained obvious errors such as zero trip times, negative trip times, duplicate records, and blank identifiers (Table 1). The second step was the layered filtering. Repeated attempts to remove records and retest distributions for fitness is computationally intensive. Therefore, the strategy used was to run the algorithm on substantially fewer records, namely, those within the lower and upper outlier 2.5 percentile. This approach effectively adds outlier records to the 95% confidence interval only if they likely contribute to an overall distribution that matches the expected distribution of the filter variable for that layer.

Previous research demonstrated that the expected distribution of trip lengths and trip times is lognormal [14–16]. Figure 1 shows the distributions of only the upper and lower 2.5 percentile of each filter variable. The algorithm then identified a cut-off point to eliminate those outlier records that likely fell outside of the overall expected distributions. For the trip duration distribution (Figure 1(a)), a significant number of records clustered very close to zero minute. A minimum slope analysis (first derivative) of that distribution identified a threshold of approximately one minute and one percentile (the arrow) below which it is unlikely that those outlier records (1% of the low outliers) belong to the overall expected distribution. The minimum slope, which was zero

in this case, corresponded to the lowest point of a parabola that fits that portion of the distribution. Hence, the algorithm automatically eliminated them. In a similar manner, the upper 2.5 percentile (Figure 1(b)) had a large number of trips with durations near 62 minutes. The outliers that followed were obvious errors because their trip durations exceeded 60,000 minutes.

An amplified view of the trip time distribution between one and five hours revealed a lognormal distribution followed by a very long tail of outlier records that represent less than 1% of the upper 2.5 percentile. The algorithm automatically eliminated records that did not contribute to the expected lognormal distribution. Possible sources for the extremely short trip duration errors may be passengers changing their mind about a trip after entering a vehicle, or electromagnetic noise interference in the trip meters. A possible source for the extremely long trip duration errors may be trip meter malfunctions that uploaded drop-off times from a memory buffer after restarting.

Applying this statistical filtering technique to the *second* filtering variable further eliminated more than 70,000 records (Table 1). They included trip records with distances of approximately zero kilometres (Figure 1(c)) and obvious outliers that extended well beyond 32 kilometres (Figure 1(d)) that did not contribute to the expected lognormal distribution. GPS signal reflections from tall structures in Dubai may be a source for the unlikely trip distances.

In the third layer of filtering, the algorithm did not eliminate records in the lower 2.5-percentile of mobility indices (Figure 1(e)) because removing them did not move the overall distribution any closer to the expected theoretical distribution. However, the algorithm did eliminate outlier records in the upper 2.5-percentile tail (Figure 1(f)) with mobility indices that distanced the overall distribution from the expected distribution for that variable. This resulted in the elimination of records that exceeded mobility indices of 92 km h^{-1}. This is a reassuring result because the highest mobility index should be less than the highest speed limit in Dubai, which is 100 km h^{-1}. Table 1 summarizes that the overall data cleaning process, including the layered filtering, eliminated approximately 5.5% of the trip records, hence retaining those within a 94.5% confidence interval (Table 1).

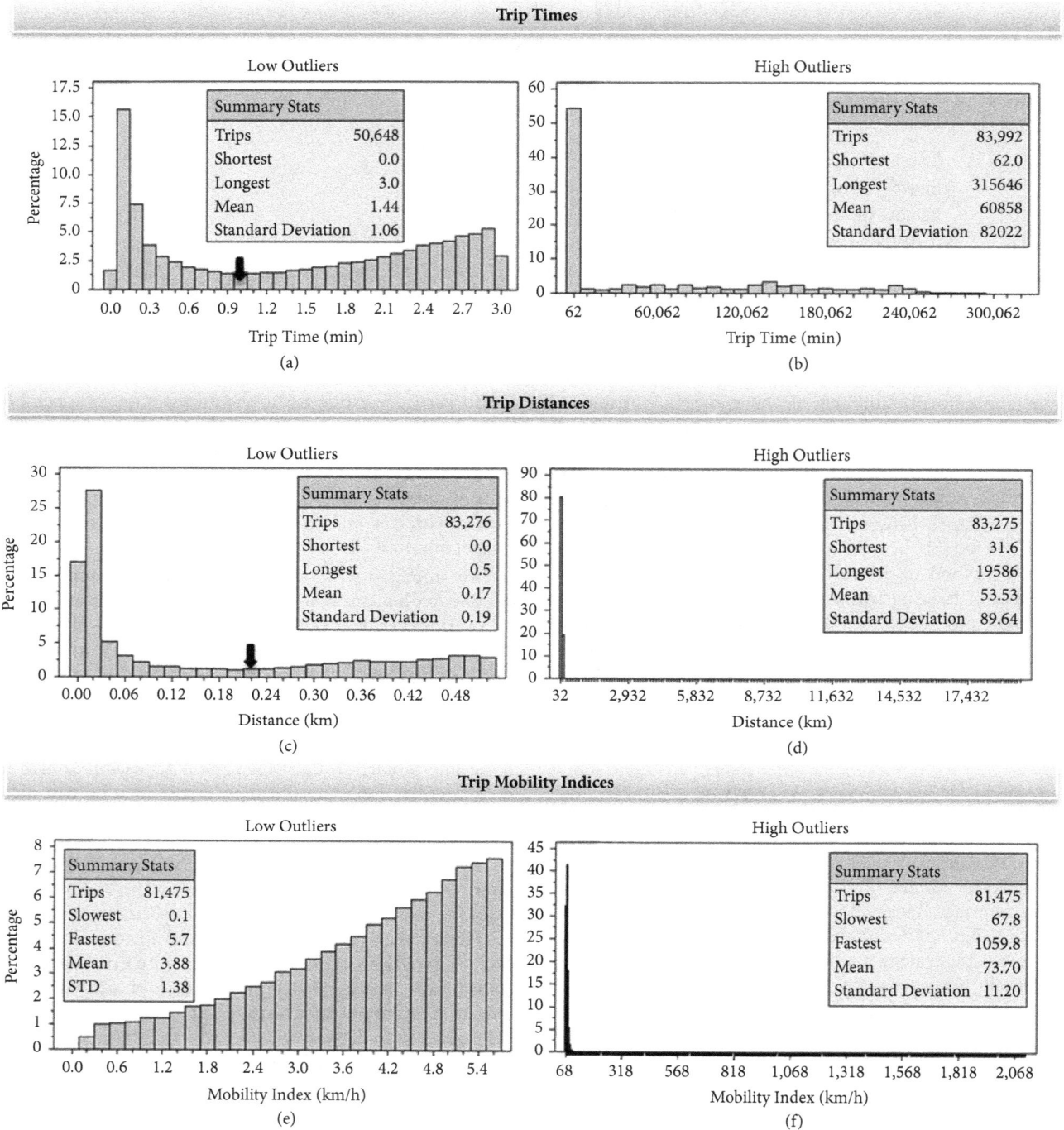

FIGURE 1: Distribution of outlier trip times, distances, and mobility indices.

3. Results

This section evaluates the effectiveness of the layered statistical filtering method by examining the degree of agreement between the distribution of the key variables of the remaining data and their expected theoretical distributions. This section also describes the results of the data mining on the cleaned dataset. A key lesson learned from this research is that the data collection, data preparation, and data cleaning efforts are far greater than those of the actual data mining. Consequently, the organization of this paper reflects those proportions.

3.1. Data Quality Evaluation. The criterion for evaluating the effectiveness of the proposed data cleaning method is the degree to which the distribution of the key variables of the

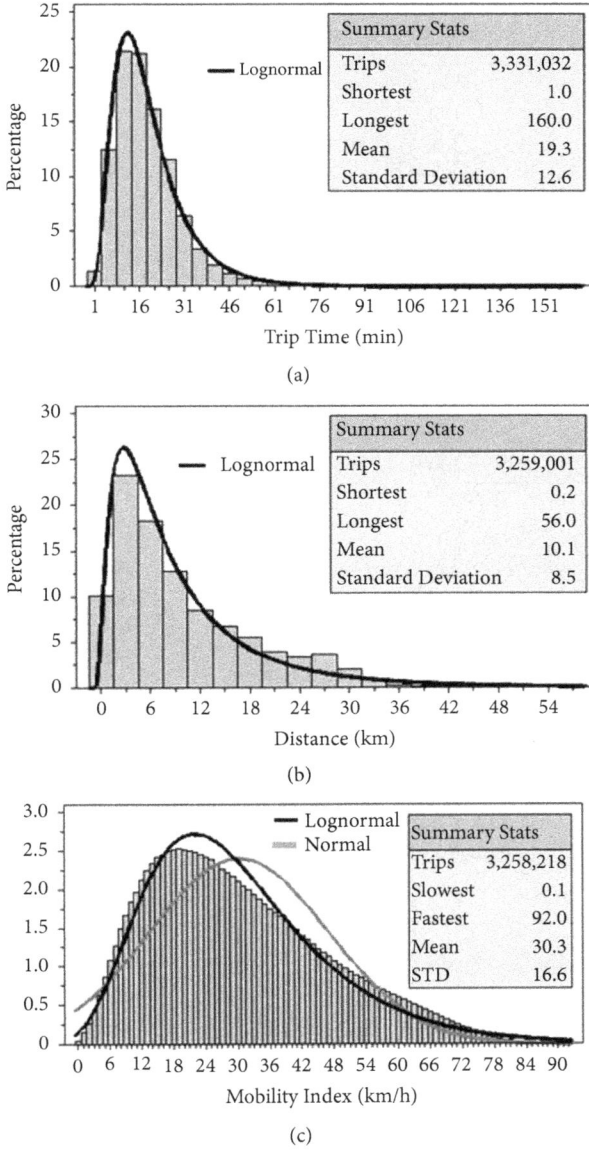

FIGURE 2: Distribution of trip times, distances, and mobility indices after cleaning.

cleaned data agrees with the expected theoretical distributions of the key variables. Figure 2 plots the distribution of the key variables of the cleaned data.

The line plot is the continuous distribution that best fits the histogram of cleaned trip times (Figure 2(a)), geodesic distances (Figure 2(b)), and mobility indices (Figure 2(c)). The iterative Levenberg-Marquardt nonlinear least squares method of curve fitting identified the parameters of the best-fit distributions [17]. The model for estimating the best-fit lognormal distribution $D_{LN}(\xi)$ is

$$D_{\ln}(\xi) = \frac{\gamma_{\ln}}{\xi\sqrt{2\pi\sigma_{\ln}^2}}\exp\left[-\frac{1}{2}\left(\frac{\ln(\xi)-\mu_{\ln}}{\sigma_{\ln}}\right)^2\right]_{\xi>0} \quad (1)$$

The constants $\gamma_{\ln}$, $\mu_{\ln}$, and $\sigma_{\ln}$ are estimates of the amplitude, mean, and standard deviation parameters, respectively. Trip distances are highly correlated to trip times; hence they distribute similarly. The mobility index is a random variable derived from the ratio of the travel distance and the travel time; therefore, it also follows a lognormal distribution. Prior knowledge establishes that the mobility index cannot be zero or infinite because neither the travel distance nor the travel time will be zero or infinite. Therefore, the mobility index is limited to a finite interval. Table 2 lists the statistics of the key variables and parameters of the distributions that best fit their histograms. The variable ΔT is in minutes and ΔL is in km.

The chi-squared goodness-of-fit test [18] indicates when there is a significant difference between the expected frequencies and the observed frequencies of the variables. The null hypothesis H_0 is that the observed distribution of the variables is the same as the candidate distribution. Failure to reject the null hypothesis will result in accepting the alternative hypothesis that there was no significant departure of the observed distribution from the candidate distribution.

The chi-squared statistic (χ^2) is

$$\chi^2 = \sum_{k=1}^{n}\frac{(O_k - E_k)^2}{E_k}. \quad (2)$$

The random variables O_k are the histogram values observed in bin k and E_k are the expected values of the hypothesized distribution. The chi-squared test rejects the null hypothesis if the χ^2-statistic exceeds the critical χ^2 value of a chi-squared distribution evaluated at degrees-of-freedom DF and a specified significance percentage. Statisticians typically set the significance value to 0.05, which represents a low probability of 5% that the test will reject the null hypothesis when in fact it is true. The alternative approach calculates the chi-squared probability values (p-values) that correspond to the observed χ^2-statistic, evaluated at the DF. The tests reject the null hypothesis when the p-values are less than the selected significance percentage.

As shown in Table 2, the chi-squared tests could not reject the null hypothesis for the distributions tested. Therefore, the tests conclude that the trip times and the trip distances of the cleaned and filtered data do not depart significantly from the expected lognormal distribution. By extension, the mobility indices do not significantly depart from the lognormal distribution because it is a dependent variable of the trip times and the trip distances. Subsequently, these tests validated the effectiveness of the proposed layered statistical filtering method of removing records that are likely erroneous.

3.2. Data Mining of Trip Production. Several different drivers can operate a taxi, and a driver can operate several different taxis. The data mining quantified the number of drivers that operated a unique taxi as the level of "taxi-split" and the number of taxis operated by a unique driver as the level of "driver-split." Given the uniqueness of the dataset, the focus of the data mining was to examine the distribution of taxi- and driver-split. Figure 3 captures the data mining results, which shows the distribution of taxi- and driver-splits by

TABLE 2: Parameters of the cleaned distributions.

Histogram	ΔT	ΔL	MI
Bins	160	29	47
Mean	19.3	10.1	30.3
STD	12.6	8.5	16.6
Min	1.0	0.2	0.1
Max	160.0	56.0	92.2
Chi-Squared	Lognormal	Lognormal	Lognormal
$\chi^2\,DF$	157	26	44
γ	101.6	234.6	214.4
μ	2.9	2.2	3.4
σ	0.6	1.3	0.7
χ^2 Critical	187.2	38.9	60.5
χ^2 Statistic	21.1	8.4	10.9
p-value (%)	100	100	100

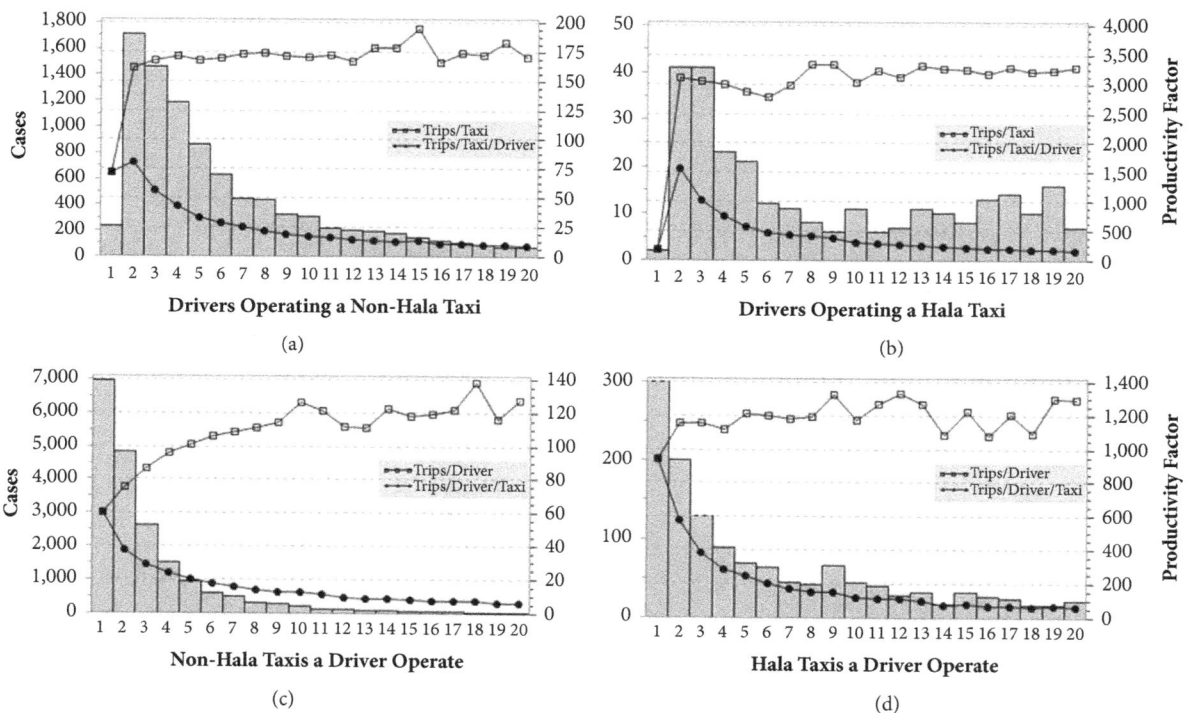

FIGURE 3: Distribution of taxi and driver splits by fleet type.

fleet type. For brevity, the figures show the frequency of *cases* for up to 20 taxis, but the maximum was actually 137. The pattern revealed was that the *productivity factor*, in terms of the number of trips-per-taxi-per-driver, peaked when only two to three drivers operated a given taxi. This two-driver taxi *case* dominated for non-Hala taxis whereas two- and three-driver taxi cases dominated equally for Hala taxis. These cases accounted for 33.5% and 15.2% of the non-Hala (Figure 3(a)) and Hala (Figure 3(b)) taxi-split cases, respectively. Scenarios of single-driver taxis accounted for only 2.5% and 0.4% of the non-Hala and Hala taxis, respectively. They were also among

the least productive of cases in terms of the trips-per-taxi productivity factor.

The data mining results also indicate that a given taxi in Dubai sustains high trip production by assigning as many drivers to them as needed to minimize their parking times (Figures 3(a) and 3(b)). However, driver changes incur significant overhead or off-duty time that reduces a taxi's trip production efficiency. The off-duty time includes time spent in depositing collected cash fares at specific bank locations, and then driving to designated locations to accommodate driver shift changes.

Using the productivity factor of trips-per-driver-per-taxi revealed that the factor peaked for cases of drivers operating a single taxi. These cases accounted for 35.3% and 20.5% of the non-Hala (Figure 3(c)) and Hala (Figure 3(d)) driver-split cases, respectively. Further inspection of the driver-split for the two-driver taxi pools (Figures 3(a) and 3(b)) revealed that their operators came from the pool of drivers of a single taxi (Figures 3(c) and 3(d)). By induction, trip productivity tends to peak when the same two drivers operate a given taxi. RTA was previously unaware that this pattern dominated. However, the pattern was not surprising because they provided two explanations for its commonness. Firstly, drivers can minimize their insurance costs by minimizing the number of different vehicles they operate. Secondly, the logistical complexity and time overhead involved with shift-changes increases with the number of drivers of a given taxi. The observed pattern along with the RTA explanation suggests that new tactics to encourage or facilitate more of the beneficial taxi assignment would likely lead to reduced overhead and enhanced trip production.

4. Discussion and Conclusions

The proliferation of shared mobility services worldwide and their growing variety has led to intense competition with traditional dispatch taxi services. Hence, the goal of this study was to mine large-scale dynamic mobility data from connected taxis to discover beneficial patterns that could inform tactics to improve the competitiveness of dispatch taxi services. However, the huge size, non-uniform composition, variable update rates, and GPS errors complicate the task of data mining. Therefore, the main *objective* and *contribution* of this research was to improve the quality of the dataset by developing an automated data cleaning and filtering method, tailored for such datasets. During the course of this research, the authors learned that the data collection, data preparation, and data cleaning efforts are far greater than that of the actual data mining effort. Therefore, the organization and relative focus of this paper reflects their relative magnitude.

The proposed *layered statistical filtering* algorithm automatically eliminated *outlier* records that contained both obvious errors and likely errors. The *main idea* of the technique was to remove records that moved the distributions further away from the known theoretical distributions of the key filter variables. Validation of the quality of the cleaned data used chi-squared statistical tests to compare the distribution of the three variables with their expected distributions. The tests determined that the overall cleaning procedure, including the filtering algorithm, removed obvious outliers and other poor quality records that represented approximately 5% of the dataset.

Subsequently, the data mining focused on examining taxi trip production as a function of taxi-driver pairing patterns. Such an analysis would not be possible without the uniqueness of the dynamic mobility dataset, which includes identifiers to distinguish individual drivers. The revealed pattern was that taxi trip production peaks for the case where only the same two to three drivers operate the same taxi. The RTA explanation was that drivers could minimize their insurance costs by minimizing the number of different vehicles that they operate. Fewer drivers per vehicle also reduce the logistical complexity and the time overhead of shift-change and cash deposit procedures. Hence, taxi companies in Dubai can use this finding to develop tactics that would encourage more of the beneficial assignment pattern. At this point, it is unknown whether similar patterns exist for dispatch taxi services in other cities of the world.

A limitation of the proposed layered data filtering method is that it relies on known statistical distributions of the selected filter variables. This necessitates the transformation of dynamic mobility data into trip records containing the timestamps and geospatial positions of trip origins and destinations.

Future research will mine the Dubai taxi data to characterize the spatial-temporal dynamics in supply and demand to guide decisions in zonal taxi allocations. The authors will also investigate various methods of predictive analysis to guide driver recruitment, fleet acquisition, network management, scheduling, and revenue management decisions.

Conflicts of Interest

The authors declare that there are no conflicts of interest regarding the publication of this paper.

Acknowledgments

The authors are grateful to the Dubai Road and Transport Authority for their review, appreciation, and feedback on the research outcome. A grant from the University of Modern Sciences, Dubai, United Arab Emirates, supported this research.

References

[1] Y. M. Nie, "How can the taxi industry survive the tide of ridesourcing? Evidence from Shenzhen, China," *Transportation Research Part C: Emerging Technologies*, vol. 79, pp. 242–256, 2017.

[2] M. W. Ulmer, L. Heilig, and S. Voß, "On the Value and Challenge of Real-Time Information in Dynamic Dispatching of Service Vehicles," *Business & Information Systems Engineering*, vol. 59, no. 3, pp. 161–171, 2017.

[3] Xiaowe Hu I, Sh An I, and Jian Wang, "Taxi Driver's Operation Behavior and Passengers' Demand Analysis Based on GPS Data," *Journal of Advanced Transportation*, vol. 2018, Article ID 6197549, 11 pages, 2018.

[4] A. W. Smith, A. L. Kun, and J. Krumm, "Predicting taxi pickups in cities: Which data sources should we use?" in *Proceedings of the 2017 ACM International Joint Conference on Pervasive*

and Ubiquitous Computing and Proceedings of the 2017 ACM International Symposium on Wearable Computers, 2017.

[5] K. Liu, T. Yamamoto, and T. Morikawa, "An analysis of the cost efficiency of probe vehicle data at different transmission frequencies," *International Journal of Intelligent Transportation Systems Research*, vol. 4, no. 1, pp. 21–28, 2006.

[6] T. Dasu and T. Johnson, *Exploratory data mining and data cleaning*, vol. 479, John Wiley & Sons, New York, NY, USA, 2003.

[7] P. D. Groves, L. Wang, and M. Ziebart, "Shadow matching: Improved GNSS accuracy in Urban canyons," *GPS World*, vol. 23, no. 2, pp. 14–29, 2012.

[8] T. Miwa, D. Kiuchi, T. Yamamoto, and T. Morikawa, "Development of map matching algorithm for low frequency probe data," *Transportation Research Part C: Emerging Technologies*, vol. 22, pp. 132–145, 2012.

[9] J. Liu, X. Yu, Z. Xu, K. R. Choo, L. Hong, and X. Cui, "A cloud-based taxi trace mining framework for smart city," *Software: Practice and Experience*, vol. 47, no. 8, pp. 1081–1094, 2016.

[10] H. J. van Zuylen, F. Zheng, and Y. Chen, "Using Probe Vehicle Data for Traffic State Estimation in Signalized Urban Networks," in *Traffic Data Collection and its Standardization*, vol. 144 of *International Series in Operations Research & Management Science*, pp. 109–127, Springer New York, New York, NY, USA, 2010.

[11] Z. Zhang, D. Yang, T. Zhang, Q. He, and X. Lian, "A study on the method for cleaning and repairing the probe vehicle data," *IEEE Transactions on Intelligent Transportation Systems*, vol. 14, no. 1, pp. 419–427, 2013.

[12] C. Yang and E. J. Gonzales, "Modeling Taxi Demand and Supply in New York City Using Large-Scale Taxi GPS Data," in *Seeing Cities Through Big Data*, P. Thakuriah, N. Tilahun, and M. Zellner, Eds., Springer Geography, pp. 405–425, Springer International Publishing, 2017.

[13] T. Vincenty, "Direct and inverse solutions of geodesics on the ellipsoid with application of nested equations," *Survey Review*, vol. 23, no. 176, pp. 88–93, 1975.

[14] H.-C. Chu, "An empirical study to determine freight travel time at a major port," *Transportation Planning and Technology*, vol. 34, no. 3, pp. 277–295, 2011.

[15] N. Wu and J. Geistefeldt, "Standard deviation of travel time in a freeway network - A mathematical quantifying tool for reliability analysis," in *Proceedings of the 14th COTA International Conference of Transportation Professionals: Safe, Smart, and Sustainable Multimodal Transportation Systems, CICTP 2014*, pp. 3292–3303, China, July 2014.

[16] E. Durán-Hormazábal and A. Tirachini, "Estimation of travel time variability for cars, buses, metro and door-to-door public transport trips in Santiago, Chile," *Research in Transportation Economics*, vol. 59, pp. 26–39, 2016.

[17] P. E. Gill, W. Murray, and M. H. Wright, *Practical Optimization*, Academic Press, n, 1981.

[18] A. Papoulis, *Probalility, Random Variables, and Stochastic Processes*, McGraw-Hill, New York, NY, USA, 4th edition, 1991.

7

Aspects of Improvement in Exploitation Process of Passenger Means of Transport

Marian Brzeziński ⓘ**, Magdalena Kijek** ⓘ**, Paulina Owczarek** ⓘ**, Katarzyna Głodowska** ⓘ**, Jarosław Zelkowski** ⓘ**, and Piotr Bartosiak**

Military University of Technology, Faculty of Logistics, Institute of Logistics, Warsaw 00908, Poland

Correspondence should be addressed to Magdalena Kijek; magdalena.kijek@wat.edu.pl

Guest Editor: Vladislav Zitricky

Effective exploitation of means of transport in transport companies is one of the most important ways of achieving competitive advantage. Mentioned problem is particularly important in the market of passenger transport services in large agglomerations, because it has a social aspect in addition to the economic dimension. In addition, most often the studies concern single objects of exploitation, while the subject of research are groups of objects of passenger transport means. The main objective of the study is to analyze and evaluate the system of exploitation of passenger transport means and to propose solutions for its improvement. On the basis of the theory of exploitation systems, quantitative utilitarian models have been built, which have been verified by applications using data obtained from Municipal Communication Company (MPK) in Wroclaw. Originality and innovation in the recognition of the research problem consist in applying to the analysis and evaluation of the Ishikawa diagram exploitation system, Pareto-Lorenzo analysis, and FMEA (Failure Mode and Effects Analysis) methods. On the other hand, a QFD (Quality Function Deployment) diagram was used to build a model of improvement of the exploitation system, with the use of which the target values of parameters for the operation of MPK passenger transport in Wroclaw were determined. The applied methods, techniques, and research tools are rarely used in the field of testing of vehicle operation systems. The work has a very practical character and built models can be used in other urban agglomerations in order to improve the operation of passenger transport means.

1. Introduction

Currently, one of the most important problems of city management is the issue of communication and development of urban transport systems [1–4]. This is mainly caused by a large number of vehicles on the road, low capacity of selected transport routes, and relatively weak condition of the linear transport infrastructure [5, 6]. In urban areas with a significant density of road infrastructure, solving transport problems by expanding the infrastructure is not very effective, because any bandwidth reserve obtained in this way is immediately used [7–9]. Among the effective methods of improving the efficiency and quality of the transport system, the use of advanced technological and organizational solutions is mentioned [10–13].

Efficient use of means of transport in any type of company is one of the major ways to achieve competitive advantage [14–17]. This problem is particularly important in the area of passenger transport market in large agglomerations, because except its social aspect it also has the economic dimension. In addition, most of the studies concern individual facilities, while the subject of research are groups of objects of means of passenger transport.

The issue of evaluating and improving the exploitation system of means of passenger transport is an extremely important issue from a practical and a theoretical point of view [15, 18, 19].

The main objective of the study is to analyze and evaluate the operation of the system of passenger transport and to propose solutions to improve it.

In the process of development of analysis and evaluation of the operating system, Ishikawa diagram, Pareto-Lorenz analysis, and FMEA (Failure Mode and Effects Analysis) [20], among others, have been used, while the QFD (Quality Function Deployment Diagram) was used to build a model for the improvement of the exploitation system, which determined

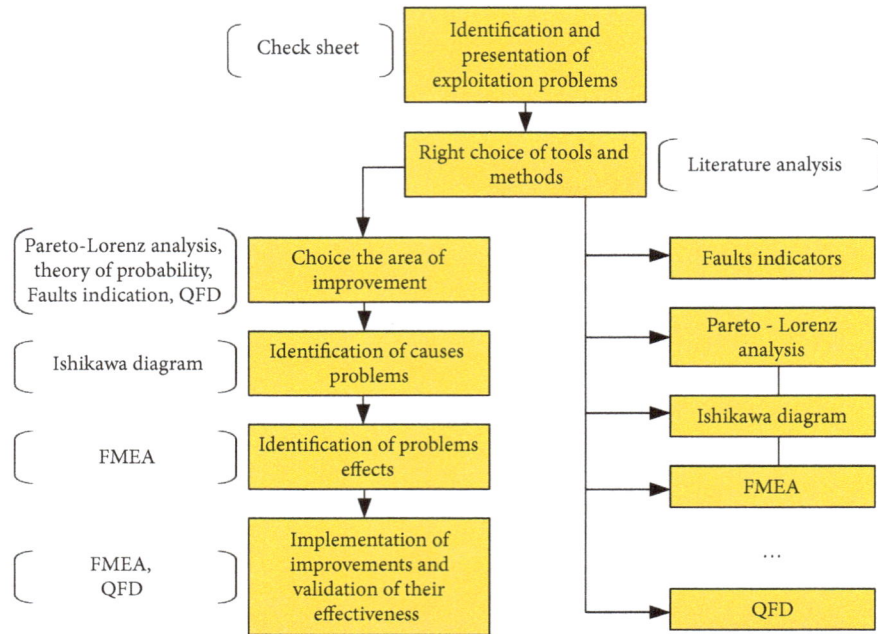

FIGURE 1: Improvement model of the exploitation process of means of passenger transport, on the example of urban buses.

the target values of the exploitation parameters of MPK passenger transport in Wroclaw [9, 21, 22]. The methods, techniques, and research tools used are rarely applied in the field of vehicle operating systems.

2. Modelling the Improvement of Exploitation Process of Means of Passenger Transport

2.1. Assumptions to Build a Model for Improving the Exploitation Process. Proper exploitation of buses contributes to the fulfilment of a certain level of quality of transport services. Operational technical parameters are used to assess these services [23]. With the development of cities and the increasing number of people in agglomerations, transport companies have to constantly search for optimal solutions. One of them is, among others, improvement of the vehicle exploitation process, which will ensure their reliability at a high level [21, 24, 25].

The article presents the model of improvement of the exploitation process and its verification on the example of urban public transport company in Wroclaw. This verification will be based on the analysis of bus damage from the year 2014. The study was carried out on the basis of selected systems failures occurring on buses, such as the braking system, electrical system, bodywork, transmission system, suspension system, steering system, engine with attachments, and chassis [22, 26, 27].

In Poland there are 12 agglomerations treated as metropolises belonging to the Union of Polish Metropolis: Bialystok, Bydgoszcz, Gdansk, Katowice, Cracow, Lublin, Lodz, Poznan, Rzeszow, Szczecin, Warsaw, and Wroclaw. Giving the city the character of a metropolis is to sanction the specifics of highly urbanized areas, stimulating the

comprehensive economic development of the whole country. There are 17 million people living in metropolitan areas, over 40% of Poles. These cities produce 42% of Poland's GDP.

Accordingly, in the workplace of the place where the analysis of urban bus damage was carried out, the City Transport Company was selected, Wroclaw. The company has a large number of vehicles, due to its high density of population and a strong urban character of the city, which has made the most reliable test results possible, hence the choice.

Infrastructure of the Municipal Communication Company in Wroclaw (MPK Wroclaw) consists of 65 bus lines: 52 daily lines and 13 night lines. Fleet of MPK Wroclaw consists of over 300 buses, including 69 city bus subcontractors-Michalczewski Sp. z o. o. Out of all the bus operators that are stationed in city car barns, an analysis was performed on the damage of 110 selected buses.

2.2. Model of Improvement in the Process of Exploitation of Passenger Means of Transport. In city transport, the key element of management is control and the provision of the highest possible quality of service. This requires constant monitoring of the current state of affairs and processes to ensure their efficiency. Figure 1 shows improvement model of the exploitation process in means of passenger transport on the example of the Urban transport company (MPK Wroclaw).

3. Application of the Model for Analysis and Evaluation of the Passenger Transportation System

The first step in the improvement model is the identification and presentation problems of exploitation of the vehicle

TABLE 1: Damages and exploitation data of city bus operator MPK Wroclaw from 2014 within 1 year.

Name of damaged bus system	Model of the bus				
	Mercedes-Benz O530 K Citaro	Mercedes-Benz O530 Citaro	Mercedes-Benz O530 G Citaro	Mercedes-Benz O530 G Citaro 2	Jelcz 120MM
Braking system	11	56	48	8	217
Electrical system	19	89	71	10	167
Bodywork	2	52	32	1	184
Engine with attachments	0	16	14	2	134
Suspension system	5	37	29	1	15
Steering system	5	18	15	3	13
Transmission system	12	60	46	8	39
Chassis	0	20	15	2	31
Number of bus failures-LU	54	348	270	35	800
Average age [years]	3	5,7	5,7	3	18,7
Total bus mileage-P [km]	65,818	4,496,745	2,709,404	54,824	395,631
Number of buses (total)	1	58	42	1	8

group under study. It was done with a check sheet for damage to individual systems of different models of city buses. In addition, the average age and mileage of vehicles are reported. The data is shown in the Table 1.

The study was conducted on 110 city buses divided into 5 types:

(i) Mercedes-Benz O530 Citaro K-Class MAXI

(ii) Mercedes-Benz Citaro O530-Class

(iii) Mercedes-Benz Citaro O530 G-Class MEGA

(iv) Mercedes-Benz Citaro O530 G 2-MEGA

(v) Jelcz 120MM-Class MAXI

The second step for improving the operation process is to choose the proper methods and tools. Based on the analysis of literature and the results of the years of research in the field of improvement of operation, test sheet, analysis of Pareto-Lorenz, indicators of damage, Ishikawa diagram, FMEA, and QFD were chosen.

The third step of the improvement model is the choice of areas of improvement, which is to determine what damage should be seen about first. For this purpose, the Pareto-Lorenz graph was used. On the left side the total number of defects is found; on the right, the percentage of defects is found.

Analysis of the graph in Figure 2 shows that 80% of all the damage is caused by 4 systems: the electrical system, brake system, bodywork, and engine with attachments. This means that they need to be addressed first. For verification of this method of work, a fault indicator of individual systems of the type was also used. The damage analysis was based on the selected operational data of the Wroclaw City Bus Operator-MPK Wroclaw. An assessment of the number of damage to a certain type of system per thousand kilometers over one year

of usage was made by using a W index that was determined from the relationship:

$$W = \frac{LU'}{P} \times 10^3 \qquad (1)$$

where LU' is number of defects of each team within 1 year, P is total mileage of buses of a given type within 1 year.

The analysis mentioned in Figure 3 takes into account vehicle mileage and determines the failure rate for each of the systems considered, for each 1000 km traveled. It can be said that Jelcz 120MM has the highest damage index in 5 out of 8 systems: bodywork, electric system, brake system, chassis, and transmission system. The Mercedes-Benz O530 K Citaro has the lowest damage rating in 3 of the 8: suspension system, steering system, and engine with attachments (Figure 3).

After defining areas for further research, one must move on to the next, fourth stage of the model, identification of causes and propose solutions to problems. To that end, examples of potential causes are presented, in the form of Ishikawa diagrams (Figure 4), and there are proposed solutions in the form of Table 2 for each of the previously specified systems, for

(i) braking system, weak braking,

(ii) electrical system, problems with starting the engine,

(iii) bodywork, corrosion of the bodywork,

(iv) engine with attachments, weak air conditioning.

The Isikawa diagram allows identification of possible causes of damage. Figure 4 shows an exemplary Ishkava diagram for the engine system with attachments divided into 6 main causes of damage: materials, machines, personnel, measurements, methods, and environment.

Table 2 presents examples of solutions of the described problems for all systems. The table has been divided into three

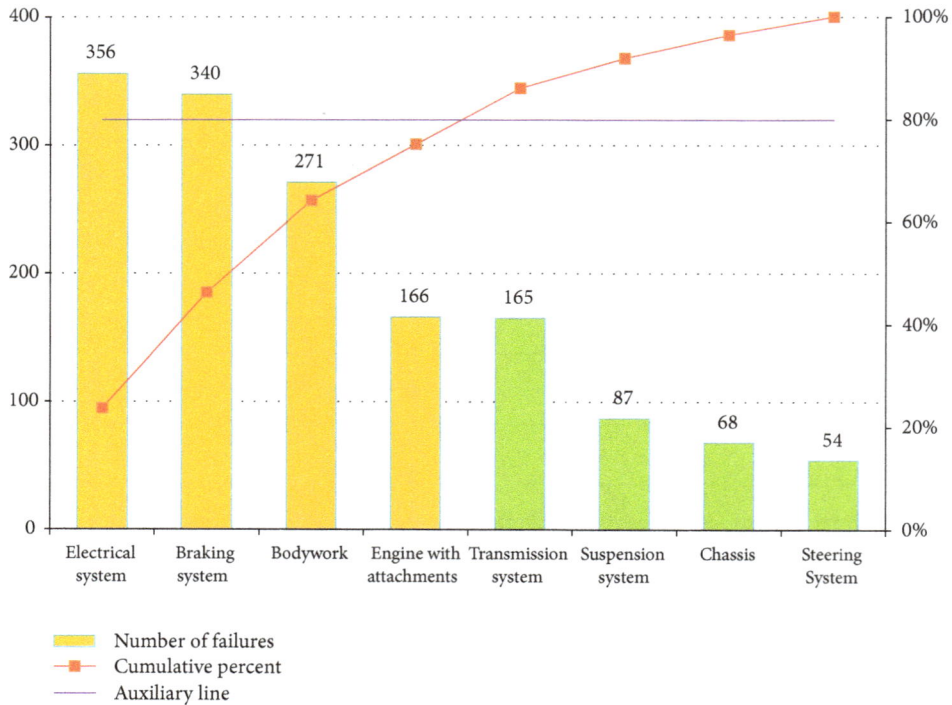

FIGURE 2: Pareto-Lorenz diagram of damage to individual bus systems based on the control sheet.

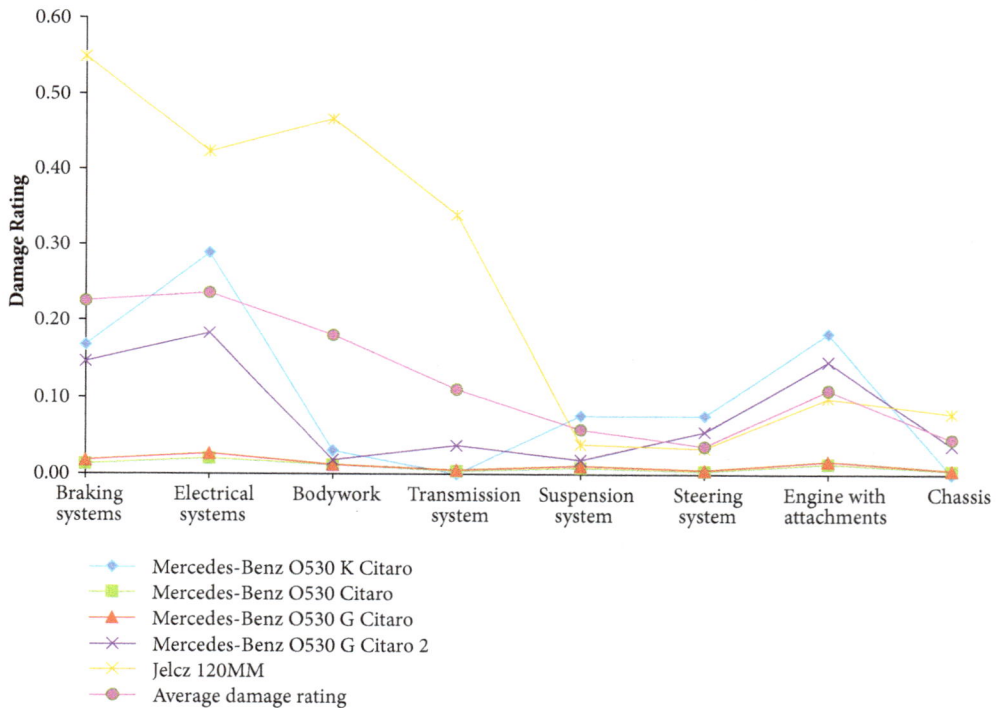

FIGURE 3: Damage indicators of individual buses of MPK operator Wroclaw.

columns: type of problem, its cause, and suggested solution. Considering the first system, the biggest problem and the cause of corrosion of the body are weak storage and improper maintenance. Therefore, actions are necessary to improve the storage and maintenance of buses by providing vehicles with heated garages (in winter), washing them frequently, and protecting them with protection measures. Problems with the engine system with attachments include inefficient air

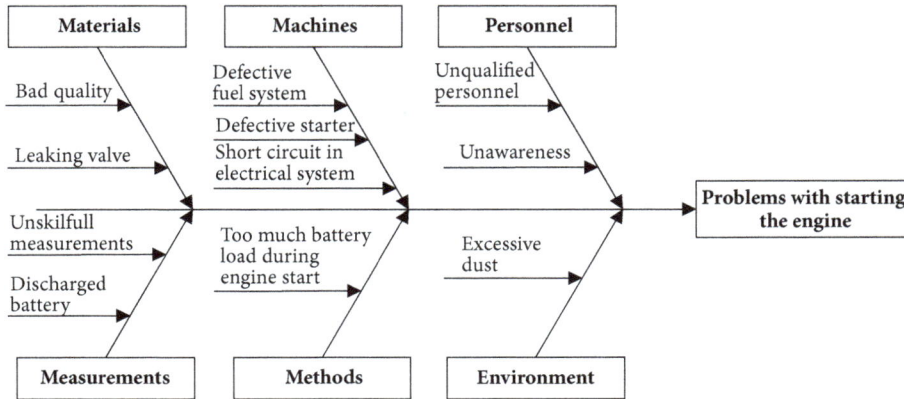

FIGURE 4: Ishikawa diagram for the electrical system.

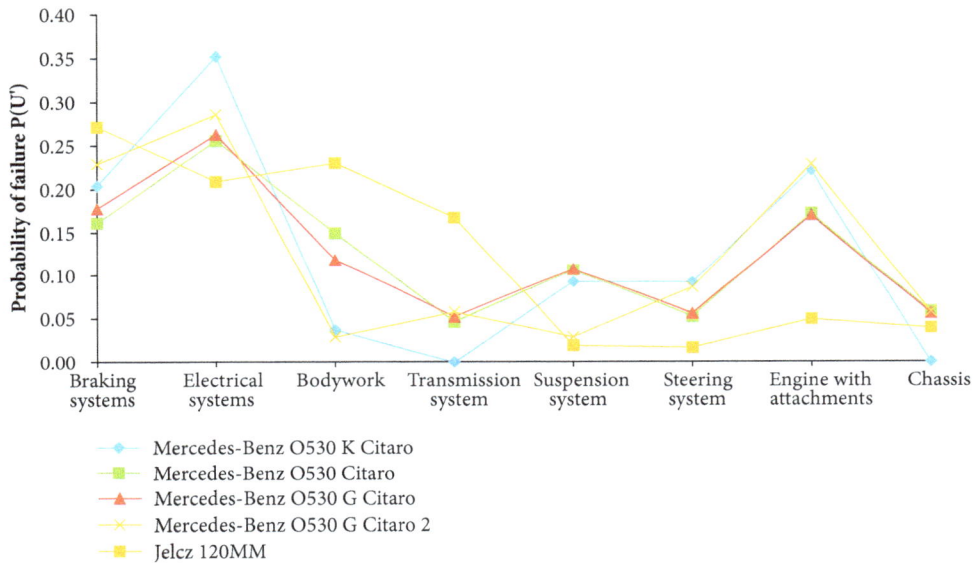

FIGURE 5: Values of the probability of failure for each of the bus systems in city buses.

conditioning. The main reasons for this issue can include primarily improper handling and use of the system. The proposed changes include starting the air conditioning system with the windows closed (securing the possibility of opening the windows to other passengers) and more frequent replacement of the refrigerant. In the case of an electrical system the most important causes are the lack of control over the battery level and improper use of vehicles by drivers. Proposed solutions include, among others, periodic training of employees increasing awareness and daily diagnostics of vehicle fleet by a qualified personnel. The last to be taken into consideration is the braking system, which is inefficient braking the vehicle. The reasons for this are the weak technical service of vehicles before leaving the depot. It is therefore necessary to daily service vehicles by checking individual fluids, condition of brake pads, and discs.

The next, fifth stage of the model is the implementation of improvements and validation of their effectiveness, implemented using FMEA method. This method consists

in calculating the Risk Priority Number (RPN), which is designated by the dependence [18]:

$$RPN = S \times O \times D \qquad (2)$$

where S is severity, O is probability of occurrence, and D is detection.

To determine the probability of failure of the analyzed system P, the following relationship is used:

$$P\left(U'\right) = \frac{LU'}{LU} \qquad (3)$$

where LU' is the number of types of damage of a given system for the type of bus under examination within 1 year and LU is total number of failures for the type of bus under examination within 1 year.

Figure 5 presents the probability of failure of each system in vehicles per 1000 kilometers driven. The highest probability of failure is related to the electrical system

TABLE 2: Proposed solutions for detected problems.

(a)

Type of problem	Causes of the problem	Suggested solutions
Bodywork		
Corrosion of the body	Weak storage of vehicles	Heating garages in winter
		Storage of vehicles in well-ventilated and dry garages
	Inadequate maintenance	Frequent washing of vehicles
		After cleaning, protective measures for varnish protection
		Increased control services in search of rust fires
	Inadequate manufacturer's corrosion protection	Inspections of vehicles for paint defects and their protection if necessary
		Additional anti-corrosion protection of closed profiles, e.g. doors

(b)

Type of problem	Causes of the problem	Suggested solutions
Engine with attachments		
Weak air supply	Defective air conditioning compressor	Do not turn on the air conditioning system with the windows open
		Before switching on the air conditioning, ventilate the vehicle
	Leaks in the air conditioning system	Periodic checking of the tightness of air conditioning system connections
		More frequent diagnosis of errors using the OBD II interface
	Too small amount of refrigerant every 2 years	Compulsory refrigerant exchanges every 1 year
	Faulty condenser	Periodic checking of the condenser for mechanical damage
Electrical systems		
Problems with starting the engine	Defective starter	Replacement of damaged starter components
	Defective fuel systems	Instructing drivers with shorter start-up time
	Uncharged battery	Periodic checking of battery voltage
		Charging the battery when stationary
	Short circuit in electrical system	Connection of diagnostic equipment and detection of faults
	Too much battery load during engine start	Disconnection of unnecessary current collectors
	Unqualified personnel	Increased number of employee training
	Unawareness	Software systems supporting analysis of driving parameters
Braking systems		
Weak braking	Aerated braking system	More frequent de-aeration of the brake system
		Control of brake fluid level by drivers
	Used brake fluid	More frequent change of brake fluid
		The use of brake fluids with better technical parameters
	Badly installed brake pads or discs	Additional training for mechanics
		Disciplining the drivers of the machine park
	Greasy brake pads or discs	Cyclic use of degreasing agents

(Mercedes-Benz O530 K Citaro 35%) and average 22% concern on failures in the brake system. The third-largest probability is the transmission system.

To define the number of defect risks the average value of all bus failures for individual systems was used, as shown in Figure 6.

For the determined average values of damage of individual systems, the appropriate number O (Table 3) is assigned. O is the probability of occurrence of a given damage in a predetermined or determined time interval [11, 18].

S is a dimensionless number denoting severity, which is an estimation of how the effects of a given damage affect

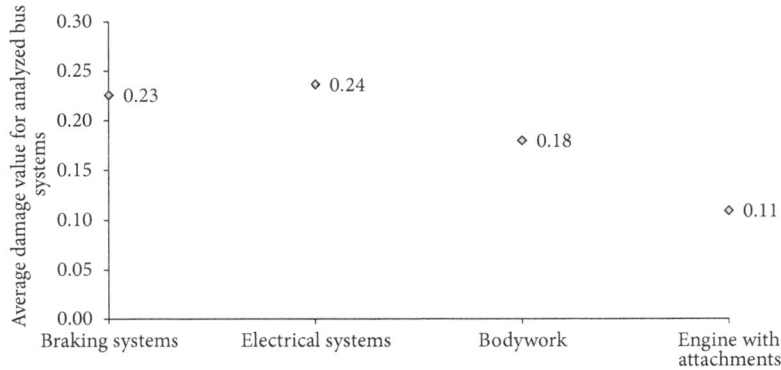

FIGURE 6: Average damage value for analyzed city bus systems.

TABLE 3: Failure mode occurrence related to frequency.

Failure mode occurrence	Frequency	Rating, O
Remote:	< 1/5000	1
Failure is unlikely		
Low:	1/5000	2
Relatively few failures	1/2000	3
Moderate:	1/1000	4
Occasional failures	1/500	5
	1/200	6
High:	1/100	7
Repeated failures	1/50	8
Very high:	1/10	9
Failure is almost inevitable	≥ 1/2	10

the use of a given system. In order to determine the severity for individual systems, the following evaluation was used (Table 4) [11, 18].

D means detection, i.e., determining the chance of identifying and eliminating a given damage before it hits the system or its use. Table 5 presents the evaluation criteria for determining the detectability for individual systems [11, 18].

After specifying the individual parameters S, O, and D, Table 6 shows the first FMEA analysis for the specified systems.

Such analysis is a valuable source of information for the manager of the operation of a public transport company, when choosing corrective measures. After the analysis, the target value of the Priority Risk Number (RPN) was determined and after analyzing the literature it was decided that the target RPN was set at RPN level of 240.

The next step is to check the effects of the proposed corrective measures, whether the RPN of each test system is less than 240; therefore four conditions must be fulfilled:

$$RPN > RPN_{\text{BRAKING SYSTEM}} \qquad (4)$$

$$RPN > RPN_{\text{ELECTRICAL SYSTEM}} \qquad (5)$$

$$RPN > RPN_{\text{ENGINE WITH ATTACHMENTS}} \qquad (6)$$

$$RPN > RPN_{\text{BODYWORK}} \qquad (7)$$

If after the first FMEA analysis the values of the individual RPN's do not meet the above inequalities, the FMEA analysis with the subsequent corrective measures shall be repeated until the desired values are achieved. Before the corrective actions, all RPN values of each system exceeded the RPN = 240, which was also done and shown by the graph in Figure 7.

After corrective actions, FMEA analysis was again performed and the RPN values of individual systems were checked to see if they went under the below target value. Figure 8 shows a graph with RPN values for individual systems after the first FMEA analysis. It can be pointed out that for three of the systems the RPN value is below the target value (240). The greatest improvement was noted at the brake system, because the RPN value dropped from 810 to 180. In the case of the electrical system, the value decreased from 432 to 144, while in case of the body system the RPN decreased from only 324 to 140. However, further corrective actions for the 4th system (engine with attachments) are necessary.

Table 7 presents the second failure mode and effects analysis (FMEA) for engine with attachments systems and corrective actions for this systems. In turn, another results are presented in Table 8.

After performing the corrective action for the engine with attachments in the second FMEA analysis, it was found that the value of RPN_{EWA} decreased to a value equal to only 98, which is the result of more than twice lower than that required. Finally, after all corrective actions, the RPN values were

$$RPN = \{98, 140, 144, 180\} \qquad (8)$$

It was found that the improvement activities had the desired effect, as illustrated in Figure 9 showing a comparison of RPN values before and after corrective measures based on FMEA analysis.

The last (sixth) stage of the improved model for the exploitation of passenger vehicles is the QFD, which shows the relationship between customer requirements and the technical characteristics of products or services (Figure 10).

TABLE 4: Failure mode severity.

Severity	Criteria	Ranking
None	No discernible effect.	1
Very minor	The defect is irrelevant and the user will hardly feel its effects (perceived by less than 25% of users).	2
Minor	The defect is irrelevant and the user will hardly feel its effects (perceived by 50% of users).	3
Very low	A defect of medium importance, causing user dissatisfaction. Seen by the majority of users (about 75%).	4
Low	A defect of medium importance, causing user dissatisfaction. The user feels its effects and is a bit dissatisfied.	5
Moderate	A defect of medium importance, causing user dissatisfaction. The user feels its effects and is dissatisfied.	6
High	A defect of great importance, resulting in reduced system performance. User very dissatisfied.	7
Very high	Inoperative system (loss of the primary function).	8
Hazardous with warning	A defect of very high importance, affecting the safety of use and/or entails failure to comply with government regulations, with warning.	9
Hazardous without warning	A defect of very high importance, affecting the safety of use and/or entails failure to comply with government regulations, without warning.	10

TABLE 5: Failure mod detection evaluation criteria.

Detection	Criteria: likelihood of detection by Design Control	Ranking
Almost certain	The inspectors will almost certainly detect a possible defect and the subsequent damage.	1
Very high	A very good chance that the inspectors will detect a possible defect and the subsequent damage.	2
High	A high chance that the inspectors will detect a possible defect and the subsequent damage.	3
Moderately high	Moderately high chance that the inspectors will detect a possible defect and the subsequent damage.	4
Moderate	A moderate chance that the inspectors will detect a possible defect and the subsequent damage.	5
Low	Low chance that the inspectors will detect a possible defect and the subsequent damage.	6
Very low	Very low chance that the inspectors will detect a possible defect and the subsequent damage.	7
Remote	A small chance that the inspectors will detect a possible defect and the subsequent damage.	8
Very remote	A very small chance that the inspectors will detect a possible defect and the subsequent damage.	9
Absolutely uncertain	Inspectors will not detect and/or cannot detect a possible defect and subsequent damage. Or no system control.	10

4. Target Values of Exploitation of Passenger Transport Parameters

The QFD method is a way of "translating" opinions and needs of customers into a technical language, understandable in the company by designers, builders, and technologists. It serves to translate market requirements into conditions that an enterprise must meet.

The use of this method is caused by reflection, that the decisive factor standing behind the financial condition of the

companies is the buyers of their products. Even if the product is correct from an engineering point of view, it does not have to provide economic success because it is determined by the consumer market, the customer.

Similar dependencies could be set for the vehicle designer and engineer having to build it. Based on the data above, a QFD diagram has been developed in the form of a "quality house", Figure 11.

The goal of a quality home is to set critical parameters and set their target values in such a way as to ensure success in the

TABLE 6: First FMEA analysis for individual systems.

Specificity defects	Effect	Cause	S	O	D	RPN	Corrective Action
Bodywork							
Corrosion of bus body	Dissatisfaction users Weakness of supporting structure of the bus	Wrong storage and maintenance of the vehicle	4	5	9	324	Improve the method maintenance and storage Make additional corrosion protection, in particular closed profiles
Engine with attachments							
Inefficient work of the air conditioning system	Too high temperature and air humidity in the bus, preventing comfortable traveling	Damaged compressor Leaks in the air conditioning system Insufficient amount of refrigerant	7	9	7	441	Regular servicing Don't turn on the air conditioning system while the windows open Replace the refrigerant every 1 year
Electrical systems							
Problems with starting the engine	Delays in commuting bus trips	Failed starter Uncharged battery Short circuit in electrical system Too thick engine oil during start-up	8	9	6	432	Replacement of damaged starter components Rechargeable batteries during stoppages The use of seasonal engine oils
Braking systems							
Inefficient braking	The bus driver is not able to brake hard enough in an emergency situation	Aerated braking system Badly fitted brake pads or discs	10	9	9	810	Frequent venting of the system and control condition of the brake fluid Additional training and disciplining their mechanics

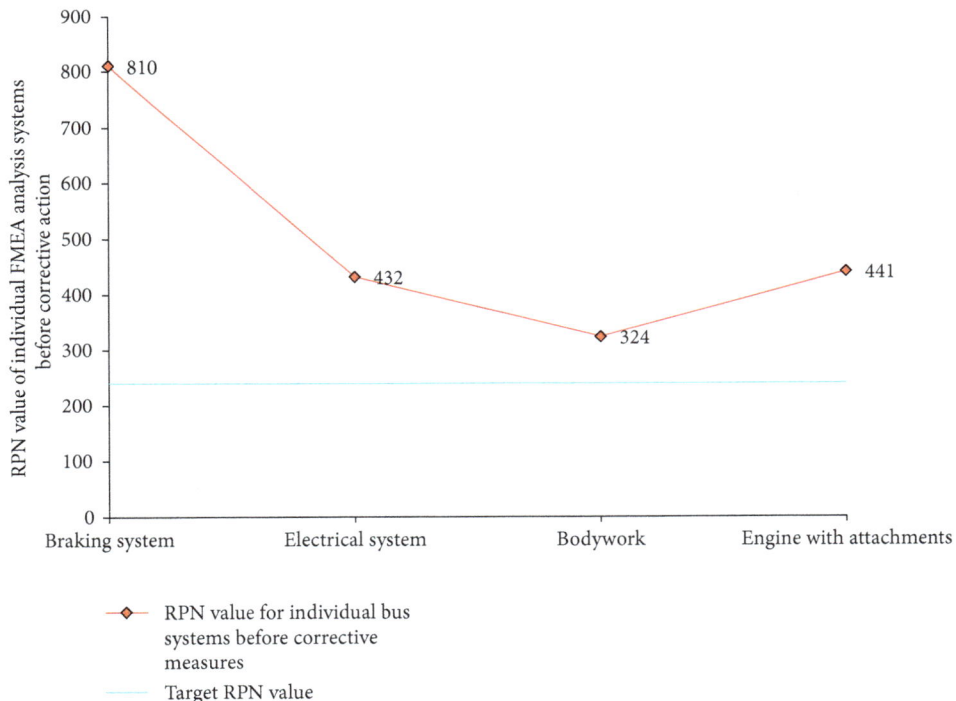

FIGURE 7: RPN value each of the bus systems before corrective action based on FMEA.

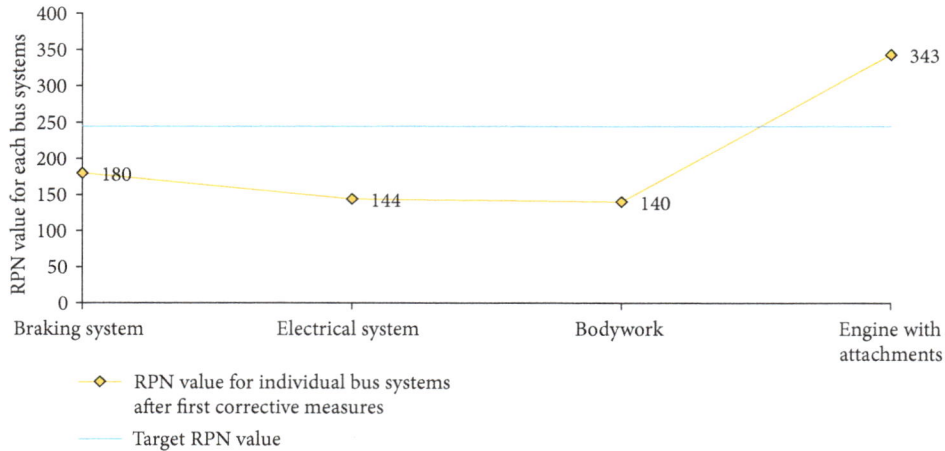

FIGURE 8: RPN value for each of the bus systems after the first phase of corrective action based on FMEA.

TABLE 7: Second FMEA analysis for the engine with attachments system.

Specificity defects	Effect	Cause	S	O	D	RPN_{EWA}	Corrective Action
Inefficient operation of the air conditioning system	Too high temperatures and humidity in the bus make it impossible to travel comfortably	Compressor damage Leaks in the air conditioning system Too little refrigerant	7	7	7	343	Ventilate the vehicle before turning on the air conditioner Periodic air tightness check

TABLE 8: Results after second FMEA analysis of the engine with attachments system.

Specificity defects	S	O	D	RPN_{EWA}	Further corrective action
Inefficient operation of the air conditioning system	7	2	7	**98**	No corrective actions

market for the services or products offered. In the presented case, 3 target parameters were defined:

(1) Servicing in the cycle 1 time per month for each of the buses.

(2) Driver training courses in the cycle 6 times per 1 year for each driver.

(3) Vehicle reliability at the average level of 1 failure per 2 weeks for each vehicle.

5. Conclusions

Based on presented study, the following conclusions were made:

(i) The process of exploitation of city buses in Wroclaw is not full correctly realized and requires improvement.

(ii) Based on the analysis of Pareto-Lorenz, it was stated that around 80% of all damage was generated by the half of investigated systems.

(iii) The proposed solutions to the problems in the various systems, based on the causes of their development

and the Ishikawa diagram have brought measurable benefits in the FMEA method.

(iv) With the FMEA analysis, a reduction in the number of RPNs was achieved in the range of 57% to 78%, most already after the first phase of corrective action.

(v) Based on the QFD method, the relationships between customer requirements and the technical parameters of city buses were determined and 3 of them were diagnosed as critical parameters: servicing, driver training courses, and vehicle reliability.

In addition, one should choose the proper operating strategy, depending on the nature of the work of the vehicle. If sudden breakdowns that prevent the vehicle from operating properly will not cause additional losses such as lost earnings and loss of image, a reactive strategy can be used to repair after the failure; otherwise preventive strategies should be chosen, much more reasonable, that is, to prevent damage. Using the QFD method that allows to improve quality causes that a company dedicated to the well-known needs of the customer has a clear advantage over its competitors. Detailed knowledge about the current needs of customers allows

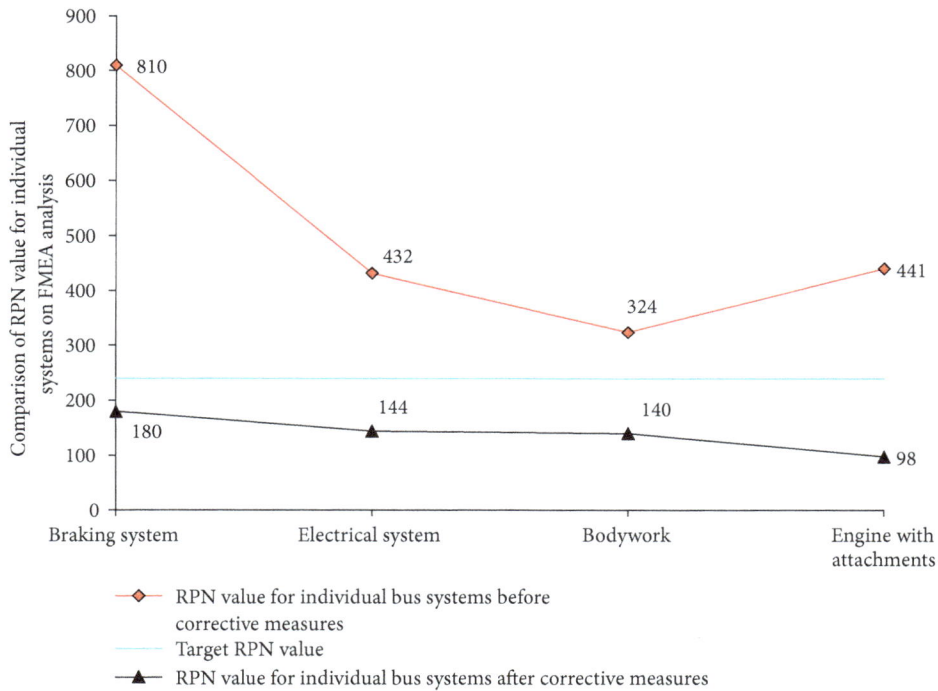

FIGURE 9: Comparison of RPN values for each of the bus systems before and after corrective actions based on FMEA analysis based on FMEA.

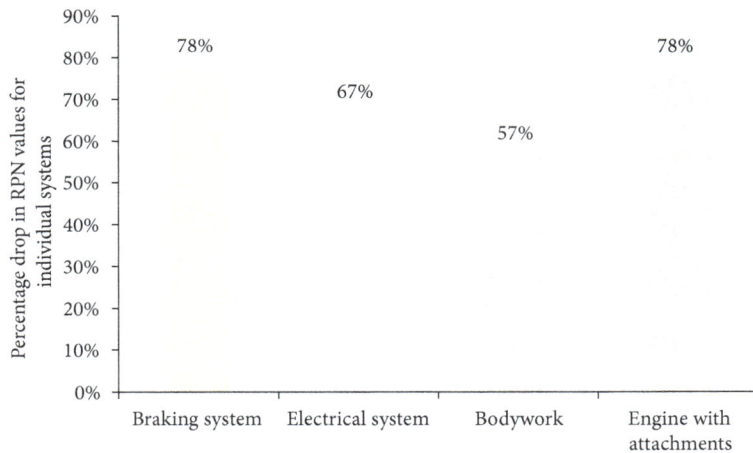

FIGURE 10: Percentage drop in RPN values for each of the bus systems.

us to take action to ensure optimal satisfaction of them. When refining and modifying the undertaken activities, it is important for marketing of the company to remember to continually show its strong sides in order to provide the company with a competitive advantage.

At present, significant development of all types of quality improvement activities of processes or services is evident, due to increasing customer demands and ever-increasing competition.

Built analysis model and evaluation of the exploitation of passenger transport have been verified with actual data

obtained from the MPK in Wroclaw. It can find practical use in other urban agglomerations, which would improve the operation of passenger transport.

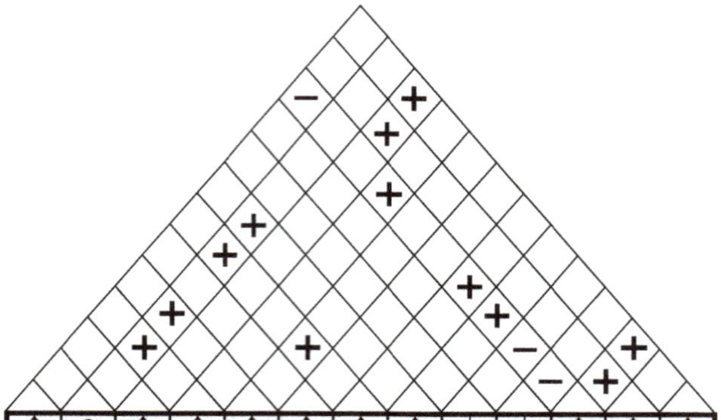

Customer Needs	Imp.	accessories of the vehicle	climatic conditions in the vehicle	servicing	drivers training courses	disciplining drivers	modern technologies	washing & cleaning vehicles	quality od mechanics work	age of the vehicle	Service life of the vehicle	reliability of the vehicle	Adjustments and maintenance	daily service	1	2	3	4	5
working interior lightning	2	3		1			1		3	2	2	2	2	2		B	A		C
pleasant smell	5	1	3	2				3	1	1	1						C	A	B
smooth driving	8			1	3	3	1		1	1	1	1	1			A		C	B
cleanliness inside & outside	8		1					3				1	1				A	B	C
efficient heating	8		2	3					3	1	1	1	1				B	A	C
efficient air conditioning	8		2	3					3	1	1	1	1			A		B	C
fast driving	5			1	3	3			1	1	1	1			B	C			A
punctuality od arrivals and	9	1		1	3	3	1					1			B	C			A
optimal temperature	7		3	2	3							3	1			A		B	C
collision-free driving	9	3		1	3	2											A	B	C
fault-free driving	8			3					3	3	3	3	3	3		A		C	B
safe diving	8	3			3	3						2				A		B	C
quiet driving	5			1					1	1	1	3				A		B	C
easy to open windows	4	1		1				1	1	1	1	1	1			C	B	A	
working validators	3			1					1				1	1	C		A	B	
Technical Evaluation		75	76	141	138	108	19	48	108	105	111	122	74	39	1	2	3	4	5
Specification Target Value		as expected of users	21°C	1/month	6/year	if necessary	as many as possible	1/week	no mistakes at work	max 5 years	max 2 mln km	1 failure/2 weeks	2/week	before & after using vehicle					
Technical Difficulty (1-Low, 5-High)		4	2	5	2	1	4	3	4	4	4	4	2	3					
Importance Rating				1.	2.								3.						

	A	B	C
	320	351	393

Color	Description
	Customer Needs
	Importance to the customer
	Engineering Characteristics
	Engineering Characteristics & Customer Needs Correlations
	Technical requirements type
	Technical Evaluation
	Engineering Characteristics Correlations
	Competitive Evaluation
	Customer perceptions Rating
	Customer perceptions Results
	Specification Target Value
	Technical Difficulty
	Importance Rating (Critical parameters)

FIGURE 11: "Quality home" the operation of city buses.

Conflicts of Interest

The authors declare that they have no conflicts of interest.

References

[1] A. Chen, H. Yang, H. K. Lo, and W. H. Tang, "A capacity related reliability for transportation networks," *Journal of Advanced Transportation*, vol. 33, no. 2, pp. 183–200, 1999.

[2] A. Chen, P. Kasikitwiwat, and C. Yang, "Alternate capacity reliability measures for transportation networks," *Journal of Advanced Transportation*, vol. 47, no. 1, pp. 79–104, 2013.

[3] G. Gentile and K. Noekel, *Modelling public transport passenger flows in the era of intelligent transport systems*, Springer International Publishing, Gewerbestrasse, Switzerland, 2016.

[4] M. Brzezinski, *Eksploatacja w logistyce wojskowej*, Bellona, Warsaw, Poland, 1996.

[5] J. Camargo Pérez, M. H. Carrillo, and J. R. Montoya-Torres, "Multi-criteria approaches for urban passenger transport systems: a literature review," *Annals of Operations Research*, vol. 226, no. 1, pp. 69–87, 2014.

[6] M. Brzezinski, M. Kijek, M. Gontarczyk et al., "Fuzzy Modeling of Evaluation Logistic Systems," in *Proceedings of the 21th International Conference, Transport Means*, pp. 377–382, 2017.

[7] S. Krawiec, B. Łazarz, S. Markusik et al., "Urban public transport with the use of electric buses – development tendencies," *Transport Problems*, vol. 11, no. 4, pp. 127–137, 2016.

[8] C. Von Ferber, T. Holovatch, Y. Holovatch, and V. Palchykov, "Public transport networks: empirical analysis and modeling," *The European Physical Journal B*, vol. 68, no. 2, pp. 261–275, 2009.

[9] D. Maritan, *Practical Manual of Quality Function Deployment*, Springer International Publishing, Cham, Switzerland, 2015.

[10] J. Zak, "The methodology of multiple criteria decision making/aiding in public transportation," *Journal of Advanced Transportation*, vol. 45, no. 1, pp. 1–20, 2011.

[11] J. Żuchowski and E. Łagowski, "Narzędzia i metody doskonalenia jakości," Tech. Rep., Publisher Technical University of Radom, Radom, Poland, 2004.

[12] K. Bergquist and J. Abeysekera, "Quality Function Deployment (QFD) - A means for developing usable products," *International Journal of Industrial Ergonomics*, vol. 18, no. 4, pp. 269–275, 1996.

[13] R. Tolley and B. J. Turton, *Transport systems, policy and planning: a geographical approach*, Routledge, 2014.

[14] J. Figurski, *Ekonomika logistyki. Cz. 2, Logistyka transportu*, MUT, 2010.

[15] L. Bertalanffy, *Ogólna teoria systemów*, PWN, 1984.

[16] P. Drucker, *Innovation and Entrepreneurship*, Routledge, New York, Ny, USA, 2015.

[17] S. Yedla, *Urban transportation and the environment – issues, alternatives and policy analysis*, Springer, New Delhi, India, 2015.

[18] EN 60812, "Analysis Techniques for system reliability – procedure for failure mode and effects analysis (FMEA)," 2006.

[19] M. Brzezinski, *Logistyka w przedsiębiorstwie*, Bellona, Warsaw, Poland, 2006.

[20] M. R. Beauregard, R. J. Mikulak, and R. E. Mcdermott, *The Basics of FMEA*, CRC Press, 2009.

[21] V. Bouchereau and H. Rowlands, "Methods and techniques to help quality function deployment (QFD)," *Benchmarking: An International Journal*, vol. 7, no. 1, pp. 8–20, 2000.

[22] Y. Fan, S. Zhang, Z. He et al., "Spatial Pattern and Evolution of Urban System Based on Gravity Model and Whole Network Analysis in the Huaihe River Basin of China," *Discrete Dynamics in Nature and Society*, vol. 2018, Article ID 3698071, 11 pages, 2018.

[23] A. Świderski, A. Jóźwiak, and R. Jachimowski, "Operational quality measures of vehicles applied for the transport services evaluation using artificial neural networks," *Eksploatacja i Niezawodnosc - Maintenance and Reliability*, vol. 20, no. 2, pp. 292–299, 2018.

[24] PN-EN 15341, "Maintenance - Maintenance Key Performance Indicators," 2007.

[25] S. Chowdhury, Y. Hadas, V. A. Gonzalez, and B. Schot, "Public transport users' and policy makers' perceptions of integrated public transport systems," *Transport Policy*, vol. 61, pp. 75–83, 2018.

[26] U. Barua and R. Tay, "Severity of urban transit bus crashes in Bangladesh," *Journal of Advanced Transportation*, vol. 44, no. 1, pp. 34–41, 2010.

[27] Y. Yin, "Multi-objective bilevel optimization for transportation planning and management problems," *Journal of Advanced Transportation*, vol. 36, no. 1, pp. 93–105, 2002.

8

Integrating Bus Holding Control Strategies and Schedule Recovery: Simulation-Based Comparison and Recommendation

Weitiao Wu ⑩,[1] Ronghui Liu ⑩,[2] and Wenzhou Jin[1]

[1]*South China University of Technology, Wushan Road, Tianhe District, Guangzhou 510641, China*
[2]*University of Leeds, Institute for Transport Studies, Leeds LS2 9JT, UK*

Correspondence should be addressed to Weitiao Wu; ctwtwu@scut.edu.cn

Academic Editor: Ondrej Stopka

In the absence of control strategies, headway fluctuation and bus bunching are commonly observed in transit operation due to the stochastic attributes such as travel time and passenger demand. Existing research on real-time control largely focused on developing operational tactics to maintain bus arrival regularity at stops without fully considering the effect of schedule recovery. This paper investigates the effect of bus driver behavior on bus holding control strategies and more specifically their effort in catching up with schedule in case of delay, i.e., schedule recovery. To this end, this paper first proposes a bus propagation model with capacity constraint to simulate the evolution of bus trajectories along a fixed route. It proceeds to explicitly incorporate both holding control actions and schedule recovery effect into the bus propagation model. Using simulation for a high-frequency bus line in Guangzhou, China, schedule- (SH) and headway-based holding (HH) control strategies are compared under various operational settings in the context of schedule recovery. These comparisons show that SH performs better under certain conditions, and SH generally benefits more from schedule recovery than HH. These results provide insights into the bus stop layout design and implementation of holding methods in the context of cruising guidance.

1. Introduction

Service reliability of public transport system is of great importance to passengers. Studies have shown that passengers value travel time reliability four times higher than they do to average travel time [1]. In the uncontrolled bus systems, buses are likely to bunch in the presence of stochastic travel time and demand, which is commonly observed in the peak hours. Bus bunching occurs when a number of buses arrive at a stop within an interval that is shorter than schedule headway or even together. Such phenomenon is undesirable for both passengers and the operator since it leads to unpredictable bus arrival times and additional waiting time at stops, which discourages passengers from choosing public transport.

A variety of operational tactics have been proposed to improve bus system performance in the literature, while holding control is the most commonly used [2]. The holding controlling approaches can be classified into three groups, including schedule-based control, headway-based control,

and optimization-based control. They works by injecting slack time into the schedule at designated stops, in which the slack time refers to the amount of time that a task in a schedule can be delayed without causing a delay to subsequent tasks. The first two methods are triggered by bus arrival time and headway deviations, respectively, while the other one optimizes holding times through formulating a mathematical programming problem based on cost or time minimization. In this sense, the slack times are predetermined static settings for schedule-based control, whereas they are determined in real-time for headway- and optimization-based control. Osuna and Newell [3] investigated the holding problem at a single point for a cyclical route, in an attempt to minimizing the overall passenger waiting time. Hickman [4] developed an analytical model to determine the optimal holding time at a control stop along a bus route considering stochastic running times. Eberlein et al. [5] formulated the holding problem as a deterministic quadratic program with the availability of real-time information. Zhao et al. [6] studied the determination of optimal slack time under schedule-based control with the aim of minimizing the expected waiting time of passengers.

Daganzo [7] proposed a headway-based control scheme, where the holding times are dynamically determined by the information of the forward headway. It was found that the proposed method could achieve a faster speed and thus lower travel time compared to the schedule-based approach. Later, Xuan et al. [8] developed a family of control strategies by combining both the forward and backward headway information. Results show that such scheme can considerably reduce slack times and enhance headway regularity. This work was extended by Argote-Cabanero et al. [9] to be generalized to evaluate a bus corridor with multiple bus lines. Daganzo and Pilachowski [10] proposed an adaptive control scheme based on vehicle-to-vehicle cooperation, in which the bus cruising speed was adjusted in a real-time manner with the information of expected demand and vehicle spacing. They reported that the scheme yields regular headways with faster travel than the earlier control strategies. Delgado [11] developed a holding control policy in combination with boarding limit in an attempt to minimizing total delays on the transit corridor. Hernandez et al. [12] presented an extended model considering multiple bus line services. Recently, Sánchez-Martínez et al. [13] formulated a holding control model with dynamic running times and demand.

The variability in travel time is one of the central sources triggering bunching. Some efforts were taken to tackle the effects of exogenous variables on the travel time variability by employing machine learning models and predictive analytics that are able to explain such variability to learn about the behavior of a given fleet of vehicles. In this way, proactive control approaches could be generated that are able to prepare the system for delays or surges in demand before they reach a critical level. For example, Yu and Yang [14] used a support vector machine to predict the arriving status in the implementation of holding control. Moreira-Matias et al. [15] integrated the bus bunching prediction model into a real-time framework to mitigate bus bunching, of which the prediction output is used to select and employ corrective actions (holding and stop skipping). Nair et al. [16] presented real-time predictive analytics for bus bunching by using the real data in Miami-Dade Transit. Recently, Andres and Nair [17] presented a predictive-control framework to reduce bus bunching, which involves hierarchically related components including headway prediction and dynamic holding control. There are also literatures that investigate the exogenous factors affecting bus bunching, such as passenger arrival patterns (Fonzone et al. [18]) and the presence of common line (Schmöcker et al. [19]). Holding strategies have also been used for transfer synchronization. As direct and transfer demand are the mainstreams of transit networks [20], such problem is usually formulated as minimizing passenger waiting time or cost accounting for these types of passengers. Recently, Wu et al. [21] incorporated schedule-based holding control with a predefined time window into the bus schedule coordination design.

With the advances in connected vehicle technology and roadside detectors development such as time control points, the information of schedule deviations can be readily collected and informed to bus drivers. This provides new opportunities for transit operators with real-time schedule adherence status. Due to the travel time variability, both early and late arrivals can occur when compared to the reference timetable at the designated time control stops. One of the operational goals of a transit agency is to maintain buses on schedule. In reality, well-experienced bus drivers constantly adjust their speeds to keep their buses on schedule [22–25]. Figure 1 shows a potential way of realizing this goal using cruising guidance with colored bars that move up and down, such that drivers are able to vary their average speed to improve the schedule adherence. According to an empirical study conducted by a transit agency in the northeastern United States, schedule recovery effort can be observed on at least half of the segments on a bus route [22]. Recently, Liu et al. [26] proposed an inter-vehicle communication scheme to achieve a planned direct transfer. Two operational tactics were employed by using real-time information: speed control and holding at transfer point, of which speed control resembles schedule recovery behavior.

Schedule-based holding control are often employed when the bus arrives earlier than the scheduled arrival time. When a bus arrival is behind schedule, schedule recovery tactics could be deployed to support/guide the driver to catch up with the schedule at the next time control point. Unlike holding control that keeps buses at stops, the schedule recovery emphasizes speed adjustment between stops. This inter-stop control action can be utilized as complementary to holding control. Thus there may exist interactions between schedule recovery and holding control. The performance of bus scheduling is closely related to the dynamic motion of buses including driving behavior described by speed and acceleration. Therefore, schedule recovery behavior should be taken into consideration in the execution of holding control strategy. However, most of the existing literatures on holding control have focused on developing and evaluating the effectiveness of different action rules, largely neglecting the inherent effect of such schedule recovery driving behavior.

On the other hand, there is a set or "library" of feasible operational tactics to be used by transit operators. Among them, speed adjustment and vehicle holding are usually employed as combinatorial strategies. Speed adjustment, when applied in bus delay scenarios, resembles schedule recovery. For example, Nesheli et al. [27] used a combination of speed change control and headway-based holding control to reduce bus bunching. On a similar combination of strategies, Milla et al. [28] integrated holding and stop-skipping control based on fuzzy rules to minimize users' travel time. While headway-based and schedule-based control are two distinctive and most commonly implemented holding control methods, no performance comparison was made between them in the combinational design of operational tactics in the literature. Understanding the combined effect could help to design proper combinational tactics in response to varying traffic conditions.

The major focus of this paper is to compare the combined effect of schedule recovery and two different holding control approaches, so as to evaluate how cruising guidance technology and the resulting schedule recovery behavior affects

Behind schedule

Ahead schedule

FIGURE 1: Cruising guidance on-board device surface (adapted from [9]).

the bus holding control strategies. This comparison should allow us to identify the scenarios for which the different holding control methods present advantages and highlight their respective strengths for further implementation in the context of cruising guidance technology. We have made an effort in this document to present both approaches with a common nomenclature. Our findings show that schedule recovery plays an important role in the design of bus holding control and that, for specific indicators, the optimal holding strategy transition will occur with certain level of schedule recovery effort and under certain conditions. We thus suggest the bus operators should select the most appropriate operation strategy that suits their operating conditions, which provides managerial insights into bus operational control. To the best of our knowledge, this is the first time that holding control strategies are compared in the context of schedule recovery.

The remainder of this paper is organized as follows. In Section 2, simulation frameworks are developed. In Section 3, performance measures are introduced to evaluate the performance of bus service. In Section 4, model experiments are performed and their practical implications are provided. Finally, the conclusions and future research directions are given in Section 5.

2. Modelling Approach

This paper is designed to investigate the effects of schedule recovery on the performance of holding control strategies, more specifically on the schedule-based and headway-based holding control. The main objectives are threefold: (a) develop an enhanced bus propagation and holding control models, which explicitly takes the schedule deviation and driving behavior into account; (b) compare the performance of the two holding control strategies under various operational settings in the context of schedule recovery; and (c)

discuss the implementation issue for holding control with schedule recovery.

2.1. Assumption and Notations. Without loss of generality, the following assumptions are made:

(A1) Passenger arrival pattern relates directly to bus head-ways ([29]). In this paper passenger arrivals at bus stops is assumed to follow Poisson distributions. This assumption is reasonable on high-frequency routes, as has already been validated and commonly used by many researchers (e.g., [14, 30, 31])

(A2) The boarding process and the variability in link travel time are attributing factors to bus bunching.

(A3) Bus overtaking maneuvers are prohibited. This is reasonable since overtaking rarely occurs under the combined effect of holding and schedule recovery. This is also a common simplification in the literature (e.g., [13]). When overtaking is allowed, the bus order may change from stop to stop. Thus allowing overtaking requires structural changes to the model which has been left for another work.

(A4) Over the study-time horizon, passenger demand is assumed to be stochastic governing Poisson distribution, while bus running times vary at results from stochastic phenomena in the network.

2.2. Bus Propagation Model with Capacity Constraint. A bus motion model is comprised of three components: departures of buses, dwell times at stops, and link travels times. The arrival time of bus i at stop j is the departure time from stop $j-1$ plus the random link travel time between stop $j-1$ and j:

$$a_{i,j} = d_{i,j-1} + t_{i,j-1} \qquad (1)$$

The uncontrolled bus departure time is determined by its arrival time and dwell time:

$$d_{i,j} = a_{i,j} + D_{i,j} \qquad (2)$$

Following Liu et al. [32], the headway between bus i and the preceding one is assumed to be the gap between bus $i-1$ leaving stop j and bus i arriving stop j.

$$h_{i,j} = a_{i,j} - \overline{d}_{i-1,j} \qquad (3)$$

where $\overline{d}_{i-1,j}$ stands for the previous departure time from the control point, which is equal to $d_{i-1,j}$ plus the corresponding holding time. $d_{i-1,j}$ is effectively equal to $\overline{d}_{i-1,j}$ without holding control.

Passenger arrival is assumed to be stochastic governing Poisson process, with the mean arrival flow equals to the product of the mean passenger arrival rate q_j and the headway $h_{i,j}$ of the bus with its leader. Therefore, the boarding demand for bus i at stop j is

$$B_{i,j} = P\left(q_j h_{i,j}\right) + l_{i-1,j} \qquad (4)$$

The alighting demand is drawn from a binomial distribution, which is related to the bus load $L_{i,j}$ before arriving at stop j and its alighting percentage ρ_j.

$$A_{i,j} = Bi\left(L_{i,j-1}, \rho_j\right) \qquad (5)$$

With vehicle capacity constraint, the actual number of boarding passengers is either the boarding demand or the remaining capacity, then we have

$$\begin{aligned}\overline{B}_{i,j} &= \min\left\{B_{i,j}, C - L_{i,j-1} + A_{i,j}\right\}\\ &= \min\left\{B_{i,j}, C - L_{i,j-1}\left(1 - \rho_j\right)\right\}\end{aligned} \qquad (6)$$

We assume that waiting passengers are loaded in a random fashion, which is appropriate when passengers mingle on waiting platforms. It is further assumed that each available space is equally likely to favoured by the waiting passengers, thus the boarding probability is the actual number of boarding passengers to the boarding demand, i.e., $\overline{B}_{i,j}/B_{i,j}$. Evidently, when the number of passengers who want to board exceed the remaining capacity, the probability is less than 1; otherwise, the probability is equal to 1. Therefore, the actual number of arriving passengers who are able to board is

$$\overline{B}'_{i,j} = \frac{\overline{B}_{i,j}}{B_{i,j}} P\left(q_j h_{i,j}\right) \qquad (7)$$

Equations (6) and (7) are used to calculate the average waiting time; see Section 3.2.

Therefore, the number of leftover passengers is the difference between total boarding demand and the actual number of boarding passengers.

$$l_{i,j} = B_{i,j} - \overline{B}_{i,j} \qquad (8)$$

The number of on-board passengers in bus i when it departs from stop j is related to the load before arriving the current stop and passenger flow at current stop.

$$L_{i,j} = L_{i,j-1} + \overline{B}_{i,j} - A_{i,j} \qquad (9)$$

When the vehicle is not crowded ($L_{i,j}/C \le \varphi$, where $0 < \varphi < 1$ is a constant), passenger boarding and alighting take place simultaneously in a front-on rear-off policy. Thus the bus dwell time at the stop is estimated as the maximum time between the boarding and alighting time, plus the open and close door time.

$$D_{i,j} = \max\left\{b\overline{B}_{i,j}, \alpha A_{i,j}\right\} + \tau \qquad (10)$$

where φ is a threshold of in-vehicle crowding degree. τ is the open and close door time. b and α represent the average boarding time and alighting time for passengers.

Note that the link travel time extracted from GPS data includes the acceleration and deceleration time.

If the vehicle is crowded, i.e., $\varphi < L_{i,j}/C \le 1$, the passengers would need more time to board and alight, and the dwell time is

$$D_{i,j} = \gamma \max\left\{b\overline{B}_{i,j}, \alpha A_{i,j}\right\} + \tau \qquad (11)$$

where γ is the crowding coefficient, $\gamma > 1$.

2.3. Bus Holding Control with Schedule Recovery. When real-time holding control is in place, the bus departure time may be modified according to the control policy. This inherently has an effect on the boarding and alighting process. In this paper, two typical control methods are investigated: the schedule-based and headway-based holding control strategies. Schedule recovery is only triggered by schedule deviation independent of control strategies. Such time deviation can be readily informed to drivers. The driver adjustment is related to the schedule adherence status of the bus. In what follows, the corresponding schedule deviations are identified and the effect of schedule recovery are incorporated.

2.3.1. Schedule-Based Holding Control (SH). Under SH, buses either depart on schedule or immediately after serving passengers if they arrive late at the time point [2]. Therefore, the scheduled departure time takes the following piecewise function:

$$d_{i,j} = \begin{cases} s_{i,j}, & a_{i,j} < s_{i,j} - D_{i,j} \\ a_{i,j} + D_{i,j}, & a_{i,j} \ge s_{i,j} - D_{i,j} \end{cases} \qquad (12)$$

where $s_{i,j} - D_{i,j}$ is the critical arrival time after which the bus has to depart later than the scheduled departure time $s_{i,j}$.

The scheduled departure time at a designated stop $s_{i,j}$ can be calculated as the scheduled departure time from the previous stop $s_{i,j-1}$ plus the scheduled link travel time. The dwell time is included in the scheduled link travel time. The reliability of bus operation under SH depends on the scheduled link travel time. The scheduled link travel time could be set as the average link travel time multiplied by a safety factor (we term it *slack ratio*) to provide time redundancy and thereby absorbs travel time randomness.

When $a_{i,j} < s_{i,j} - D_{i,j}$, the early arriving bus will be held until time $s_{i,j}$. The schedule deviation, and therefore schedule recovery, arises when the bus arrives at a designated stop later than the critical arrival time, i.e., when $a_{i,j} \geq s_{i,j} - D_{i,j}$, then the delay experienced by bus i at stop j is $d_{i,j} - s_{i,j}$. Similar to Chen et al. [22] and Yan et al. [25], the driver's adjustment between the current time point and the preceding time point is assumed to be proportional to the schedule deviation at the current point. Therefore, the driver adjustment can be estimated as $\beta_{i,j}(d_{i,j} - s_{i,j})$. As a result, by modifying (1), the arrival time with schedule recovery is

$$a_{i,j+1} = d_{i,j} + t_{i,j} - \beta_{i,j}\left(d_{i,j} - s_{i,j}\right) \qquad (13)$$

where $\beta_{i,j}$ represents the adjustment factor between stop j and $j + 1$ for bus i. In practice, this adjustment parameter can be estimated from historical trip information and is a stochastic variable following a specific distribution will be discussed in Section 2.3.3.

2.3.2. Headway-Based Holding Control (HH).
HH approach is usually triggered by headway deviation. In line with SH control, schedule recovery in HH works when arrival delay arises. In this study, we use a heuristic HH similar to that proposed by Sánchez-Martínez et al. [13]. The rational is that hold bus i at control point j to ensure preceding headways are never less than a prescribed design headway. In order to attain the desired headway, when the current headway is smaller than the desired headway, the vehicle should be held at the stop; otherwise, the schedule recovery should be employed. The recovery time is based on the headway deviation. The scenarios are specified as follows.

Scenario I. When the headway is shorter than the design headway, i.e., $h_{i,j} \leq H$, hold bus i at stop j for time $H - h_{i,j}$, thus the arrival at the next stop is simply expressed as

$$a_{i,j+1} = \overline{d}_{i,j} + t_{i,j} \qquad (14)$$

Scenario II. When the headway is larger than the design headway, i.e., $h_{i,j} > H$, the bus should depart immediately and schedule recovery starts, and the arrival at the next stop is

$$a_{i,j+1} = d_{i,j} + t_{i,j} - \beta_{i,j}\left(h_{i,j} - H\right) \qquad (15)$$

where $h_{i,j} - H$ represents the schedule deviation under headway-based holding control, which can be informed to the bus driver for schedule recovery instruction immediately when bus i completes serving passengers at stop j.

2.3.3. Calibration of Schedule Recovery Factor.
As discussed, bus drivers tend to actively pursue schedule recovery if the bus is delayed. Naturally, driver's behavior is highly dependent on the his/her experience, which may vary considerably from scenario to scenario, fleet to fleet, or even from vehicle to vehicle. Therefore, it is reasonable to assume that the schedule recovery factor follows a specific distribution. Based on automatic vehicle location (AVL), automatic passenger counting (APC) data, and the timetable, the delay at stops could be calculated and then correlated with the travel time deviation on the next leg of the journey to the next downstream stop. Consequently, using the historical trip data, one can get the distribution of the adjustment factor $\beta_{i,j}$ along the route.

According to the empirical study by Chen et al. [22], the average adjustment factors vary within a range between -0.5 and 0.5 on most segments. Since the early arrival will be compensated by holding control, here we assume the adjustment factor is nonnegative; that is, bus drivers are always trying to recover the schedule deviation at the preceding time control point.

2.4. Solution Algorithms for Bus Trajectories Evolution.
Instead of using simulation tools such as multiagent approach or discrete event-based simulation software (e.g., [33]), in this study we develop bus propagation models to simulate the evolution of bus motion and evaluate the bus holding methods. Unlike the disaggregated models in which system dynamics are explicitly simulated by individual travel behavior, in our simulation framework the passengers' activity is modelled in an aggregated way. The advantage of aggregate models is their greater computational efficiency, which facilitates repeated simulation.

With the above formulations, Algorithm 1 outlines the general simulation framework in which alighting process, capacity constraint, and leftover passengers are incorporated. The algorithm is made up of three components: calculating, respectively, the departures of buses, link travel times, and dwell times. We consider a unidirectional bus route where vehicles depart from one terminal to another. Although extension from one-way line to the general bus route with bidirectional traffic would be straightforward, the modelling a unidirectional route avoids considering traffic continuity at terminal stations and fleet size limitation problem as with the cyclic route where inbound and outbound headways are correlated. Since late or early arrivals at terminals can occur due to travel time variability [34], modelling cyclic route may also result in departure headway fluctuations from the terminals. Such effect is a special case of travel time variability. Let M denote the fleet size of the modelled bus line and N the number of bus stops on the corridor served by the bus line. To discourage bunching at the beginning of the simulation, headways are set deterministic and at a uniform headway; thereafter headways become stochastic. In order to make the system evolve to be chaotic enough for bus bunching to appear, the number of buses M is set sufficiently large in each run of simulation. In this regard, one may consider the simulation of the first few buses in the system as a "warm-up" period.

In order to avoid bus overtaking phenomenon, the bus headway $h_{i,j}$ is required to be larger than a value. When the preceding bus is caught by the next incoming bus during the

```
Initialization: Set input parameters and the counter of simulations
Procedure:
Step 1: Generate the departure times for all trips from the terminal
for bus i=1: M do
    Compute the departure time for the bus line, satisfying d_{i,1} := d_{1,1} + (i − 1)H
end
Step 2: Generate the stochastic bus link travel time
for bus i=1: M do
    for stop j=2: N do
        Compute the bus link travel time t_{i,j−1} from a truncated normal distribution.
    end
end
Step 3: Generate the full trajectories of the first bus
for stop j=2: N do
    Compute the arrival time of bus 1 at stop j, satisfying a_{1,j} := d_{1,j−1} + t_{1,j−1}
    Compute the departure time of bus 1 from stop j, satisfying d_{1,j} := a_{1,j} + P(q_jH)/b
    Compute the number of on-board passengers, satisfying L_{1,j} := L_{1,j−1}(1 − ρ_j) + P(q_jH)
        Let the leftover demand, l_{1,j} := 0
end
Step 4: Generate the trajectories for the remaining trips of the bus line
for stop j=2: N do
    for bus i=2: M do
        Compute the trajectory and passenger flows of bus i at stop j using Eqs.(1)-(11) subject to Eqs.(16)
        Apply holding control and schedule recovery where necessary, and update the departure time according to Eqs.(12)-(15).
    end
end
```

ALGORITHM 1: Bus trajectories evolution algorithm.

simulation, i.e.,$h_{i,j} < 0$, let the preceding bus restart after a delay time δ, which we call it *minimum safety interval*, i.e.,

$$a_{i,j} = \overline{d}_{i-1,j} + \delta \qquad (16)$$

3. Performance Measure

To evaluate the performance of different control strategies, we use a number of performance measures to take into account the views of different stakeholders: passengers and operator. The headway variability is the major concern of both passengers and operator, since uneven headway is the main cause of spatially uneven loads and thus bus bunching. The average waiting time reflects the level of service and appears to be uppermost for passengers. In addition, the operator may concern the travel time reliability, since it is crucial for schedule design and operation costs. A smoother and more robust operation and planning at terminals requires lower variability of travel time. Multiple simulation runs are conducted, from which we generate distributions of performance measures.

3.1. Headway Variability Coefficient (HVC). Similar to Turnquist and Bowman [35], Liu and Sinha [36], and Wu et al. [37], we use the headway variability coefficient (HVC) to measure the reliability of the observed headways, which is defined as the ratio of the standard deviation to the mean headway. This coefficient is the coefficient of variation as known in statistics and probabilities.

3.2. Average Waiting Time. As mentioned previously, in this study, the passengers' activity is modelled in an aggregated way; thus it would be difficult to obtain the waiting time of each individual passenger. The average passenger waiting time could be achieved by (6)-(8).

The waiting passengers at a station are divided into two groups: those who are able to board and those who are left behind due to capacity constraint. The former, of which the number is $\overline{B}'_{i,j}$ (see (7)), arrives randomly during time window $[0, h_{i,j}]$; thus their expected waiting time can be approximated to be half of the headway $h_{i,j}/2$ (Sánchez-Martínez et al. [13]; Salek and Machemehl [30]). The passengers who are left behind, $l_{i-1,j}$, have to wait for the next bus, and their additional waiting time is the entire headway$h_{i,j}$. Summing up these two groups we have the total waiting time expressed as $(1/2)\sum_i \sum_j \overline{B}'_{i,j} h_{i,j} + \sum_i \sum_j l_{i-1,j} h_{i,j}$.

Dividing the total waiting time by the total number of boarding passengers $\overline{B}_{i,j}$(see (6)), we have the average waiting time per passenger as follows:

$$\overline{w} = \frac{(1/2)\sum_i \sum_j \overline{B}'_{i,j} h_{i,j} + \sum_i \sum_j l_{i-1,j} h_{i,j}}{\sum_i \sum_j \overline{B}_{i,j}}$$
$$= \frac{\sum_i \sum_j \left(\overline{B}'_{i,j} + 2l_{i-1,j}\right) h_{i,j}}{2\sum_i \sum_j \overline{B}_{i,j}} \qquad (17)$$

As shown in Section 2.2, the number of waiting passengers and the headways are interdependent. The expected

FIGURE 2: Bus route 87. The direction of the bus route discussed in this article is from the bottom to the up.

average waiting time $E(\overline{w})$ can be drawn from multiple simulation runs.

3.3. Average Travel Time. While bus holding control could improve the service regularity, it prolongs the terminal-to-terminal running time for the vehicles. The average travel time for buses is an important operational performance measure for transit operators since it is related to the fleet size. In each run of simulation, an average travel time is derived as the arithmetic mean over all buses simulated. Then an expected average travel time, derived from multiple simulation runs, is used as the performance measure.

3.4. Load Variation. Headway variability and bus bunching lead to uneven load. Such spatially heterogeneous demand is one of the major sources triggering crowding effect since discomfort happens at high load factors. Therefore, a more balanced load factor across buses yields a more comfortable experience to users. In view of this, we introduce load variation to evaluate the performance of bus control strategies, which is defined as the standard deviation of all bus loads across all segments. Then the expected load variation can also be drawn from multiple simulation runs.

4. Comparative Results on a Simulated Real-Life Route

In this section, we compare the performance of two bus holding controls with and without schedule recovery (SR) effort: the simple schedule-based holding (SH) and headway-based holding (HH) and SH with schedule recovery (SHSR) and HH with schedule recovery (HHSR).

These four control methods are applied to bus route 87 in the city of Guangzhou, China. The bus route (shown in Figure 2) has a total distance of 14.7 km. It connects Yijin Cuiyuan Terminal and Airport Road Terminal in the city. The passenger demand and link travel time data are provided by a local bus company. We use the data during the morning peak hour (9:00-10:00 am) and in the direction from Yijin Cuiyuan

Terminal to Airport Road Terminal in the city. The scheduled headway of the route is taken as 8 minutes.

Following Liu et al. [32], we assume that passengers boarding at a station will evenly alight at the downstream stops; thus the stop-specific average alighting rate ρ_j can be obtained from the boarding rate. Table 1 shows the empirical data on passenger arrival rate and the derived alighting distributions. The link travel time data are obtained from on-board GPS tracking devices, from which the mean and standard deviation of travel time between stops are calculated and listed in Table 2. At present, the bus route operates in an unscheduled and uncontrolled manner.

The minimum safety interval is set as δ = 0.3 *min*, the average boarding and alighting time are set as α = 2*s* and b = 4*s*, respectively, and vehicle capacity is set as C = 100 *pax/veh*. The open and close door time is taken as τ = 4*s*. The threshold of the in-vehicle crowding degree φ is set to be 0.8. The crowding coefficient is set as γ = 1.5, and the number of buses M is set as 20. In order to highlight the relative effect of holding and schedule recovery, all intermediate stops are considered key time control stops. Buses are set to depart from the terminal on time in the base case. The link riding times are drawn from a truncated normal distribution with means and standard deviations as listed in Table 2. The simulation is run 1000 times.

The detailed output in a typical simulation includes vehicle trajectories, the vehicle load, and the number of leftover passengers. In what follows, we test the effectiveness of the simulation model under various operating settings. To represent differences in driver behavior and their impact on schedule recovery, in each simulation the schedule recovery factor β_{ij} is randomly generated for each segment following a uniform distribution.

4.1. Slack Ratio for Schedule-Based Holding Control. The slack time is a crucial predetermined setting for SH control. To investigate, we define slack ratio as the multiplier of average link travel times, and we test the system performance for a range of slack ratios: {0.6, 0.7, 0.8, 0.9, 1, 1.1, 1.2, 1.3, 1.4, 1.5}.

TABLE 1: Observed passenger arrival flow and derived alighting proportion on 87 Route.

stop	1	2	3	4	5	6	7	8	9	10	11	12	13
Arrival rate (pax/min)	0.34	0.22	0.17	0.23	0.25	0.27	0.45	0.91	0.64	0.99	0.56	0.74	0.25
Alighting Proportion (%)	0	4.2	4.3	4.5	4.8	5	5.3	5.6	5.9	6.3	6.7	7.1	7.7
Stop	14	15	16	17	18	19	20	21	22	23	24	25	
Arrival rate (pax/min)	0.79	0.26	0.38	0.35	0.27	0.29	0.30	0.11	0.14	0.08	0.06	0.03	
AlightingProportion(%)	8.3	9.1	10	11.1	12.5	14.3	16.7	20	25	33.3	50	100	

TABLE 2: Observed mean and standard deviation of link travel times on 87 Route (unit: min).

stop	1	2	3	4	5	6	7	8	9	10	11	12	13
Mean	1.46	2.05	0.89	1.87	1.66	1.65	1.63	4.41	0.82	0.79	0.83	1.35	0.2
STD	2	0.7	1.6	0.47	0.27	0.68	0.63	2.63	0.51	0.23	0.26	0.67	0.03
Stop	14	15	16	17	18	19	20	21	22	23	24	25	
Mean	3.27	2.72	3.04	2.90	1.53	2.21	2.94	1	2.64	2.56	0.74	1.6	
STD	1.07	0.75	1.22	1.36	0.48	1.03	1.06	0.28	0.73	0.66	0.27	0.28	

Moreover, for a given slack ratio, we introduce two levels of schedule recovery effort: a low level uniformly distributed in an interval [0.1, 0.2] and a high level in an interval [0.4, 0.5]. The results are shown in Figure 3. A summary of findings and their implications is listed as follows.

(a) In aspect of HVC, the HH control generally outperforms the SH under various slack ratios (Figure 3(a)). However, in the provision of schedule recovery, the HH is not always better than SH control in passenger waiting time $E(\overline{w})$, when the slack ratio lies between 0.6 and 1.2 (Figure 3(b)). The average travel time under HH control is shorter than that of SH when the slack ratio is larger than 2 (Figure 3(c)). When the slack ratio ranges from 0.6 to 1.2, the SH with high level of schedule recovery presents the less variability and most uniform pattern in bus loads (Figure 3(d)).

(b) The load variation (Figure 3(d)) under SH first decreases and then increases with the increasing slack ratio. There are two possible reasons for this. First, 1.2 is already a high slack ratio to mitigate travel time variability; any improvement of headway stability could become more difficult through increasing the slacks. Second, higher slacks lead to less frequent service and greater accumulated boarding demand, which could in turn result in high crowding at some stops.

(c) Performance improvements for all indicators are observed at more sophisticated holding strategies with SR. The improvement is more significant when more schedule recovery effect is made. This is because schedule recovery compensates holding times vehicle spending. For SH control, schedule recovery could improve performance by a greater degree when the slack ratio is smaller. The reason is that when the slack time is sufficiently large, most of the travel time randomness and resulting delay have been mitigated, such that schedule recovery takes less effect. Therefore, one can see that when the slack ratio is relatively small, the benefit of HVC, waiting time, and travel time gained by schedule recovery is greater with SH control.

Naturally, a longer slack time for SH will lead to better schedule adherence, but at an expense of the negative effect of less frequent service and greater mean headways. As shown in Figure 3(b), such negative effect would overweight the reduction of headway variation when the slack time reaches a threshold (about 1); thereafter the waiting time increases instead. This suggests that sufficient holding times is not productive, and the operator should make a trade-off between the headway stability and efficiency in the planning. Hence, we analyse in the following the performance of holding control strategies with a reasonable level of slack ratio, at 1. Optimizing the slack time has been left for future study. In addition, to highlight the effect of schedule recovery effort, from now on the schedule recovery factor is set as a high level, i.e., uniformly distributed in an interval [0.4, 0.5]. The control policies are compared under the same operational settings, except where they are the subject of a test. This is approximated in the interest of presenting the incremental improvement, though it might be possible to improve performance further by optimizing headway.

4.2. Sensitivity to Travel Time Variability. In this section, we analyse the performance improvement from schedule recovery under various levels of travel time variability. The performance improvement is calculated as relative performances between with and without SR controls: $(SH-SHSR) \times 100/SHSR$ or $(HH-HHSR) \times 100/HHSR$. The travel time variability is reflected by the standard deviation of truncated normal distributed link travel time. We amplify the standard deviation by factors 1.2, 1.4, 1.6, 1.8, and 2. The results are presented in Figure 4.

When the travel time variability increases, more performance improvement could be achieved by schedule recovery. While schedule recovery benefits both of the holding control strategies, the performance improvement of SHSR is much greater than that of HHSR under the same level of travel time variability. Since schedule recovery acts as the contributor to service recovery, these results suggest that SH is more sensitive to schedule recovery than HH in the presence of the travel time variability. In other words, SH method is less able to stabilize headways but could benefit more from schedule

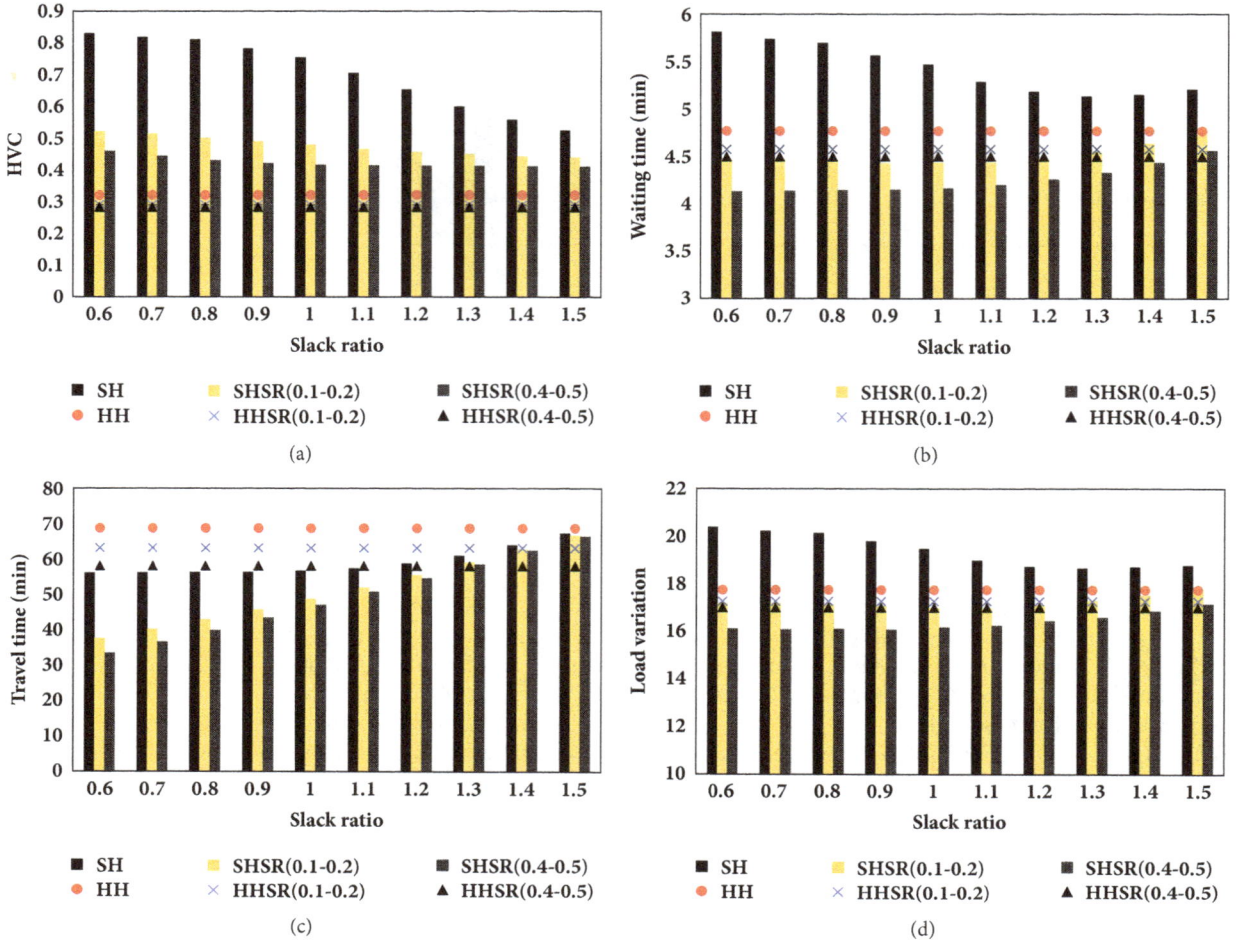

FIGURE 3: Performance measures under different control policies: (a) HVC; (b) $E(\overline{w})$; (c) average travel time; (d) load variation.

recovery. This implies that, in practice, cruising guidance technology has more effect in SH rather than HH under the same schedule recovery effort.

4.3. Sensitivity to Total Demand.

Demand is one of the most important factors affecting the performance of the control strategy. To analyse the independent effect of demand congestion on the holding control performance, in this section we vary the demand levels without changing the headway and vehicle capacity. Figure 5 shows performance measures with varying demand levels.

SH appears to take no effect in terms of HVC in high demands (Figure 5(a)). A possible reason is that the effect of SH method is closely related to dwell times; when the demand volume grows (unless exceeding the capacity), buses are likely to stay longer at stops and depart later than the given scheduled departure time. Such effect may propagate along the downstream route. Consequently, the schedule departure times in SH takes less effect as the demand volume grows. This suggests that operators should pay attention to the level of passenger demand when using SH method.

HH is more sensitive to demand variation than SH in terms of average waiting time (Figure 5(b)). HH works well only in low demand; when the demand increases, the control performance deteriorates quickly in the average waiting time. In particular, SH outperforms HH in terms of the average waiting time when the demand ratio is larger than 2. This is because, as discussed previously, SH takes less effect with higher demand, whereas HH is always in effect regardless of the demand level. In this sense, the service frequency under HH will be reduced as the demand increases as opposed to SH.

Bus load variation for both SH and HH increases with the demand ratio but at a decreasing rate (Figure 5(d)). This is because when the demand level reaches a threshold, the bus capacity can only meet the transport demand of the first several bus stops, after which buses tend to be full of passengers at the following stops. Schedule recovery takes more effect in smoothing bus loads under SH compared to HH. As the demand ratio reaches 2, the effect of schedule recovery is trivial for HH.

4.4. Summary of the Key Findings and Their Practical Implications.

In this section, we highlight the key findings from the sensitivity analysis of the proposed holding control with schedule recovery and discuss their operational implications.

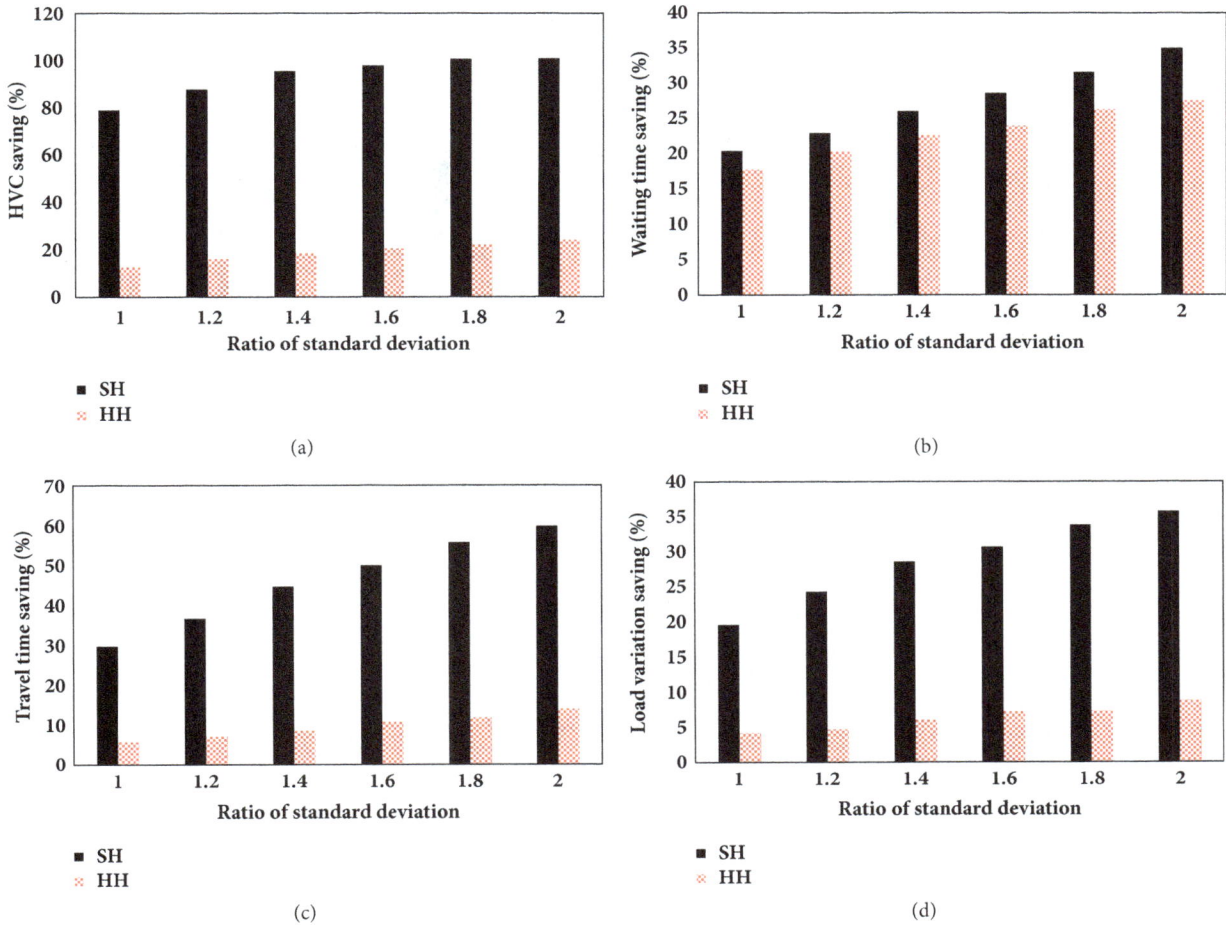

FIGURE 4: Percentage reductions in (a) HVC; (b) $E(\overline{w})$; (c) average travel time; (d) load variation under different travel time variation levels.

First, HH method generally outperforms the SH method with respect to HVC (Figure 3(a)). When schedule recovery (SR) effort is included; however, the relative performance of HHSR and SHSR would depend on the slack time. HHSR outperforms SHSR only with large slack ratios.

Second, we have shown that bus performance will always improve with SR effort, and SH control benefits more than HH control with SR, particularly in the presence of travel time variability (Figure 4). SR has less effect with larger slack ratios (Figure 3), and its effect in smoothing bus loads is greater under SH than under HH (Figure 5(d)).

Third, due to the inherent control mechanism, the SH takes less effect as the demand grows, whereas HH is always in effect. This may result in the phenomenon that SH outperforms HH in aspect of waiting time when the demand is sufficiently high (Figure 5(b)).

Based on the key findings described above, the following practical insights and recommendations can be drawn.

(a) *Bus Stop Layout (Re)design.* The above results show that cruising guidance and schedule recovery could improve the reliability of bus system. To facilitate the en route driver guidance and schedule recovery effort, some strategies can be introduced. For example, fewer stop activities and passenger flow control should be encouraged. According to the previous empirical analysis, bus stop consolidation has no significant effects on passenger activity [38]. Therefore, transit planners should make a trade-off between stop spacing and passengers' access to service in the design or redesign of bus route and possibly introduce wider stop spacing through the removal or consolidation of existing stops. Since the bus service time at bus stop areas usually occupies a large proportion of the total on-road bus operational time, it would also be helpful to invest in a quicker fare collection technology, such as building enclosed areas to allow off-bus fare collection for rapid boarding like the bus rapid transit system. In addition, for those bus routes traversing the center and suburb, such as Express Route 2 of Suzhou City in China, the stops in the suburban area with low demand could be converted into request stops (or flag stops), at which the vehicle will stop only on request.

(b) *Determination of Slack Times.* Including slack times in the schedule could stabilize headways, whereas too much slacks reduce service frequency. Transit operator need to consider the trade-off between the headway stability and efficiency in the planning. In the presence of schedule recovery, schedule-based method could achieve the best compromise

(a)

(b)

(c)

(d)

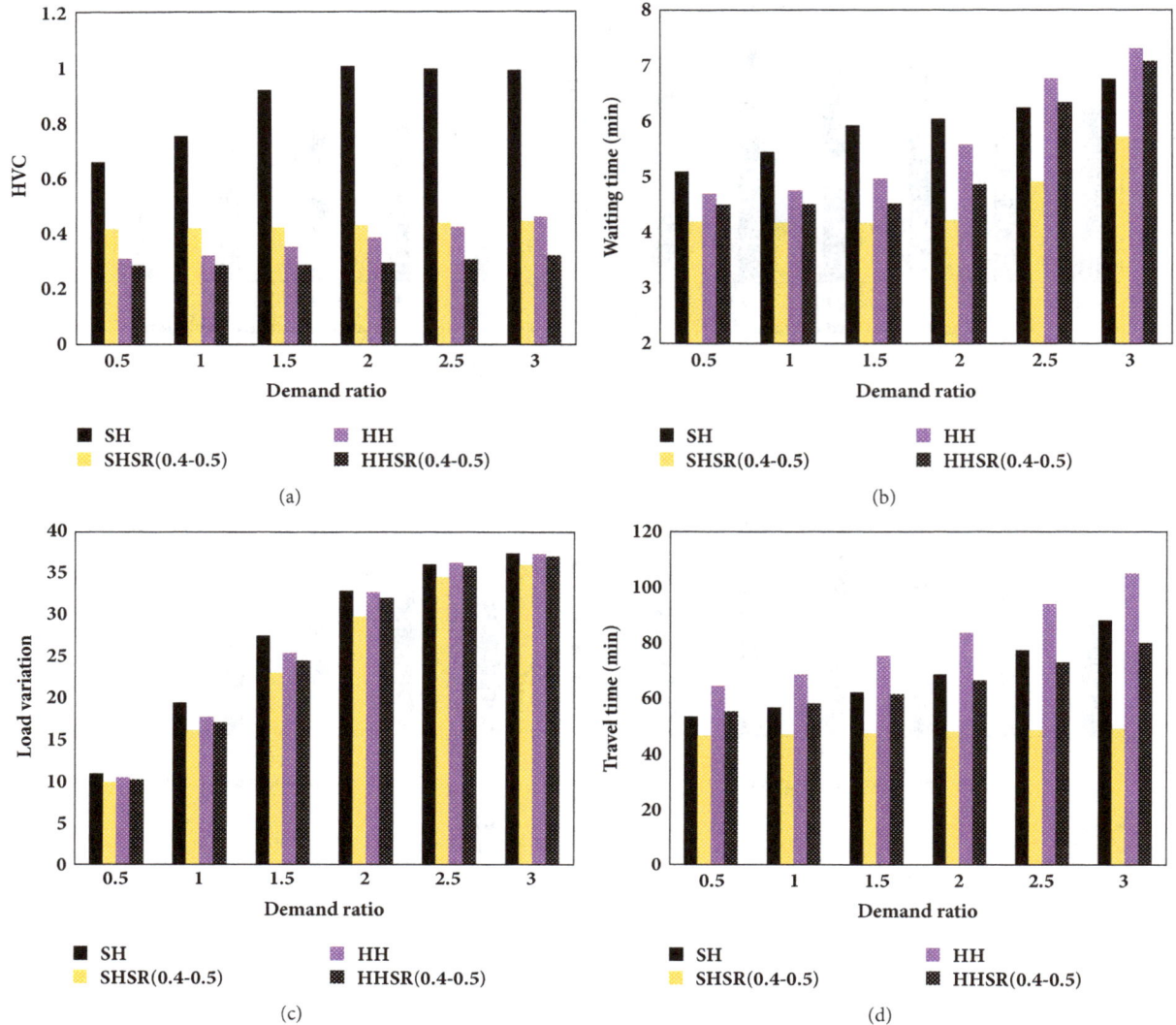

FIGURE 5: Performance measures for different total demand: (a) HVC; (b) $E(\overline{w})$; (c) average travel time; (d) load variation.

between headway regularity and efficiency with reasonable amounts of slack times, making it the preferable method for routes to reduce the waiting time.

(c) *Selection of Bus Holding Control Method.* Simulation results show that SH could benefit more from schedule recovery compared to HH under the same level of schedule recovery effort. Such property can be utilized to choose proper tactics in response to different operational conditions. Since there exist potential safety issues for schedule recovery, such as speeding and pushing traffic lights when there is little time left for the driver to catch up with the arrival schedule, the SH is preferable in the case of dedicated bus lanes, exclusive right of ways, or wide stop spacing where cruising guidance and schedule recovery could be more easily implemented. On the other hand, the HH may be recommended when the cruising guidance technology is absent or restricted, such as in normal lanes exposed to exogenous conditions. This is also the case during peak

hours and traffic congestion, which implies that HH might be converted to SH in the transitional period from peak to off-peak hours, vice versus. Therefore, in practice, transit agencies should be mindful of choosing bus control methods in the context of cruising guidance technology. In addition, since SH will take less effect with higher demand as opposed to HH, transit operators should pay attention to the level of passenger demand when choosing control method.

5. Concluding Remarks

This study was conducted to understand the impact of schedule recovery on real-time holding control strategies for a fixed bus route. Such schedule recovery behavior is readily available in the context of cruising guidance technology but so far has been neglected in the holding control analysis and evaluation.

To capture the stochastic nature of travel times and demand, we first developed a bus propagation model with vehicle capacity constraint, which was extended by combining with the real-time holding control strategies (i.e., schedule-based and headway-based) and the effect of schedule recovery. The recovery time is proportional to the arriving delay time and the corresponding adjustment factor. Such stop-specific factor could be calibrated with the help of AVL-APC data.

The combined effects of holding control and schedule recovery were tested through a case study for a simulated real-life bus route in Guangzhou, China. We analysed and compared how the schedule recovery behavior affects the system performance of two holding control methods under different operating settings. We found that schedule recovery acts as the contributor to service reliability, and that schedule-based holding method is less able to stabilize headways but could benefit more from schedule recovery compared to headway-based holding method. The findings are specific to a real-life simulated bus route. The comparative results can help provide supporting tool for different bus control options and bus stop design in the provision of emerging technologies.

Future research may continue to develop an extended list of operational tactics and compare the combined effects under various operational settings. There exist common-line corridors where several bus lines serve the same stops [39]. The further study will also extend to considering multiple bus lines and investigate the impact of common-line stops on the system performance.

Notations

i: The subscript of vehicle

j: The subscript of bus stop

C: The vehicle capacity

q_j: The passenger arrival rate at stop j (pax/min)

ρ_j: The alighting proportion at stop j (%)

b: The average boarding time for passengers (s)

α: The average alighting time for passengers (s)

H: The design headway of a bus line (min)

$\beta_{i,j}$: The adjustment factor between stop j and $j+1$ for bus i

δ: Minimum safety interval (min)

$h_{i,j}$: The headway between bus i and the preceding bus at stop j (min)

$a_{i,j}$: The arrival time of bus i at stop j (min)

$d_{i,j}$: The departure time of bus i at stop j (min)

$t_{i,j}$: The travel time between stop j and $j+1$ for bus i (min)

$B_{i,j}$: The total number of waiting passengers for bus i at stop j during the headway (pax)

$\overline{B}_{i,j}$: The actual number of boarding passengers for bus i at stop j during the headway (pax)

$\overline{B}'_{i,j}$: The actual number of arriving passengers at stop j who are able to board bus i during the headway (pax)

$A_{i,j}$: The number of alighting passengers of bus i at stop j (pax)

$D_{i,j}$: The dwell time for bus i at stop j (min)

$s_{i,j}$: The scheduled departure time of bus i at stop j for schedule-based holding control (min)

$l_{i,j}$: The number of leftover passengers of bus i when it departs from stop j (pax)

$L_{i,j}$: The number of on-board passengers of bus i between stop j and $j+1$ (pax).

Conflicts of Interest

The authors declare that they have no conflicts of interest.

Acknowledgments

This work is supported by the National Science Foundation of China (Project nos. 61703165 and 61473122), the China Postdoctoral Science Foundation (Project no. 2016M600653), the Fundamental Research Funds for the Central Universities (Project no. D2171990), and the Royal Academy of Engineering (Project UK-CIAPP\286).

References

[1] Y. Hollander and R. Liu, "Estimation of the distribution of travel times by repeated simulation," *Transportation Research Part C: Emerging Technologies*, vol. 16, no. 2, pp. 212–231, 2008.

[2] W. Wu, R. Liu, and W. Jin, "Modelling bus bunching and holding control with vehicle overtaking and distributed passenger boarding behaviour," *Transportation Research Part B: Methodological*, vol. 104, pp. 175–197, 2017.

[3] E. E. Osuna and G. F. Newell, "Control strategies for an idealized public transportation system," *Transportation Science*, vol. 6, no. 1, pp. 52–72, 1972.

[4] M. D. Hickman, "An analytic stochastic model for the transit vehicle holding problem," *Transportation Science*, vol. 35, no. 3, pp. 215–237, 2001.

[5] X. J. Eberlein, N. H. M. Wilson, and D. Bernstein, "The holding problem with real-time information available," *Transportation Science*, vol. 35, no. 1, pp. 1–18, 2001.

[6] J. Zhao, M. Dessouky, and S. Bukkapatnam, "Optimal slack time for schedule-based transit operations," *Transportation Science*, vol. 40, no. 4, pp. 529–539, 2006.

[7] C. F. Daganzo, "A headway-based approach to eliminate bus bunching: Systematic analysis and comparisons," *Transportation Research Part B: Methodological*, vol. 43, no. 10, pp. 913–921, 2009.

[8] Y. Xuan, J. Argote, and C. F. Daganzo, "Dynamic bus holding strategies for schedule reliability: Optimal linear control and performance analysis," *Transportation Research Part B: Methodological*, vol. 45, no. 10, pp. 1831–1845, 2012.

[9] J. Argote-Cabanero, C. F. Daganzo, and J. W. Lynn, "Dynamic control of complex transit systems," *Transportation Research Part B: Methodological*, vol. 81, pp. 146–160, 2015.

[10] C. F. Daganzo and J. Pilachowski, "Reducing bunching with bus-to-bus cooperation," *Transportation Research Part B: Methodological*, vol. 45, no. 1, pp. 267–277, 2001.

[11] F. Delgado, J. C. Munoz, and R. Giesen, "How much can holding and/or limiting boarding improve transit performance?" *Transportation Research Part B: Methodological*, vol. 46, no. 9, pp. 1202–1217, 2012.

[12] D. Hernández, J. C. Muñoz, R. Giesen, and F. Delgado, "Analysis of real-time control strategies in a corridor with multiple bus services," *Transportation Research Part B: Methodological*, vol. 78, pp. 83–105, 2015.

[13] G. Sánchez-Martínez, H. N. Koutsopoulos, and N. H. M. Wilson, "Real-time holding control for high-frequency transit with dynamics," *Transportation Research Part B: Methodological*, vol. 83, pp. 1–19, 2016.

[14] B. Yu and Z. Yang, "A dynamic holding strategy in public transit systems with real-time information," *Applied Intelligence*, vol. 31, no. 1, pp. 69–80, 2009.

[15] L. Moreira-Matias, O. Cats, J. Gama, J. Mendes-Moreira, and J. F. De Sousa, "An online learning approach to eliminate Bus Bunching in real-time," *Applied Soft Computing*, vol. 47, pp. 460–482, 2016.

[16] R. Nair, E. Bouillet, Y. Gkoufas et al., "Data as a resource: real-time predictive analytics for bus bunching," in *Proceedings of the Annual Meeting of the Transportation*, Washington, D.C, 2014.

[17] M. Andres and R. Nair, "A predictive-control framework to address bus bunching," *Transportation Research Part B: Methodological*, vol. 104, pp. 123–148, 2017.

[18] A. Fonzone, J.-D. Schmöcker, and R. Liu, "A model of bus bunching under reliability-based passenger arrival patterns," *Transportation Research Part C: Emerging Technologies*, vol. 59, pp. 164–182, 2015.

[19] J.-D. Schmöcker, W. Sun, A. Fonzone, and R. Liu, "Bus bunching along a corridor served by two lines," *Transportation Research Part B: Methodological*, vol. 93, pp. 300–317, 2016.

[20] B. Yu, Z. Z. Yang, P. H. Jin, S. H. Wu, and B. Z. Yao, "Transit route network design-maximizing direct and transfer demand density," *Transportation Research Part C: Emerging Technologies*, vol. 22, pp. 58–75, 2012.

[21] W. Wu, R. Liu, and W. Jin, "Designing robust schedule coordination scheme for transit networks with safety control margins," *Transportation Research Part B: Methodological*, vol. 93, pp. 495–519, 2016.

[22] M. Chen, X. Liu, and J. Xia, "Dynamic prediction method with schedule recovery impact for bus arrival time," *Transportation Research Record*, no. 1923, pp. 208–217, 2005.

[23] Y. Ji, L. He, and H. M. Zhang, "Bus drivers' responses to real-time schedule adherence and the effects on transit reliability," *Transportation Research Record*, vol. 2417, pp. 1–9, 2014.

[24] W.-H. Lin and R. L. Bertini, "Modeling schedule recovery processes in transit operations for bus arrival time prediction," *Journal of Advanced Transportation*, vol. 38, no. 3, pp. 347–365, 2004.

[25] Y. Yan, Q. Meng, S. Wang, and X. Guo, "Robust optimization model of schedule design for a fixed bus route," *Transportation Research Part C: Emerging Technologies*, vol. 25, pp. 113–121, 2012.

[26] T. Liu, A. Ceder, J. Ma, and W. Guan, "Synchronizing pub-

lic transport transfers by using intervehicle communication scheme: Case study," *Transportation Research Record*, vol. 2417, pp. 78–91, 2014.

[27] M. M. Nesheli, A. Ceder, and V. A. Gonzalez, "Real-Time Public-Transport Operational Tactics Using Synchronized Transfers to Eliminate Vehicle Bunching," *IEEE Transactions on Intelligent Transportation Systems*, vol. 17, no. 11, pp. 3220–3229, 2016.

[28] F. Milla, D. Sáez, C. E. Cortés, and A. Cipriano, "Bus-Stop Control Strategies Based on Fuzzy Rules for the Operation of a Public Transport System," *IEEE Transactions on Intelligent Transportation Systems*, 2012.

[29] J. B. Ingvardson, O. A. Nielsen, S. Raveau, and B. F. Nielsen, "Passenger arrival and waiting time distributions dependent on train service frequency and station characteristics: A smart card data analysis," *Transportation Research Part C: Emerging Technologies*, vol. 90, pp. 292–306, 2018.

[30] M. D. Salek and R. B. Machemehl, "Characterizing bus transit passenger waiting times," SWUTC Research Peport 99, Austin, Tex, USA, 1999.

[31] E. Hans, N. Chiabaut, L. Leclercq, and R. L. Bertini, "Real-time bus route state forecasting using particle filter and mesoscopic modeling," *Transportation Research Part C: Emerging Technologies*, vol. 61, pp. 121–140, 2015.

[32] Z. Liu, Y. Yan, X. Qu, and Y. Zhang, "Bus stop-skipping scheme with random travel time," *Transportation Research Part C: Emerging Technologies*, vol. 35, pp. 46–56, 2013.

[33] W. Wu, L. Shen, X. Ji, and W. Jin, "Analysis of freeway service patrol with discrete event-based simulation," *Simulation Modelling Practice and Theory*, vol. 47, pp. 141–151, 2014.

[34] B. Z. Yao, P. Hu, X. H. Lu, J. J. Gao, and M. H. Zhang, "Transit network design based on travel time reliability," *Transportation Research Part C: Emerging Technologies*, vol. 43, pp. 233–248, 2014.

[35] M. A. Turnquist and L. A. Bowman, "The effects of network structure on reliability of transit service," *Transportation Research Part B: Methodological*, vol. 14, no. 1-2, pp. 79–86, 1980.

[36] R. Liu and S. Sinha, *Modelling urban bus service and passenger reliability. Paper presented at the International Symposium on Transportation Network Reliability*, The Hague, Netherlands, 2007.

[37] W. Wu and W. Jin, "Comparative analysis of bus holding control strategies with the effect of schedule recovery," in *Proceedings of the 97th Annual Meeting of the Transportation Research Board*, Washington, DC, USA.

[38] A. M. El-Geneidy, J. G. Strathman, T. J. Kimpel, and D. T. Crout, "Effects of bus stop consolidation on passenger activity and transit operations," *Transportation Research Record*, no. 1971, pp. 32–41, 2006.

[39] B. Yu, W. H. K. Lam, and M. L. Tam, "Bus arrival time prediction at bus stop with multiple routes," *Transportation Research Part C: Emerging Technologies*, vol. 19, no. 6, pp. 1157–1170, 2011.

9

Public Transit Loyalty Modeling Considering the Effect of Passengers' Emotional Value

Shi-chao Sun (ID)

College of Transportation Engineering, Dalian Maritime University, Dalian, 116026, China

Correspondence should be addressed to Shi-chao Sun; sunshichao1988@hotmail.com

Academic Editor: Luigi Dell'Olio

To better sustain passengers' loyalty towards bus service, this paper addressed the modeling of the public transit loyalty by the use of structural equation model. As a novel hypothesis, the emotional value was considered to have effects on the perceived value of bus services in this study, which reflected the degree of passengers' emotional dependence on the public transit. Specifically, in order to better assess the loyalty, seven unobserved variables were measured to construct the structural model, namely, "service guarantee," "operational services and efficiency," "emotional value," "perceived value," "expectation," "satisfaction," and "loyalty." The goodness-of-fit of the model was estimated and evaluated by using the survey data harvested from Xiamen, China. Besides, the index score of variables was also computed to help determine targeted approaches to better improve the level of bus service. The results indicated that the time cost and the monetary cost actually had no effects on the perceived value of users in the case study. At the same time, however, it also proved that passengers' emotional value towards the public transit indeed affected passengers' perception of the service value. In addition, whether users' perceived value was as expected determined how much passengers satisfied with the service. Regarding the index score of variables, it indicated a great dissatisfaction of passengers towards the current bus service. Unexpectedly, the score of loyalty even still retained a relatively high level, which reflected continue-to-use willingness of passengers. It implied that being subject to economic conditions and other factors, passengers were captive and had to continue relying on the public transit, in spite of their dissatisfaction. As for the improvement direction of bus services, targeted approaches should be determined to improve the quality of bus service, regarding the aspects of "condition of facilities in the bus," "driving stability and comfort," "vehicle speed," and "safety."

1. Introduction

Public transportation (PT) systems including common buses and metro systems have been long considered as an effective and sustainable way to alleviate traffic jams and promote the quality of trips. The rapid growth in the number of private cars led to a very serious traffic congestion problem in most metropolis, especially in China where the level of urbanization has been developed significantly in the past decade. To this end, the local governments have been hard working for years to construct an efficient "Transit Metropolis with a Bus Priority," where most residents were expected to prefer and rely on the public transit. In this context, developing the bus priority does not simply rely on the improvement or new constructions of public transit facilities on the supply side

but also needs a relatively competitive improvement of bus services to satisfy the passengers' demand. Thus, in order to evaluate the effectiveness of public transit improvements, an integrated assessment method should be established to monitor the dynamic development of bus service quality. However, the current practice in China usually employs several objective operational indicators to measure the quality of bus service, such as travel time, time-reliability, and waiting time. It lacks assessing the bus services from the perspective of the demand side and also neglects to discuss how the bus services quality are perceived by passengers. This limitation has led to a common phenomenon of public transportation systems in China, which can be summarized as "sufficient facilities with no high-quality service." Therefore, the quality assessment method of bus services in China needs to be improved, and passengers' perception of bus service quality should be measured with the "public transit loyalty" as the core [1–3].

In marketing, the definition of customer's "satisfaction" is different from "loyalty." When a customer feels satisfied, it mostly reflects a high degree of conformity between one's expectation of the product/service utility and the actual value perceived by him/her in a short term. However, customers' loyalty can be regarded as a long-term combination of behaviors and willingness of patronage [4]. Although the satisfaction is usually treated as a variable affecting the loyalty directly and significantly, the latter one seems to be more complex since the long-term accumulation of personal emotions/attitudes plays an important role in its development progress. The methodology of customer loyalty and also the assessment of perceived service quality have been widely applied in the area of public transportation [3, 5–12]. In this context, the relationship between passengers' perceived value of bus services, the level of satisfaction, and the loyalty has recently received much attention, since understanding what affects the public transit loyalty can be used to develop targeted approaches to retain existing passengers and increase the ridership [11]. To this end, when regarding this issue, it should account for how to assess the perceived quality of bus services, the definition of public transit loyalty, and the mechanism of related variables interacting with each other.

In the context of perceived quality of bus services, a number of previous studies have focused on the quality measurement [13]. Susniene [14] employed the framework of SERQUAL model and derived the related factors to assess the bus service quality, such as reliability, tangibility, and assurance. Sezhian et al. [2] adopted a discriminant analysis method to study the passengers' expectations by analyzing attribute-based perceptual mapping in a bus company in India, and the quality of bus services was measured by using 18 indicators, such as seat comfortable, cleanliness, driver behavior, stopping at correct place, and obeying traffic rules. Similarly, de Ona et al. [6] measured the bus service quality by using 12 indicators, among which punctuality, cleanliness, safety, courtesy, and accessibility were considered. While the number of parameters evaluating the service quality was reduced to six in the work of dell'Olio et al. [8], a discrete choice model was adopted to assess the effectiveness and influence of related factors. Kendall's algorithm and the Delphi method were, respectively, employed by Cafiso et al. [15, 16] to study the safety issues in the management of bus operation, and factors related to drivers' skills and behaviors, the maintenance of vehicles, and traffic issues on roads were taken into account for assessing bus service quality. Gonzalez-Diaz and Montoro-Sanchez [17] segmented the indicators into three groups to evaluate the bus services, respectively, including (1) quality of service outside the vehicle, (2) quality of vehicle, and (3) fares and schedules. Although the combination of indicators used to assess the bus service quality varied from studies to studies in the literature, the general evaluation methods and the aspects of related indicators selection were relatively consistent. Nevertheless, it was also necessary to use statistical analysis techniques to better explore and confirm the correlations of variables in the practice.

Regarding the definition of public transit loyalty, TCRP Report 49 indicated "customer loyalty is reflected by a combination of attitudes and behaviors. It is usually driven by customer satisfaction, yet also involves a commitment on the part of the customer to make a sustained investment in an ongoing relationship with transit services (p. 18)" [18]. However, there still seems to be no standardized or even common way to assess the loyalty. Chou and Kim [5] adopted a structural equation model (SEM) to examine the relationship between corporate image, quality of bus services, complaints, satisfaction, and loyalty through an empirical test by using the data harvested from Taiwan and South Korea. The typical characteristics of loyal passengers were defined as repeat patronage, willingness of recommending to others, and high price tolerance. However, Minser and Webb [19] defined the public transit loyalty based on the likeliness of continuous service usage and willingness to recommend. They analyzed the empirical data collected from the Chicago Transit Authority by using a SEM and found that the quality of service, perceived value of service, satisfaction, and corporate image were the key factors affecting passengers' loyalty. Both of the above studies defined a unique "public transit loyalty," but the variable which could reflect passengers' attitudinal preference to continue to use the service was not demonstrated in the model. To this end, some researches extended the definition of loyalty by including the consideration of intended future use and other variables [3, 12, 20, 21].

Nevertheless, in the context of public transit loyalty, passengers may generate a positive or negative emotion/impression towards the service after taking a bus, and how users felt about the mode would be accumulated with the increase in the number of trips [3]. The perception/impression would gradually affect the degree of passengers' emotional dependence on the public transit as well as the loyalty. Thus, it was additionally somewhat different from the conception reflected by the discrete choice model, in which the decision-making process was not only dependent on the utilities but also required emotional attitude shifting for a period of time. This was what the emotional value implied and how it worked. However, the above existing works still neglected to account for the influence of long-term accumulation of personal emotions/attitudes on the perceived value and eventually affecting the level of loyalty. This paper aimed to bridge the gap and enriched the literature. Two main hypotheses were proposed and intended to be examined in the case study of this paper:

(i) The passengers' emotional value towards the public transit would affect how the value of bus services perceived by the users.

(ii) The perceived value combined with user's expectation would have a main effect on his/her satisfaction and eventually influence the loyalty.

Moreover, it was also attempted in the current paper to adopt the index score of variables to better determine the

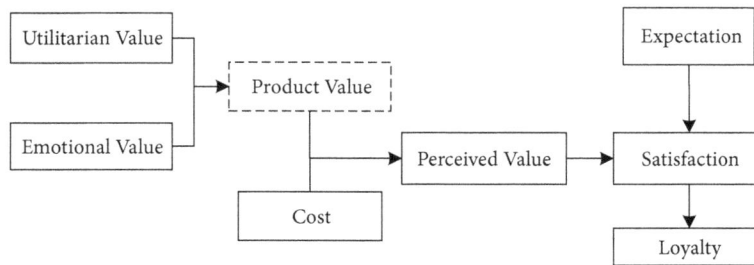

FIGURE 1: The hypothesized path diagram of the public transit loyalty.

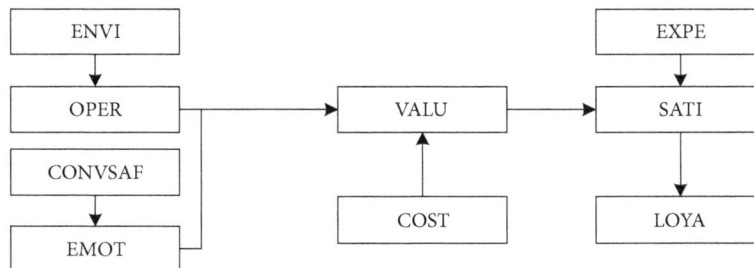

FIGURE 2: The structural model of public transit loyalty.

direction of improvement on bus services and retain the loyalty by mining related factors.

2. Methods

2.1. SEM and Factor Analyses Methods. The methodology of SEM has been widely applied in statistics and other disciplines. It is a combination technique including factor analysis, path analysis, and regression models. SEM can be used to better identify, estimate and verify the relationship between unobserved variables, and also benefit examining the correlations of an appropriate set of observed variables to measure the latent variable. Thus, there are two kinds of submodels included in SEM: measurement models and structural models. The structural model examines and verifies the links between unobserved variables, in which several measurement models should be determined for unobserved variable measures.

Exploratory factor analysis (EFA) and confirmatory factor analysis (CFA) are two common methods of factor analyses to determine/verify the relationships between variables in SEM. EFA is commonly used to find the relationship between variables, since researchers may not be able to have the empirical knowledge in this context previously [13]. In contrast, however, CFA could only be applied to verify whether the relationships between the variables are as expected. Since the main goal of this study was to examine and verify the proposed hypotheses, CFA was employed here to assess the extended loyalty model derived from the literature, by additionally considering the effect of emotional value on the target variable.

2.2. Hypotheses and Variables. In marketing, the product/ service value was usually considered to consist of two parts: the utilitarian value and the emotional value. In this paper, the utilitarian value referred to the subjective measures of bus service quality that corresponded to the objective utilities, while the emotional value denoted the overall feelings or impressions of passengers towards the mode, reflecting the degree of emotional dependence. It could be positive or negative and derived from a long-term accumulation of passengers' bus service experience. It was assumed that the value of bus services perceived by passengers was affected by a combination of the utilitarian value, the emotional value, and also the service cost [3]. However, the perceived value was not considered as the only factor that affected users' satisfaction. It depended on whether the perceived value was as expected. Thus, the satisfaction was determined by the consistency between passenger's perceived value and his/her expectation and eventually affected the loyalty. The hypothesized path diagram was illustrated in Figure 1.

The path diagram could be regarded as a conceptual framework of the structural model, which was derived from the existing literature but additionally taking account of passengers' emotional values (Figure 2). Specifically, the utilitarian value consisted of three unobserved variables, namely, "travel environment and facilities" (ENVI), "operational service and efficiency" (OPER), and "convenience and safety" (CONVSAF). It was assumed that facility conditions may have a direct effect on the bus operation management, and additionally how passengers felt about the mode was more sensitive to the travel safety [12]. Thus, the relationship between unobserved variables could be denoted as shown in Figure 2, which were intended to be confirmed in this study.

The latent variables in the structural model shown in Figure 2 could not be directly observed. Thus, each latent variable needs to be measured by an appropriate set of observed variables. To this end, the measurement model could be estimated and verified through CFA by examining the correlations of corresponding observed variables. The observed variables involved in each initial measurement model were as follows:

(1) Utilitarian Value. The observed variables for measuring the utilitarian value should be, respectively, in terms of environment, facilities, operational efficiency, safety, and also convenience [13].

(1-1) Travel Environment and Facilities (ENVI). In the context of travel environment, the degree of crowdedness, cleanness, suitable temperature, and odor were, respectively, concerned, while on the side of facilities, the overall condition of facilities in the buses and at the stations was taken into consideration.

(1-2) Operational Services and Efficiency (OPER). Drivers' behavior (driving stability and comfort) and the courtesy of crew members were treated as the main indicators for evaluating the service quality inside the bus in this paper [2, 6]. At the same time, regarding the service quality outside the bus, the efficiency of operations, vehicle speed, punctuality, and waiting time were concerned.

(1-3) Convenience and Safety (CONVSAF). As for the convenience of bus services, walking distance reflected the coverage of public transit networks, while the number of transfers indicated the degree of matching between the public transportation route setting and passengers' actual travel need. In addition, the safety of vehicles was measured in terms of security risks caused by human factors and the insufficient attention to the maintenance of the vehicle.

(2) Emotional Value. How to measure the emotional value of consumers had received much attention in marketing. Therefore, according to the measurement model proposed by Babin et al. [1], nine most related observed variables were selected out of twelve to measure passengers' emotions towards bus services, as shown in Table 1.

(3) Costs. The total cost of bus services mainly considered the time cost and also the fare, in which the time cost included waiting time, transfer time, and travel time.

(4) Perceived Value. In this paper, the perceived value was in terms of "the level of service on the basis of the current fare" and "satisfaction of fare at the current level of service," according to Chinese Customer Satisfaction Index Model (CCSI).

(5) Expectation. According to American Customer Satisfaction Index Model (ACSI), passengers' expectation towards bus services could be described by considering both their expectation of overall bus service quality and the expectation of services that would meet their personalized demand.

(6) Satisfaction. Regarding the satisfaction of bus services, Shiftan et al. [3] extended the concept firstly proposed by Oliver [22] to better account for passengers' satisfaction. Other than Oliver [22] using a path analysis method, he constructed a measurement model by taking account of three observed variables, namely, "overall satisfaction towards bus services," "whether the quality of service was as expected," and "how far is the gap from the ideal level of bus services." Such measures of passengers' satisfaction were also employed in the current study.

(7) Loyalty. In marketing, Oliver [22] proposed a four-stage formation theory of loyalty, which had received much attention of researchers. The theory was then extended, and each stage of loyalty was, respectively, measured by using four observed variables [4]. Shiftan et al. [3] applied the hypothesis to the public transit and examined the relationship between the four stages of passengers' loyalty. The results indicated that it was not necessary to segment the formation of loyalty into four separate measurement models, since it could be regarded as a combination. Therefore, the measurement model assessing passengers' loyalty was employed in this paper, which reflected passengers' intention of continuing to use the bus services in the near future.

3. Data Collection and Analysis

3.1. Survey Designs and Data Preparation. According to the appropriate parameters for measuring each latent variable described above, forty observed variables were finally selected and converted into the content of the questionnaire (Table 1). Each question was set as a choice format in a five-level Likert scale, and it required the interviewee to feedback his/her view towards each question, from "disagree/dissatisfy" to "agree/satisfy." The survey was implemented in a central business district of Xiamen, China, on the 29th of July, 2017. The interviewees were selected by random sampling from their workplaces, and a total of 900 questionnaires were issued. Eventually, 664 valid questionnaires were collected with completed replies. The number of valid samples met the requirements of sample size under 95% confidence levels and 4% maximum permissible errors.

Some descriptive statistics of valid samples were addressed as shown in Table 2, which was in line with characteristics of census data in 2016.

3.2. Data Reliability. The reliability of survey data was evaluated by using the value of Cronbach's alpha, which could reflect the internal consistency of observed variables in each measurement model. Generally, if the coefficient was greater than 0.7, it could be regarded as a good construct reliability. According to the results reported in Table 3, the data reliability was considered to be acceptable to measure the unobserved variable.

3.3. Data Validity. In this paper, Kaiser-Meyer-Olkin (KMO) statistics combined with the results of Bartlett's spherical test were employed to assess the data validity. The former one took values from 0 to 1, and the closer to 1, the higher correlations

TABLE 1: The designed content of questionnaires in the survey.

Latent Variable	Questions in the survey/Observed variables	Notation
ENVI	(1) Do you think it is very crowded in the bus during peak-hours	X_1
	(2) Are you satisfied with the cleanness and tidy condition in the bus	X_2
	(3) Are you satisfied with temperature and odor conditions in the bus	X_3
	(4) Are you satisfied with overall conditions of facilities in the bus	X_4
	(5) Are you satisfied with overall conditions of facilities at the station	X_5
OPER	(1) Are you satisfied with the driving stability and comfort	X_6
	(2) Are you satisfied with the vehicle speed	X_7
	(3) Are you satisfied with the courtesy of the crew	X_8
	(4) Are you satisfied with the average waiting time at stations	X_9
	(5) Are you satisfied with the punctuality of vehicle at stations	X_{10}
CONVSAF	(1) Are you satisfied with the overall condition of driving safety	X_{11}
	(2) Are you satisfied with the convenience of transfers	X_{12}
	(3) Are you satisfied with the walking distance to stations	X_{13}
	(4) Are you satisfied with the frequency of encountering breakdown	X_{14}
EMOT	(1) Travelling by bus could have a positive effect on your mood	X_{15}
	(2) Do you enjoy the fun of traveling by bus	X_{16}
	(3) Do you ever used to think about shifting to another transport mode	X_{17}
	(4) Are you feeling really relax and comfortable when taking a bus	X_{18}
	(5) Are you a member of bus priority supporters	X_{19}
	(6) Are you proud to be a public transit rider	X_{20}
	(7) Are you satisfied with the living environment	X_{21}
	(8) Sometimes whether taking a bus is subject to other factors	X_{22}
	(9) The setup of routes and timetable changes are what you concerned	X_{23}
COST	(1) Do you think the fare for bus service is reasonable	X_{24}
	(2) Do you think the time cost for bus service is reasonable	X_{25}
VALU	(1) Are you satisfied with the service on the basis of the current fare	Y_1
	(2) Are you satisfied with the fare at the current level of service	Y_2
EXPE	(1) Overall expectation of bus services	X_{26}
	(2) Expectation of the efficiency of bus operations	X_{27}
	(3) Expectation of bus services that would meet personalized needs	X_{28}
SATI	(1) Do you satisfied with the bus services	Y_3
	(2) Do you think the quality of service is as you expected	Y_4
	(3) Do you think the current service quality is not far from the ideal level	Y_5
LOYA	(1) Are you sure PT will still be your first choice in future	Y_6
	(2) Are you sure PT will still be your first choice even if the fare rise	Y_7
	(3) Are you sure PT will still be the first choice even if the quality retains	Y_8
	(4) Do you think traveling by bus is very suitable for you	Y_9
	(5) Do you think PT will still be your choice even if you can afford a car	Y_{10}
	(6) Do you think the bus service quality can be comparable to that of a car	Y_{11}
	(7) Have you always had positive attitudes towards bus services	Y_{12}

TABLE 2: Descriptive statistics of valid samples.

Characteristics	Statistics
Gender	Male (47.7%), Female (52.3%)
Age (years)	<=20 (3.1%), 21-30 (49.6%), 31-40 (22.0%), 41-50 (12.6%), 51-60 (9.9%), ≥60 (2.8%)

TABLE 3: Results of survey data reliability analysis.

Unobserved variable	Number of observed variables	Cronbach's alpha
EMOT	5	0.735
OPER	5	0.755
ENVI	4	0.759
CONVSAF	9	0.801
COST	2	0.787
VALU	2	0.873
EXPE	3	0.755
SATI	3	0.731
LOYA	7	0.702

TABLE 4: Results of data validation analysis.

Unobserved variable	KMO statistics	Bartlett's spherical test	Contribution rate of accumulative variance
EMOT	0.823 >0.7	0.000 (p<0.001)	58.222
OPER	0.794 >0.7	0.000 (p<0.001)	57.793
ENVI	0.772 >0.7	0.000 (p<0.001)	52.235
CONVSAF	0.831 >0.7	0.000 (p<0.001)	66.980
COST	Fixed value of 0.5	0.000 (p<0.001)	57.568
VALU	Fixed value of 0.5	0.000 (p<0.001)	60.340
EXPE	0.713 >0.7	0.000 (p<0.001)	58.969
SATI	0.845 >0.7	0.000 (p<0.001)	61.665
LOYA	0.793 >0.7	0.000 (p<0.001)	42.800

TABLE 5: Variables that need to be removed.

Unobserved variable	Variables need to be removed	Factor load
EMOT	X_{19}	.368
	X_{20}	.126
	X_{21}	.413
	X_{22}	.341
	X_{23}	.216
ENVI	X_5	.284
CONVSAF	X_{14}	.106
EXPE	X_{28}	.285
LOYA	Y_{10}	.374
	Y_{11}	.238
	Y_{12}	-.197

between variables were. Generally, KMO statistics equal to or greater than 0.7 was considered to have a strong correlation between variables (a fixed value of 0.5 if only two variables were considered). Moreover, if the result of Bartlett's sphere test reported that the contribution rate of accumulative variance was greater than 40% and the p value was less than the significance level; this implied that the unobserved variable could be better explained by the extracted common factor. As shown in Table 4, the results indicated that the survey data satisfied all above conditions for common factor extraction, under the configuration of each measurement model in SEM. The observed variables in each measurement model were strongly correlated and could be well explained by the corresponding unobserved variables. Thus, the validity of survey data could be confirmed.

In addition, the factor load of each variable in the component matrix should generally be greater than 0.5; otherwise, it could be removed from the corresponding measurement model to improve the data validity. Therefore, the selected observed variables in each measurement model still needed to be adjusted based on the above criterion, as shown in Table 5. Then, the adjusted measurement model could be used to establish the initial SEM for assessing public transit loyalty.

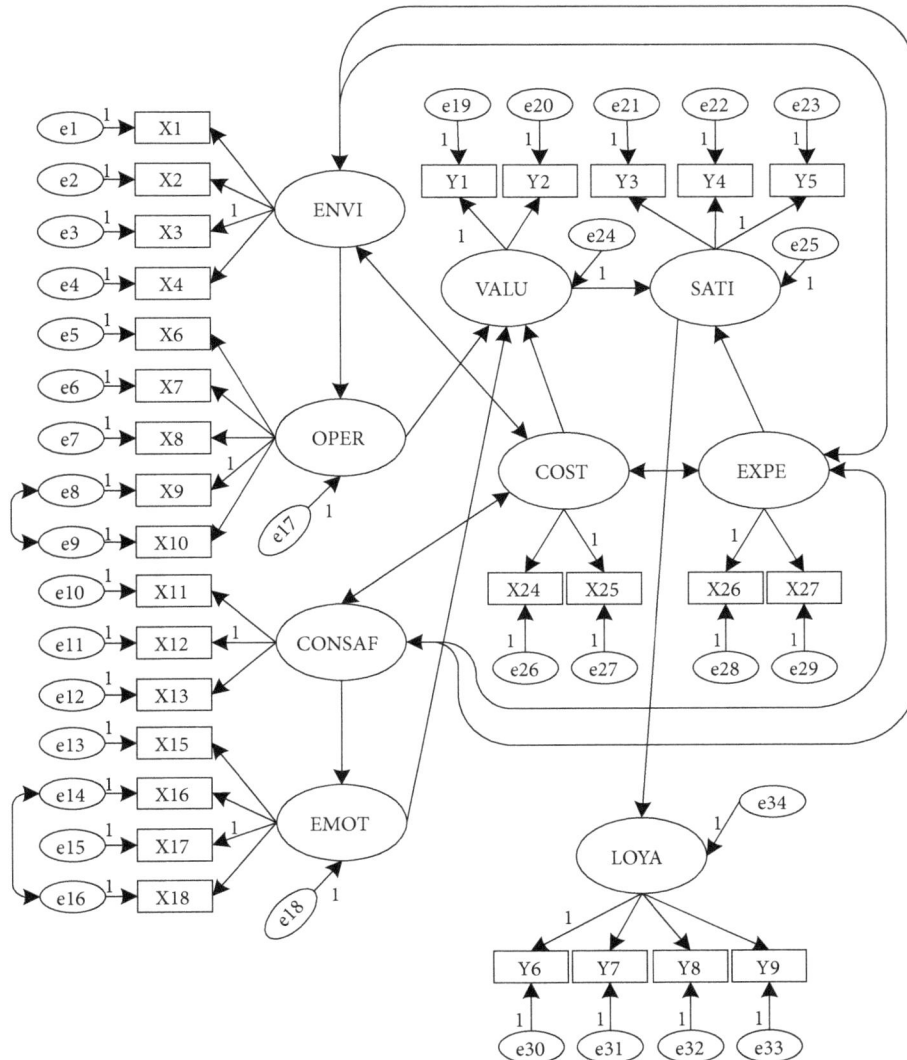

FIGURE 3: The initial model for assessing passengers' loyalty.

4. Findings

4.1. Modeling and Results Analysis. The initial model was built by using AMOS 21.0, as shown in Figure 3. The relationship between unobserved variables was derived from the conceptual framework proposed above, and corresponding adjusted measurement models were also adopted.

As for the goodness-of-fit of the model, five indicators were employed to assess it in the aspects of "Absolute Fitness Index," "Value-Added Fitness Index," and "Simplified Fitness Degree Index." The standard for each goodness-of-fit index was listed as shown in Table 6.

The path coefficients were estimated based on the survey data (Figure 4), while the estimation results indicated that the goodness-of-fit of the model was not well, since the index "GFI" was less than 0.9 not meeting with its criterion. Thus, the initial model could not be accepted and needed to be adjusted. According to the output of the path coefficients, the effect of "COST" (-0.08) on the perceived value was not

TABLE 6: Standards for evaluating goodness-of-fit.

Index	Goodness-of-fit standard
GFI	>0.90
RMSEA	<0.05
CMINDF	[1, 3]
CFI	>0.90
CN	>200

significant. In addition, the correlation coefficient between "ENVI" and "CONVSAF" was very high (0.925), which implied that another higher-order common factor could be extracted by fusing the two unobserved variables.

Regarding the adjustment of the initial model, since the average travel distance in Xiamen, China, was moderate with short travel time and the fare was relatively low compared with local income, the travel time cost combined with the

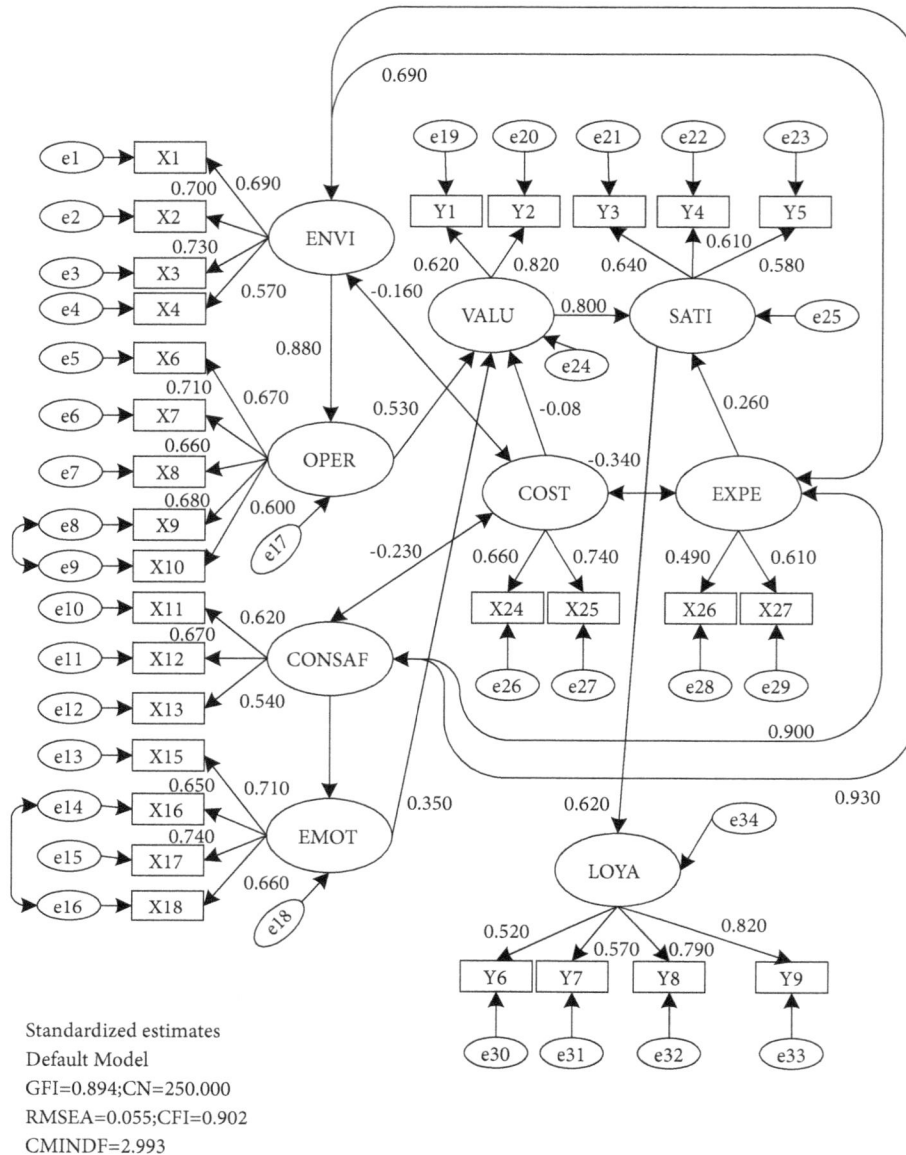

FIGURE 4: Results of the initial model.

monetary cost may had little effect on passengers' perception of service. Thus, the insignificant unobserved variable "COST" and its corresponding measurement model tended to be removed. Secondly, "ENVI" and "CONVSAF," which were strongly correlated, tended to be merged to extract a new common factor. Since the new factor, respectively, accounted for the travel environment, facilities, vehicle safety, and also the convenience of bus services, it could be named as "Service Guarantee" (SERGUA). The estimation results of the adjusted SEM were shown in Figure 5. The target variable "LOYA" could be well assessed, since R^2 test value was 0.672 and the goodness-of-fit indexes were fine. Thus, the adjusted model was considered to be acceptable.

4.2. The Index Score of Variables. Index score of variables could reflect the degree of interviewees' agreement or satisfaction towards the issue. The more it scored, the more

passengers were satisfied with the issues reflected by the variable. Thus, it was usually adopted in marketing to improve the level of customers' satisfaction. In this study, the index score was employed to determine the direction of bus services improvement, and the formula was as following:

$$Index\ Score = \frac{\sum_{i=1}^{k} w_i \bar{x}_i - \sum_{i=1}^{k} w_i}{\sum_{i=1}^{k} w_i \left(\max\{x_i\} - \min\{x_i\} \right)} \times 100 \quad (1)$$

In the formula, each variable scored from 0 to 100 points and

w_i denoted the regression coefficient of each observed variable;

$\bar{x}_i$ denoted the mean value of the ith variable;

$\max\{x_i\}$ denoted the full score of the ith variable, namely, 5 in this study;

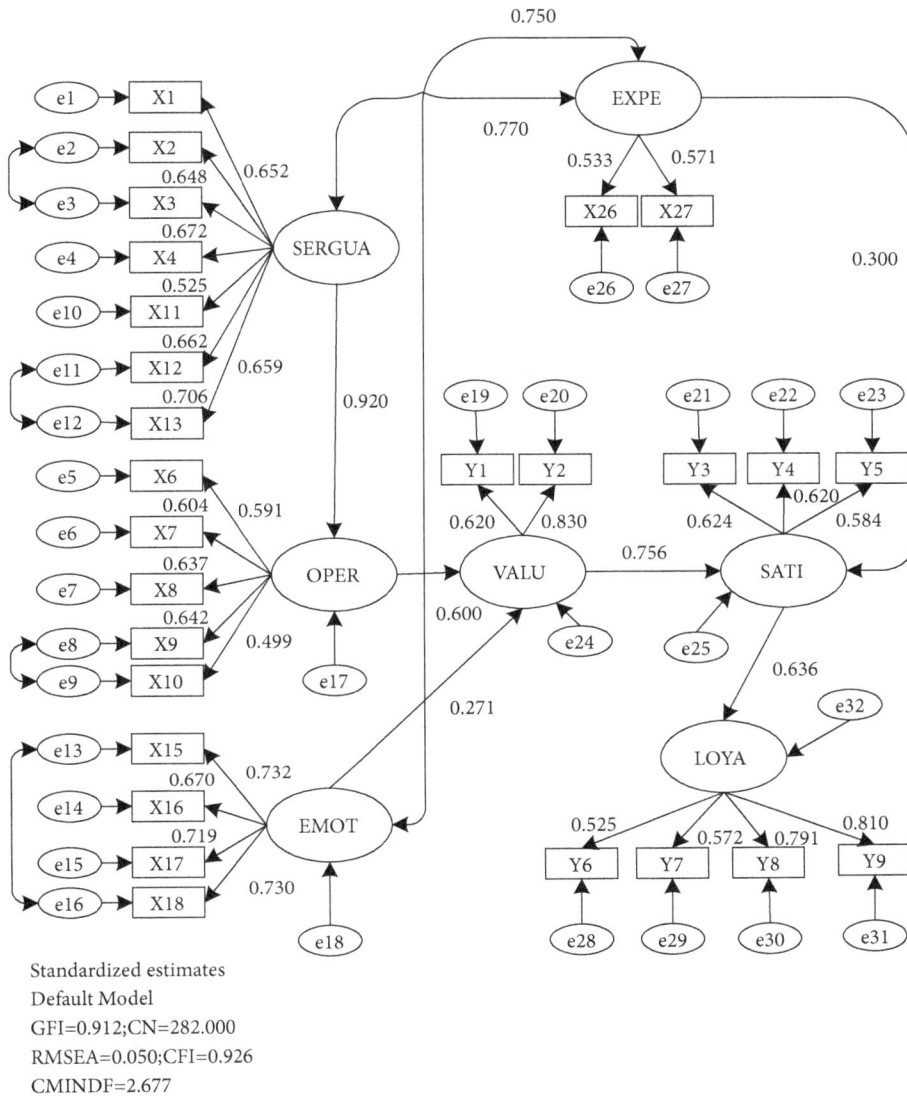

Standardized estimates
Default Model
GFI=0.912;CN=282.000
RMSEA=0.050;CFI=0.926
CMINDF=2.677

FIGURE 5: Results of the adjusted model.

min{x_i} denoted the lowest score of the ith variable, namely, 1 in this study.

Then, index score of variables was given including observed variables and corresponding unobserved variables, reported in Table 7.

4.3. Discussions. Based on the estimated path coefficients of the adjusted model and the index score of variables, some findings were discussed in the following. Firstly, "SERGUA" was found to have a very significant effect on "OPER" (0.920), as shown in Figure 5. Since "SERGUA" mostly accounted for travel environment, facility conditions, vehicle safety, the convenience of bus services, and so on, it certainly influenced the output related to the operational service and efficiency, as expected. It was also in line with the results in the related works [3, 11]. Secondly, "OPER" was considered to be the main factor (0.600) directly influencing the perceived value of passengers towards bus services. However,

regarding the emotional value which reflected the degree of passengers' emotional dependence on bus services, with each unit increase of "EMOT," passengers' perceived value would be promoted by 0.271 unit. It proved the hypothesis that a long-term accumulation of emotional attitude was also a part of perceived value in addition to the utilitarian value. Thus, it was necessary to take account of how users felt about the public transit when assessing the loyalty, since the formation process of loyalty was not only dependent on the objective utilities but also affected by an accumulation of emotional attitude for a period of time. Different from the existing works, this paper concerned and confirmed the effect of passengers' emotion/impression on the service value perceived by users, enriching the literature. Moreover, passengers' perceived value (0.756) was not the only factor that affected the satisfaction directly and significantly. The results indicated that passengers' satisfaction depended on whether the value of bus services perceived by passengers

TABLE 7: The index scores of variables in SEM.

Unobserved variable	Observed variable	Standardized weight	Scores of observed variables	Scores of unobserved variables
VALU	Y1	0.618	59.95	56.54
	Y2	0.829	51.50	
SATI	Y3	0.624	52.68	46.20
	Y4	0.62	40.85	
	Y5	0.584	44.57	
LOYA	Y6	0.525	63.55	57.90
	Y7	0.572	58.15	
	Y8	0.791	46.40	
	Y9	0.81	61.60	
SERGUA	X1	0.652	52.70	49.75
	X2	0.648	51.95	
	X3	0.672	55.68	
	X4	0.525	37.13	
	X11	0.662	46.45	
	X12	0.706	51.85	
	X13	0.659	53.65	
OPER	X6	0.591	44.73	49.75
	X7	0.604	51.58	
	X8	0.637	53.13	
	X9	0.642	53.25	
	X10	0.499	55.00	
EMOT	X15	0.732	53.35	46.88
	X16	0.67	48.95	
	X17	0.719	47.70	
	X18	0.73	37.70	
EXPE	X26	0.533	35.20	44.74
	X27	0.571	51.43	

was as expected, which was in line with the hypothesis. Furthermore, the satisfaction had a direct effect on the public transit loyalty, which was consistent with the conception in marketing and also the results derived from other related works [3, 10–12, 22].

Regarding the index score of variables, passengers' satisfaction on public transit of Xiamen only scored less than half (46.20 points). It implied that the overall level of bus services needed to be improved, because the passengers had already shown a great dissatisfaction with the public transport system. In addition, although the passengers' satisfaction was low, the score of loyalty was actually the highest among all the unobserved variables (57.9 points). Since the public transit loyalty here was defined as the indicator reflecting continue-to-use willingness, it indicated that passengers were captive and had to keep relying on public transit in the near future, in spite of the dissatisfaction. The score related to the operational indicators reflected that the quality of bus services was still low ("SERGUA" and "OPER" were both 49.75 points). As reported in Table 7, it was necessary to focus

on the core content of improving bus service quality, which were in terms of "condition of facilities in the bus," "driving stability and comfort," "vehicle speed," and "safety."

5. Conclusion

This paper addressed the public transit loyalty modeling based on SEM. As a main hypothesis, the effect of passengers' emotional value on their perception of bus services was taken into account. Specifically, for assessing the loyalty, seven unobserved variables and their corresponding relationships were eventually considered in the structural model, namely, "service guarantee," "operational services and efficiency," "emotional value," "perceived value," "expectation," "satisfaction," and "loyalty." In addition, twenty-seven observed variables were involved in the measurement models. The goodness-of-fit of the model was estimated and evaluated by using the empirical data harvested from Xiamen, China. Index score of variables was also calculated to help determine the direction of bus services improvement. The results indicated the following:

(i) Due to the short average travel distance in Xiamen, China, and the relatively low fare, the travel time cost combined with the monetary cost had little effect on passengers' perception of service.

(ii) Passengers' emotional value had a significant effect on perception value of bus services. In addition to utility value, the long-term accumulation of emotional attitude also contributed to loyalty formation.

(iii) Passengers' satisfaction towards the bus services was affected by whether users' perceived value were as expected.

(iv) The overall level of bus services needed to be improved, since the passengers had already shown a great dissatisfaction with the service. However, the score of loyalty even still retained a relatively high level. Since the public transit loyalty here was defined as the indicator reflecting continue-to-use willingness, it implied that passengers were captive and had to rely on public transit, in spite of the dissatisfaction. It could also be concluded that the cost of shifting to private motorization was still high in China, subject to economic or other factors.

(v) According to the index score of variables in the case study of Xiamen, the performances of service quality were not well, in terms of "condition of facilities in the bus," "driving stability and comfort," "vehicle speed," and "safety." The results could be used to determine the core direction of improving bus services efficiently in practice.

The model of public transit loyalty proposed in this paper could account for the internal mechanism of travel behavior and attitude, and grasp the improvement requests of passengers. In further studies, the passengers can be segmented by their spatial and temporal travel patterns or other characteristics. Multiple causality models then can be used to examine the direct and indirect factors that influenced the loyalty of respective subdivided interviewees.

Conflicts of Interest

The author declares that there are no conflicts of interest in this paper.

Acknowledgments

This research is supported by "the Fundamental Research Funds for the Central Universities (3132016301)."

References

[1] B. J. Babin, W. R. Darden, and M. Griffin, "Work and/or fun: measuring hedonic and utilitarian shopping value," *Journal of Consumer Research*, vol. 20, no. 4, pp. 644–656, 1994.

[2] M. V. Sezhian, C. Muralidharan, T. Nambirajan, and S. G. Deshmukh, "Attribute-based perceptual mapping using discriminant analysis in a public sector passenger bus transport company: a case study," *Journal of Advanced Transportation*, vol. 48, no. 1, pp. 32–47, 2014.

[3] Y. Shiftan, Y. Barlach, and D. Shefer, "Measuring passenger loyalty to public transport modes," *Journal of Public Transportation*, vol. 18, no. 1, pp. 1–16, 2015.

[4] L. C. Harris and M. M. H. Goode, "The four levels of loyalty and the pivotal role of trust: A study of online service dynamics," *Journal of Retailing*, vol. 80, no. 2, pp. 139–158, 2004.

[5] J.-S. Chou and C. Kim, "A structural equation analysis of the QSL relationship with passenger riding experience on high speed rail: an empirical study of Taiwan and Korea," *Expert Systems with Applications*, vol. 36, no. 3, pp. 6945–6955, 2009.

[6] J. de Oña, R. de Oña, L. Eboli, and G. Mazzulla, "Perceived service quality in bus transit service: a structural equation approach," *Transport Policy*, vol. 29, pp. 219–226, 2013.

[7] L. dell'Olio, A. Ibeas, and P. Cecín, "Modelling user perception of bus transit quality," *Transport Policy*, vol. 17, no. 6, pp. 388–397, 2010.

[8] L. Dell'Olio, A. Ibeas, and P. Cecin, "The quality of service desired by public transport users," *Transport Policy*, vol. 18, no. 1, pp. 217–227, 2011.

[9] S. A. Figler, P. S. Sriraj, E. W. Welch, and N. Yavuz, "Customer Loyalty and Chicago, Illinois, Transit Authority Buses Results from 2008 Customer Satisfaction Survey," *Transportation Research Record*, no. 2216, pp. 148–156, 2011.

[10] P. J. Foote, D. G. Stuart, and R. Elmore-Yalch, "Exploring customer loyalty as a transit performance measure," *Transit Planning, Intermodal Facilities, and Marketing*, no. 1753, pp. 93–101, 2001.

[11] D. van Lierop and A. El-Geneidy, "Enjoying loyalty: The relationship between service quality, customer satisfaction, and behavioral intentions in public transit," *Research in Transportation Economics*, vol. 59, pp. 50–59, 2016.

[12] J. Zhao, V. Webb, and P. Shah, "Customer loyalty differences between captive and choice transit riders," *Transportation Research Record*, vol. 2415, pp. 80–88, 2014.

[13] S. Jomnonkwao and V. Ratanavaraha, "Measurement modelling of the perceived service quality of a sightseeing bus service: An application of hierarchical confirmatory factor analysis," *Transport Policy*, vol. 45, pp. 240–252, 2016.

[14] D. Susniene, "Quality approach to the sustainability of public transport," *Transport-Vilnius*, vol. 27, no. 1, pp. 102–110, 2012.

[15] S. Cafiso, A. Di Graziano, and G. Pappalardo, "Road safety issues for bus transport management," *Accident Analysis & Prevention*, vol. 60, pp. 324–333, 2013.

[16] S. Cafiso, A. Di Graziano, and G. Pappalardo, "Using the Delphi method to evaluate opinions of public transport managers on bus safety," *Safety Science*, vol. 57, pp. 254–263, 2013.

[17] M. González-Díaz and Á. Montoro-Sánchez, "Some lessons from incentive theory: Promoting quality in bus transport," *Transport Policy*, vol. 18, no. 2, pp. 299–306, 2011.

[18] Transportation Research Board, *A Hand Book for Measuring Customer Satisfaction and Service Quality*, TCRP Report, Washington, DC, USA, 1999.

[19] J. Minser and V. Webb, "Quantifying the benefits: Application of customer loyalty modeling in public transportation context," *Transportation Research Record*, no. 2144, pp. 111–120, 2010.

[20] R. Carreira, L. Patrício, R. Natal Jorge, and C. Magee, "Understanding the travel experience and its impact on attitudes, emotions and loyalty towards the transportation provider-A quantitative study with mid-distance bus trips," *Transport Policy*, vol. 31, pp. 35–46, 2014.

[21] W.-T. Lai and C.-F. Chen, "Behavioral intentions of public transit passengers-The roles of service quality, perceived value, satisfaction and involvement," *Transport Policy*, vol. 18, no. 2, pp. 318–325, 2011.

[22] R. L. Oliver, "Whence consumer loyalty?" *Journal of Marketing*, vol. 63, pp. 33–44, 1999.

10

Research on Taxi Driver Strategy Game Evolution with Carpooling Detour

Wei Zhang, Ruichun He ⓘ, Changxi Ma ⓘ, and Mingxia Gao

School of Traffic and Transportation, Lanzhou Jiaotong University, Lanzhou 730070, China

Correspondence should be addressed to Ruichun He; tranman@163.com

Academic Editor: Giulio E. Cantarella

For the problem of taxi carpooling detour, this paper studies driver strategy choice with carpooling detour. The model of taxi driver strategy evolution with carpooling detour is built based on prospect theory and evolution game theory. Driver stable strategies are analyzed under the conditions of complaint mechanism and absence of mechanism, respectively. The results show that passenger's complaint mechanism can effectively decrease the phenomenon of driver refusing passengers with carpooling detour. When probability of passenger complaint reaches a certain level, the stable strategy of driver is to take carpooling detour passengers. Meanwhile, limiting detour distance and easing traffic congestion can decrease the possibility of refusing passengers. These conclusions have a certain guiding significance to formulating taxi policy.

1. Introduction

Taxi carpooling mode has the characteristics of improving the transportation efficiency, easing the traffic pressure, and reducing environmental pollution [1, 2]. In recent years many cities in China have begun to implement taxi carpooling policy to solve the serious traffic problems. Taxi carpooling operation is in the initial stage, and the carpool policy is imperfect. It is very necessary to study taxi carpooling problem in the present situation.

Many scholars have carried out researches on the problem of carpooling. At present, most researches on carpooling are focused on the problem of carpooling matching and path optimization [3–6]. Aissat and Oulamara [7] proposed a method of optimizing starting point position of driver and passenger based on the heuristic algorithm, and the results can minimize the travel cost and detour distance. Santi et al. [8] proposed a method of carpooling strategy optimization based on network, which can shorten the passenger travel time and improve the utilization rate of vehicles. Shinde and Thombre [9] solved the problem of carpooling path optimization based on genetic algorithms. He et al. [10] obtained the optimal path by mining GPS positioning data. Nourinejad and Roorda [11] designed a centralized and decentralized

optimization algorithms based on binary integer programming and dynamic auction-based multiagent for the problem of matching passengers and drivers. Huang et al. [12] proposed the genetic-based carpooling route and matching algorithm for the multiobjective optimization problem. Xia et al. [13] built an optimization model of carpooling matching and designed heuristic algorithm to solve it. Chiou and Chen [14] proposed a dynamic matching method for carpooling, which can be used in mobile devices via ad hoc Wi-Fi networks. Boukhater et al. [15] presented a GA with a customized fitness function that searches for the solution with minimal travel distance, efficient ride matching, timely arrival, and maximum fairness, which can be used in carpooling system. Xiao et al. [16] studied passengers matching problem based on clustering and pattern recognition methods. The above researches establish the model of carpooling matching optimization and then find the optimal matching mode by heuristic algorithm or data mining algorithms, which solve the problem of the carpooling matching and path optimization and provide theory support for carpooling implementation.

Detour problem is a common phenomenon of taxi carpooling. In general detour passengers will get more cost discount than usual carpooling to make up for the losses caused by detour. But from the perspective of the drivers,

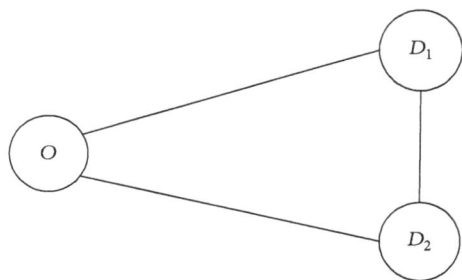

FIGURE 1: Travel routes of carpooling passengers.

drivers' income will be affected. Driver may be more resistant to carpooling detour, which makes carpooling detour impossible in reality. So it is necessary to study the problem of driver strategy with carpooling detour. The study has an important significance to the implementation of carpooling system. However few researches involve the problem of carpooling detour. Therefore this paper establishes a model of driver strategy evolution based on prospect theory and evolutionary game theory and studies the problem of driver strategy with carpooling detour.

2. Problem Description and Research Framework

2.1. Problem Description. Suppose two passengers depart from the same location O. The destination of passenger 1 is D_1 and that of passenger 2 is D_2. The best path for passenger 1 is OD_1 and passenger 2 is OD_2. OD_1 and OD_2 do not coincide, as shown in Figure 1. There are three road sections in Figure 1, where OD_1 is section 1, OD_2 is section 2, D_1D_2 is section 3, and $|OD_1| = l_1$, $|OD_2| = l_1$, $|D_1D_2| = l_3$. The traffic status of each section is uncertain. The traffic status of the section i is represented as j_i. When the traffic status of the section i is normal, $j_i = 1$; When the status is congestion, $j_i = 2$. The traffic congestion rate of the section i is represented as φ_i. The arrival time of the section i with normal traffic is represented as t_i. The waiting time of the section i with congestion is represented as r_i. Now the two passengers ride the same taxi through path OD_1D_2. That is, passenger 1 is sent to his destination first, and then passenger 2. Passenger 1 will pay a part of the fare, and payment ratio is θ. Passenger 2 will get more discount than passenger 1 because he makes a detour. The driver faces with two strategies: taking the two passengers or refusing them. This paper will study the problem of drivers' stability strategy under the above conditions.

2.2. Research Framework. The evolution game theory is an effective method to study human stability strategy. The theory analyzes problem on the premise of the bounded rationality. It believes that the human behaviors reveal bounded rationality, and human strategies reach stability gradually through constant adjustment [17, 18]. Evolutionary game theory studies stable strategies of human behaviors, which solves the problem of full rational analysis divorcing from

reality [19–21]. The evolution game theory has become one of the most important fields of modern game theory.

The evolutionary game model has been applied into analysis of stable strategies with bounded rationality, but the theory model still shows the defect, which has characteristic of full rationality. In the payoff matrix of the evolutionary game model, revenue value is usually defined as direct benefits of each actor under different strategies. That is, the fundamental condition of evolutionary game is that actor can accurately obtain revenue values of different strategies, which coincides with traditional full rational game. But the actors whose cognitive ability, computational ability, and judgment ability cannot achieve such a high requirement are not full rational. So the actors are not able to obtain accurate revenue values. Then the fundamental condition of evolutionary game can not be satisfied, and the subsequent evolutionary analysis will lose significance.

In this paper, prospect theory is introduced into evolutionary game model to solve above problem. Prospect theory is a cross research theory of psychology and behavioral science, which reveals the human psychology of decision-making. This theory describes people's decision psychology rules in uncertain environments [22], and reflects more truly people's behavior tendency. According to prospect theory, strategy prospect value describes people's perception results of revenue for the strategy with respect to personal psychological expectations [23, 24]. So strategy prospect value is more close to people's judgment result, reflecting bounded rationality [25–27]. Therefore, this paper combines prospect theory and evolutionary game theory and establishes driver strategy evolution model based on prospect value to study driver stable strategy under the condition of carpooling detour, with strategic prospect value instead of strategic revenue value of game model. The model includes two parts: driver strategy prospect model and evolutionary game model. Firstly, driver strategy prospect model is established. The prospect value of each strategy is obtained through the established model. Then evolutionary game model is built based on strategy prospect value, using strategy prospect value as strategic revenue. Driver stable strategy is analyzed by replicated dynamic mechanism. This method improves the traditional evolutionary game model and makes up for the full rationality shortcoming of payoff matrix in the traditional evolutionary game model, which makes research work closer to reality.

This paper studies driver stable strategy based on the improved model. The study finds out that refusing behaviors appear easily with carpooling detour. So passenger complaint mechanism is introduced next. Driver stable strategy is studied with passenger complaint mechanism. Eventually, we confirm that passenger complaint mechanism is an effective method to avoid refusing behaviors. We also find out related regulations.

3. Modeling

3.1. Driver Strategy Prospect Value. Suppose taxi charging standard stipulates that initiate fee is f_s¥/d km, f_r¥ per kilometer more than d kilometers, and waiting fee is f_q¥/min.

Then when travel distance is l and waiting time is r, cost $Z(l, r)$ is

$$Z(l, r) = \begin{cases} f_s + r f_q & \text{if } l \le d \\ (l - d) f_r + f_s + r f_q & \text{if } l > d. \end{cases} \quad (1)$$

If two passengers take the same taxi, cost of passenger 1 is

$$C_1 = \begin{cases} \theta Z(l_1, 0) & \text{if } j_1 = 1 \\ \theta Z(l_1, r_1) & \text{if } j_1 = 2. \end{cases} \quad (2)$$

The travel route causes detour for passenger 2. So it is necessary to reduce the cost of passenger 2 to make up for his time loss. Considering that the more the detour time the lower the detour cost, the cost of passenger 2 is inversely proportional to his nondetour time, following relational expression given.

$$C_2 = \begin{cases} \theta Z(l_2, 0) \dfrac{t_2}{t_1 + t_3} & \text{if } j_1 = 1, \ j_3 = 1 \\ \theta Z(l_2, 0) \dfrac{t_2}{t_1 + t_3 + r_3} & \text{if } j_1 = 1, \ j_3 = 2 \\ \theta Z(l_2, 0) \dfrac{t_2}{t_1 + t_3 + r_1} & \text{if } j_1 = 2, \ j_3 = 1 \\ \theta Z(l_2, 0) \dfrac{t_2}{t_1 + t_3 + r_1 + r_3} & \text{if } j_1 = 2, \ j_3 = 2. \end{cases} \quad (3)$$

If driver takes the two passengers in carpooling mode, starting from O, through D_1 before arriving in D_2, the driver's income is the sum of the two passengers' costs.

$$I = C_1 + C_2. \quad (4)$$

If driver takes only a passenger, starting from O, through D_1 before arriving in D_2, the driver's income is

$$I^0 = \begin{cases} Z(l_1 + l_3, 0) & \text{if } j_1 = 1, \ j_3 = 1 \\ Z(l_1 + l_3, r_3) & \text{if } j_1 = 1, \ j_3 = 2 \\ Z(l_1 + l_3, r_1) & \text{if } j_1 = 2, \ j_3 = 1 \\ Z(l_1 + l_3, r_1 + r_3) & \text{if } j_1 = 2, \ j_3 = 2. \end{cases} \quad (5)$$

The driver's income I_0 when he takes only a passenger which is regarded as driver reference point; then the driver's gain relative to reference point is

$$y^D = I - I^0. \quad (6)$$

According to prospect theory, value function is defined as follows:

$$V(y^D) = \begin{cases} (y^D)^\alpha & \text{when } y^D \ge 0 \\ -\lambda(-y^D)^\beta & \text{when } y^D < 0, \end{cases} \quad (7)$$

where α and β ($0 \le \alpha, \beta \le 1$) are risk attitude coefficients. The bigger α or β is, the more adventurous decision maker tends to. λ ($\lambda > 1$) is loss aversion coefficient, which reflects

that decision maker is more sensitive to loss. Value function shows the decreasing sensitivity in the two directions of gain and loss. The result of the value function reflects the psychological revenue more realistically. The parameters values $\alpha = 0.89$, $\beta = 0.92$, $\lambda = 2.25$ are more consistent with psychological characteristic of decision maker [22].

Path OD_1D_2 includes section 1 and section 3. Traffic congestion rate of the path is defined as φ_{13}.

$$\varphi_{13} = \begin{cases} (1 - \varphi_1)(1 - \varphi_3) & \text{if } j_1 = 1, \ j_3 = 1 \\ (1 - \varphi_1)\varphi_3 & \text{if } j_1 = 1, \ j_3 = 2 \\ \varphi_1(1 - \varphi_3) & \text{if } j_1 = 2, \ j_3 = 1 \\ \varphi_1\varphi_3 & \text{if } j_1 = 2, \ j_3 = 2. \end{cases} \quad (8)$$

The probability of traffic congestion which is perceived by passenger is different from actual probability according to prospect theory. The perceived probability of traffic congestion is

$$W(\varphi_{13}) = \begin{cases} \dfrac{(\varphi_{13})^\chi}{\left((\varphi_{13})^\chi + (1 - \varphi_{13})^\chi\right)^{1/\chi}} & \text{if } y^D \ge 0 \\ \dfrac{(\varphi_{13})^\delta}{\left((\varphi_{13})^\delta + (1 - \varphi_{13})^\delta\right)^{1/\delta}} & \text{if } y^D < 0. \end{cases} \quad (9)$$

The parameter values most reflecting individual behavior of decision maker are $\chi = 0.61$, $\delta = 0.69$ [22]. The transformation reflects that people tend to overestimate small probability events and underestimate medium and large probability events, while people are relatively insensitive to the intermediate stage. Transformed probability is more close to people's subjective probability.

If the driver selects taking carpooling passenger strategy, the prospect value EV_1^D is

$$EV_1^D = \sum_{v=1}^{2}\sum_{u=1}^{2} W(\varphi_{13}(j_1 = u, \ j_3 = v)) V \quad (10)$$
$$\cdot (y^D(j_1 = u, \ j_3 = v)),$$

where $\varphi_{13}(j_1 = u, \ j_3 = v)$ is traffic congestion probability under the condition of $j_1 = u$ and $j_3 = v$. $y^D(j_1 = u, \ j_3 = v)$ is driver gain under the condition of $j_1 = u$ and $j_3 = v$.

Because driver reference point is the gain when he takes only a passenger, the prospect value is 0 if the driver takes only a passenger; the prospect value is the result of $I = 0$ condition if the driver does not take any passenger.

3.2. Evolutionary Analysis of Driver Carpooling Strategy. Driver has two strategy choices: taking carpooling detour passengers (strategy 1) and refusing carpooling detour passengers (strategy 2). If the driver selects strategy 1, and the passengers select sharing the taxi, carpooling succeeds; then the driver's prospect value is represented as EV_1^D; If the driver selects strategy 2, and the passengers select riding the taxi lonely, the driver takes only a passenger; then the prospect value is 0. If the driver's strategy and passengers' strategy are not different, the driver does not take any passenger and the prospect value is represented as EV_2^D.

Suppose the percentage of passengers choosing carpooling detour strategy is p_1, and the percentage of drivers choosing carpooling detour strategy is q. Expected gains of driver choosing strategy 1 and strategy 2 respectively are as follows.

$$u_1 = EV_1^D p_1 + EV_2^D (1 - p_1)$$
$$u_2 = EV_2^D p_1. \tag{11}$$

Average expected gain is

$$\bar{u} = u_1 q + u_2 (1 - q)$$
$$= EV_1^D p_1 q + EV_2^D (1 + p_1 (1 - 2q)). \tag{12}$$

Replicator dynamic equation of driver percentage is

$$F(q) = \frac{dq}{dt} = q(u_1 - \bar{u})$$
$$= q(1 - q)\left(\left(EV_1^D - 2EV_2^D\right)p_1 + EV_2^D\right). \tag{13}$$

Suppose $F(q) = dq/dt = 0$; then $q = 0$, $q = 1$, $p_1^* = EV_2^D / (2EV_2^D - EV_1^D)$.

According to the condition, we know $EV_2^D < 0$. The stability analysis is as follows:

(1) When $EV_1^D - 2EV_2^D > 0$,

① if $p_1 = p_1^*$, then $F(q) = 0$, $F'(q) = 0$; here any strategy may be stable;

② if $p_1 > p_1^*$, then $F'(0) > 0$, $F'(1) < 0$; here $q = 1$ is stable strategy. That is, when the percentage of passengers choosing carpooling is more than p_1^*, driver will choose strategy 1;

③ if $p_1 < p_1^*$, then $F'(0) < 0$, $F'(1) > 0$; here $q = 0$ is stable strategy. That is, when the percentage of passengers choosing carpooling is low than p_1^*, driver will choose strategy 2.

(2) When $EV_1^R - 2EV_2^R < 0$, inevitably $p_1 > p_1^*$; here $q = 0$ is stable strategy, and driver will choose strategy 2.

Through the above analysis, driver stable strategy is taking passengers only if $EV_1^D - 2EV_2^D > 0$ and $p_1 > p_1^*$; otherwise stable strategy is refusing passengers.

4. Researches on Driver Strategy Evolution with Complaint Mechanism

Driver income will be affected when detour occurs in carpooling. Drivers are inclined to refusing passengers. Next passenger complaint mechanism is introduced to further analyze driver strategy change with complaint mechanism.

4.1. Prospect Value of Driver Strategy with Complaint Mechanism. Under the condition of passenger complaint mechanism, driver will make a choice carefully after considering the consequences of refusing passengers. Driver has two

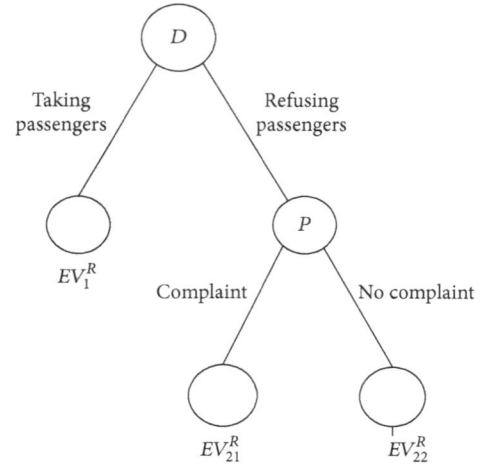

FIGURE 2: Driver strategies and prospect values.

choices: taking passengers and refusing passengers. When the behavior of refusing passengers occurs, passengers will choose the complaint strategies or no complaint strategies. The specific strategies are shown in Figure 2. If driver selects taking carpooling detour passengers, the prospect value is still EV_1^D; if driver selects refusing passengers, and passengers select complaint, the prospect value is EV_{21}^D; if driver selects refusing passengers, and passengers select no complaint, the prospect value is EV_{22}^D.

When the driver selects refusing strategy (assuming that only passenger 1 is taken at this time), and the passengers select complaint strategy, the prospects value of the driver is

$$EV_{21}^D = \sum_{u=1}^{2} W\left(\varphi_{13}\left(j_1 = u\right)\right)V\left(-g\right), \tag{14}$$

where g is punishment for refusing behavior after complaint.

When the driver selects refusing strategy, and the passengers select no complaint strategy, the prospects value of the driver is

$$EV_{22}^D = 0. \tag{15}$$

4.2. Stability Strategy Analysis. Suppose the percentage of passengers choosing complaint strategy is p_2, and the percentage of drivers choosing carpooling detour strategy is q. Expected gains of driver choosing strategy 1 and strategy 2, respectively, are as follows.

$$u_{11} = EV_1^D$$
$$u_{12} = EV_{21}^D p_2 + EV_{22}^D (1 - p_2). \tag{16}$$

Average expected gain is

$$\bar{u}_1 = u_{11} q + u_{12} (1 - q)$$
$$= EV_1^D q + \left(EV_{21}^D p_2 + EV_{22}^D (1 - p_2)\right)(1 - q). \tag{17}$$

Replicator dynamic equation of driver percentage is

$$F(q) = \frac{dq}{dt} = q(u_{11} - \bar{u}_1) = q\big(EV_1^D$$
$$- \big(EV_1^D q + \big(EV_{21}^D p_2 + EV_{22}^D (1 - p_2)\big)(1 - q)\big)\big) \quad (18)$$
$$= q(1 - q)\big(EV_1^D - EV_{21}^D p_2 - EV_{22}^D(1 - p_2)\big).$$

Suppose $F'(q) = 0$; then $q = 0$, $q = 1$, $p_2^* = (EV_{22}^D - EV_1^D)/(EV_{22}^D - EV_{21}^D)$. Due to $EV_{22}^D = 0$, $p_2^* = EV_1^D/EV_{21}^D$.

According to the condition, we know $EV_{21}^D < 0$. The stability analysis is as follows:

(1) When $EV_1^D < 0$,

① if $p_2 = p_2^*$, then $F(q) = 0$, $F'(q) = 0$; here any strategy may be stable.

② if $p_2 > p_2^*$, then $F'(0) > 0$, $F'(1) < 0$; here $q = 1$ is stable strategy. That is, when the percentage of passengers choosing carpooling is more than p_2^*, driver will choose strategy 1.

③ if $p_2 < p_2^*$, then $F'(0) < 0$, $F'(1) > 0$; here $q = 0$ is stable strategy. That is, when the percentage of passengers choosing carpooling is lower than p_2^*, driver will choose strategy 2.

(2) When $EV_1^D > 0$, inevitably $EV_1^D/EV_{21}^D < 0$, $p_1 > p_1^*$; here $q = 1$ is stable strategy, and driver will choose strategy 1.

Through the above analysis, drivers will select taking carpooling passengers strategy unless $EV_1^D < 0$ and $p_2 > p_2^*$.

5. Driver Evolution Strategies Comparison under the Different Mechanisms

(1) When $EV_1^D > 0$. Only to meet the condition that the initial percentage of passengers choosing carpooling detour strategy is more than p_1^* is drivers' stable strategy taking carpooling detour passengers under the absence of complaint mechanism; however, drivers' stable strategy is taking carpooling detour passengers under the compliant mechanism.

(2) When $EV_1^D < 0$

① If $EV_1^D - 2EV_2^D < 0$. Drivers' stable strategy is refusing carpooling passengers under the absence of complaint mechanism; however drivers' stable strategy is taking carpooling detour passengers under the compliant mechanism as long as the condition that the initial percentage of passengers choosing complaint strategy is more than p_2^* is satisfied.

② If $EV_1^D - 2EV_2^D > 0$. If the initial percentage of passengers choosing carpooling detour strategy is more than p_1^*, drivers' stable strategy is taking carpooling passengers under the absence of complaint mechanism; if the initial percentage of choosing complaint strategy is more than p_2^*, drivers' stable strategy is taking carpooling passengers under the compliant mechanism.

Under the above conditions, prove $p_1^* > p_2^*$ as follows.

Proof.

$$\frac{EV_2^D}{2EV_2^D - EV_1^D} - \frac{EV_1^D}{EV_{21}^D}$$
$$= \frac{EV_2^D EV_{21}^D - (2EV_2^D - EV_1^D)EV_1^D}{(2EV_2^D - EV_1^D)EV_{21}^D}$$
$$= \frac{EV_2^D(EV_{21}^D - 2EV_1^D) + (EV_1^D)^2}{(2EV_2^D - EV_1^D)EV_{21}^D} \quad (19)$$
$$= \frac{(EV_1^D - EV_2^D)^2 + EV_2^D(EV_{21}^D - EV_2^D)}{(2EV_2^D - EV_1^D)EV_{21}^D}.$$

Due to $EV_1^D - 2EV_2^D > 0$, $EV_{21}^D < 0$, $EV_2^D < 0$ and $EV_{21}^D < EV_2^D$, we can know $EV_2^D/(2EV_2^D - EV_1^D) - EV_1^D/EV_{21}^D > 0$, that is $p_1^* > p_2^*$. □

Through the above analysis, the complaint mechanism makes the conditions of driver stabilizing at taking carpooling detour strategy easier to be satisfied. The complaint mechanism can reduce the rejection rate of drivers.

6. Example Analysis

Suppose taxi charging standard stipulates that initiate fee is 10¥ per 3 km, 1.4¥ per kilometer more than 3 kilometers, and waiting fee is 1.2¥ per 2.5 min. Two passengers set out from the same location O to different destinations. The destination of passenger 1 is D_1 and that of passenger 2 is D_2. They intend to ride the same taxi. It is known that $l_1 = l_2 = 5$ and $l_3 = 1$. Travel speed is 30 km per hour. Travel time is twice as normal condition when traffic congestion occurs. The traffic congestion rate of each section is $\varphi = 0.5$, and carpooling payment ratio is $\theta = 0.7$.

We can know $p_1^* = EV_2^R/(2EV_2^R - EV_1^R) = 0.6$ under the absence of complaint mechanism. $q = 1$ is drivers' stable strategy when $p > 0.6$. That is, drivers' stable strategy is taking passengers if the initial percentage of passengers choosing carpooling detour strategy is more than 60%.

Suppose driver's punishment is $g = 50$ under the complaint mechanism. We can know $p_2^* = (EV_{22}^R - EV_1^R)/(EV_{22}^R - EV_{21}^R) = 0.17$ under the condition. $q = 1$ is drivers' stable strategy when $p > 0.17$. That is, drivers' stable strategy is taking passengers if the initial percentage of passengers choosing complaint strategy is more than 17%.

So under the complaint mechanism, the conditions that cause driver to select carpooling strategy are more easily satisfied, and drivers tend to take carpooling passengers at this time.

Thresholds p_1^* and p_2^* are influenced by the factors, such as detour distance, penalty, payment ratio, and traffic congestion rate. The influence analysis is as follows.

Figure 3 shows the influences of detour distance on the threshold under the two different mechanisms. Figure 3(a) is the influences analysis under the absence of complaint mechanism, and Figure 3(b) is the influences analysis under

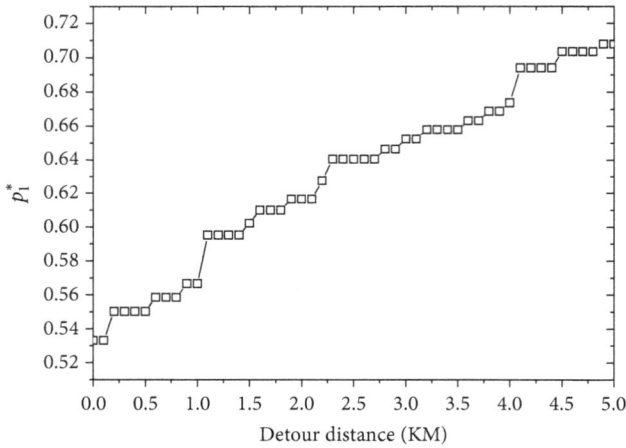

(a) Influence of detour distance on threshold under the absence of complaint mechanism

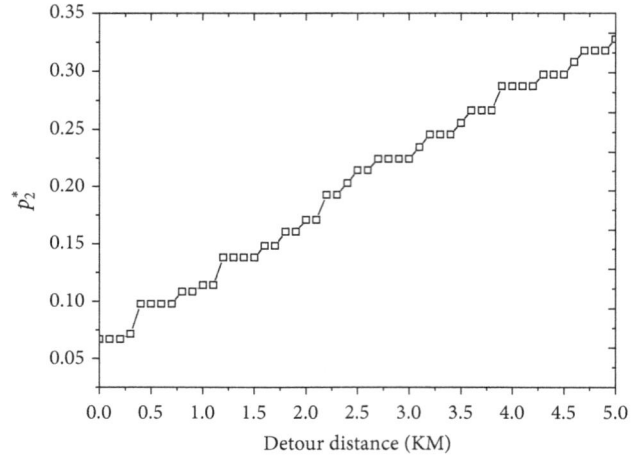

(b) Influence of detour distance on threshold under the complaint mechanism

FIGURE 3: Influence of detour distance on threshold.

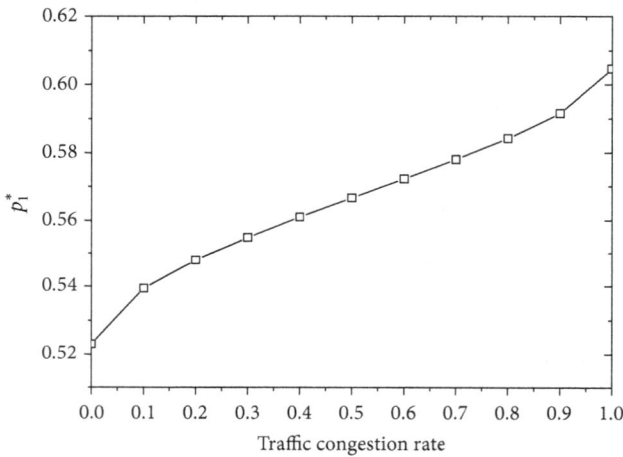

(a) Influence of traffic congestion rate on threshold under the absence of complaint mechanism

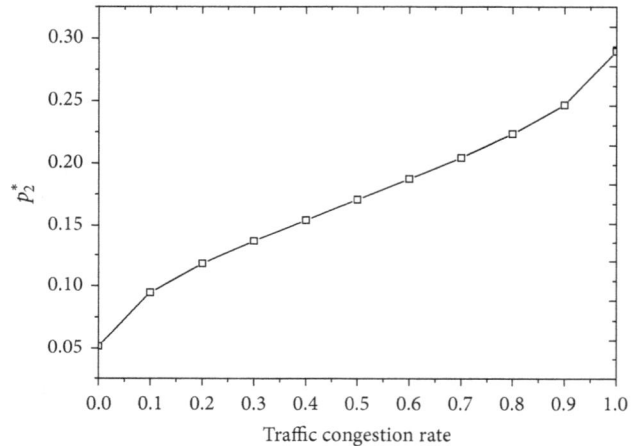

(b) Influence of traffic congestion rate on threshold under the complaint mechanism

FIGURE 4: Influence of traffic congestion rate on threshold.

the complaint mechanism. We can see that the threshold increases gradually with detour distance increases. Under the absence of complaint mechanism, the initial percentage of passengers choosing carpooling strategy needs to reach 53% when detour distance is 0; the initial percentage needs to reach 71% when detour distance is 5 KM. However, under the complaint mechanism, the initial percentage of passengers choosing complaint strategy needs to only reach 7% when detour distance is 0; the initial percentage needs to reach 33% when detour distance is 5 KM.

Figure 4 shows the influences of traffic congestion rate on the threshold under the two different mechanisms. Figure 4(a) is the influences analysis under the absence of complaint mechanism, and Figure 4(b) is the influences analysis under the complaint mechanism. We can see that

the threshold increases gradually with traffic congestion rate increases. Under the absence of complaint mechanism, the initial percentage of passengers choosing carpooling strategy needs to reach 52% when traffic is certainly normal; the initial percentage needs to reach 60% when traffic congestion occurs inevitably. However, under the complaint mechanism, the initial percentage of passengers choosing complaint strategy needs to only reach 5% when traffic is certainly normal; the initial percentage needs to reach 28% when traffic congestion occurs inevitably.

Figure 5 shows the influences of passenger payment ratio on the threshold under the two different mechanisms. Figure 5(a) is the influences analysis under the absence of complaint mechanism, and Figure 5(b) is the influences analysis under the complaint mechanism. We can see that the

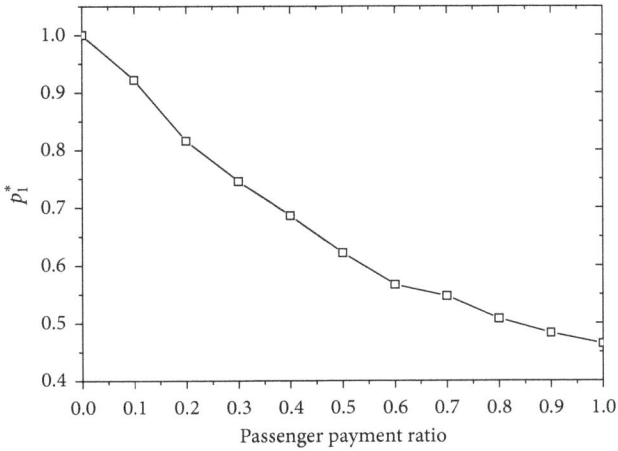

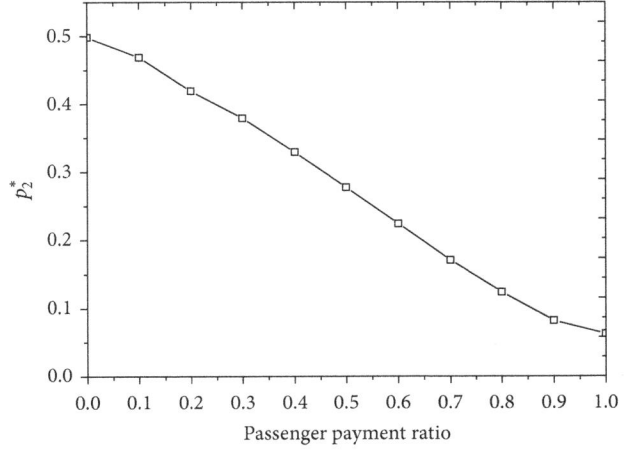

(a) Influence of payment ratio on threshold under the absence of complaint mechanism

(b) Influence of payment ratio on threshold under the complaint mechanism

FIGURE 5: Influence of payment ratio on threshold.

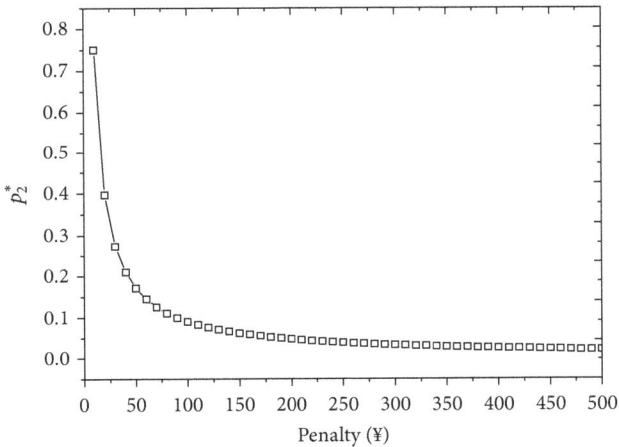

FIGURE 6: Influence of penalty on threshold.

threshold decreases gradually with passenger payment ratio increases. The threshold of the absence of complaint mechanism is lower obviously than that of complaint mechanism.

The influences of penalty on the threshold are shown in Figure 6. The threshold decreases gradually with penalty increases, and the rate of decline slows gradually. The threshold is about 75% when the penalty is less than 20¥. The threshold remains at about 5% when the penalty is over 200¥, and the continuing increase of the penalty has little effect on the threshold. There is no sense to set excessive penalty, and too low penalty can not affect drivers' strategy choice. So it is necessary to set the appropriate penalty under the complaint mechanism.

In summary, the thresholds that can ensure drivers to select taking carpooling strategy are more easily satisfied under the complaints mechanism. Complaint mechanism can reduce the rate of drivers refusing passengers and makes detour possible.

7. Conclusions

This paper studies the problem of taxi driver's strategy choice under the condition of carpooling detour. The model combines prospect theory and evolutionary game theory and improves traditional evolutionary game model by replacing the gain with prospect value, which makes the model more fit in with the actual psychology of human beings. Through the researches and analysis, the following conclusions are obtained:

(1) The behavior of drivers refusing passengers under the condition of carpooling detour can easily happen. The passenger complaints mechanism is an effective measure to reduce the driver refusing behavior.

(2) Whether or not drivers choose the strategy of taking carpooling detour passengers depends on the initial percentage of passengers choosing carpooling strategy or complaint strategy. Under the absence of complaint mechanism, drivers' stable strategy is taking passengers if the percentage of passengers choosing carpooling strategy is more than the threshold p_1^*; Under the complaint mechanism, drivers' stable strategy is taking passengers if the percentage of passengers choosing complaint strategy is more than the threshold p_2^*.

(3) The thresholds relate to detour distance, traffic congestion rate, payment ratio, and penalty. The thresholds increase gradually with detour distance or traffic congestion rate increases. The thresholds decrease gradually with payment ratio or penalty increases. It is an effective way to avoid driver refusing behavior to limit detour distance, ease traffic congestion, and select the appropriate payment ratio and penalty.

Therefore, the impacts of the factors, such as detour distance and traffic congestion, have to be considered in the implementation process of taxi carpooling system. It is necessary to provide a simple and effective way of complaints to encourage passengers complaining actively in order to control refusing behavior.

Conflicts of Interest

The authors declare that they have no conflicts of interest.

Acknowledgments

This research was funded by National Natural Science Foundation of China (61364026, 51408288, and 71661021) and Youth Science Foundation of Lanzhou Jiaotong University (2015032). The authors express their thanks to all who participated in this research for their cooperation.

References

[1] D. Zhang, T. He, Y. Liu, S. Lin, and J. A. Stankovic, "A carpooling recommendation system for taxicab services," *IEEE Transactions on Emerging Topics in Computing*, vol. 2, no. 3, pp. 254–266, 2014.

[2] R. Javid, A. Nejat, and M. Salari, "The environmental impacts of carpooling in the United States," in *Proceedings of the Transportation, Land and Air Quality Conference*, 2016.

[3] J. Jamal, A. E. Rizzoli, R. Montemanni, and D. Huber, "Tour planning and ride matching for an urban social carpooling service," in *Proceedings of the 5th International Conference on Transportation and Traffic Engineering (ICTTE '16)*, Switzerland, July 2016.

[4] W. Shen, C. V. Lopes, and J. W. Crandall, "An online mechanism for ridesharing in autonomous mobility-on-demand systems," in *Proceedings of the 25th International Joint Conference on Artificial Intelligence*, pp. 475–481, New York, USA, 2016.

[5] S. Galland, L. Knapen, A.-U.-H. Yasar et al., "Multi-agent simulation of individual mobility behavior in carpooling," *Transportation Research Part C: Emerging Technologies*, vol. 45, pp. 83–98, 2014.

[6] S.-K. Chou, M.-K. Jiau, and S.-C. Huang, "Stochastic set-based particle swarm optimization based on local exploration for solving the carpool service problem," *IEEE Transactions on Cybernetics*, vol. 46, no. 8, pp. 1771–1783, 2016.

[7] K. Aissat and A. Oulamara, "Dynamic ridesharing with intermediate locations," in *Proceedings of the 2014 IEEE Symposium on Computational Intelligence in Vehicles and Transportation Systems (CIVTS '14)*, pp. 36–42, USA, December 2014.

[8] P. Santi, G. Resta, M. Szell, S. Sobolevsky, S. H. Strogatz, and C. Ratti, "Quantifying the benefits of vehicle pooling with shareability networks," *Proceedings of the National Acadamy of Sciences of the United States of America*, vol. 111, no. 37, pp. 13290–13294, 2014.

[9] T. Shinde and B. Thombre, "An effective approach for solving carpool service problems using genetic algorithm approach in cloud computing," *International Journal of Advance Research in Computer Science and Management Studies*, vol. 3, no. 12, pp. 29–33, 2015.

[10] W. He, K. Hwang, and D. Li, "Intelligent carpool routing for urban ridesharing by mining GPS trajectories," *IEEE Transactions on Intelligent Transportation Systems*, vol. 15, no. 5, pp. 2286–2296, 2014.

[11] M. Nourinejad and M. J. Roorda, "Agent based model for dynamic ridesharing," *Transportation Research Part C: Emerging Technologies*, vol. 64, pp. 117–132, 2016.

[12] S.-C. Huang, M.-K. Jiau, and C.-H. Lin, "A genetic-algorithm-based approach to solve carpool service problems in cloud computing," *IEEE Transactions on Intelligent Transportation Systems*, vol. 16, no. 1, pp. 352–364, 2015.

[13] J. Xia, K. M. Curtin, W. Li, and Y. Zhao, "A new model for a carpool matching service," *PLoS ONE*, vol. 10, no. 6, Article ID e0129257, 2015.

[14] S.-Y. Chiou and Y.-C. Chen, "A mobile, dynamic, and privacy-preserving matching system for car and taxi pools," *Mathematical Problems in Engineering*, vol. 2014, Article ID 579031, 10 pages, 2014.

[15] C. M. Boukhater, O. Dakroub, F. Lahoud, M. Awad, and H. Artail, "An intelligent and fair GA carpooling scheduler as a social solution for greener transportation," in *Proceedings of the 2014 17th IEEE Mediterranean Electrotechnical Conference (MELECON '14)*, pp. 182–186, Lebanon, April 2014.

[16] Q. Xiao, R.-C. He, W. Zhang, and C.-X. Ma, "Algorithm research of taxi carpooling based on fuzzy clustering and fuzzy recognition," *Journal of Transportation Systems Engineering and Information Technology*, vol. 14, no. 5, pp. 119–125, 2014.

[17] G. Mallard, *Bounded rationality and behavioral economics*, Routledge Press, London, UK, 2015.

[18] P. van den Berg and F. J. Weissing, "Evolutionary Game Theory and Personality," in *Evolutionary Perspectives on Social Psychology*, Evolutionary Psychology, pp. 451–463, Springer International Publishing, Cham, 2015.

[19] M. S. Eid, I. H. El-Adaway, and K. T. Coatney, "Evolutionary stable strategy for postdisaster insurance: Game theory approach," *Journal of Management in Engineering*, vol. 31, no. 6, Article ID 04015005, 2015.

[20] C. Wu, Y. Pei, and J. Gao, "Evolution game model of travel mode choice in metropolitan," *Discrete Dynamics in Nature and Society*, vol. 2015, Article ID 638972, 11 pages, 2015.

[21] C. S. Gokhale and A. Traulsen, "Evolutionary multiplayer games," *Dynamic Games and Applications*, vol. 4, no. 4, pp. 468–488, 2014.

[22] A. Tversky and D. Kahneman, "Advances in prospect theory: cumulative representation of uncertainty," *Journal of Risk and Uncertainty*, vol. 5, no. 4, pp. 297–323, 1992.

[23] J. Wang and T. Sun, "Fuzzy multiple criteria decision making method based on prospect theory," in *Proceedings of the Proceeding of the International Conference on Information Management, Innovation Management and Industrial Engineering (ICIII '08)*, vol. 1, pp. 288–291, Taipei, Taiwan, December 2008.

[24] P. A. de Castro, A. . Barreto Teodoro, L. I. de Castro, and S. Parsons, "Expected utility or prospect theory: which better fits agent-based modeling of markets?" *Journal of Computational Science*, vol. 17, no. part 1, pp. 97–102, 2016.

[25] W. Zhang and R. C. He, "Dynamic route choice based on prospect theory," in *Proceedings of the*, vol. 138, pp. 159–167.

[26] M. Abdellaoui, H. Bleichrodt, O. L'Haridon, and D. van Dolder, "Measuring loss aversion under ambiguity: a method to make prospect theory completely observable," *Journal of Risk and Uncertainty*, vol. 52, no. 1, pp. 1–20, 2016.

[27] L. A. Prashanth, J. Cheng, F. Michael, M. Steve, and S. Csaba, "Cumulative prospect theory meets reinforcement learning: prediction and control," in *Proceedings of the 33rd International Conference on Machine Learning*, pp. 2112–2121, New York, NY, USA, 2016.

11

Passenger Travel Regularity Analysis Based on a Large Scale Smart Card Data

Qi Ouyang ⓘ, Yongbo Lv, Yuan Ren, Jihui Ma ⓘ, and Jing Li

School of Traffic and Transportation, Beijing Jiaotong University, Beijing 100044, China

Correspondence should be addressed to Qi Ouyang; 14114203@bjtu.edu.cn

Guest Editor: Javier Sánchez-Medina

Analysis of passenger travel habits is always an important item in traffic field. However, passenger travel patterns can only be watched through a period time, and a lot of people travel by public transportation in big cities like Beijing daily, which leads to large-scale data and difficult operation. Using SPARK platform, this paper proposes a trip reconstruction algorithm and adopts the density-based spatial clustering of application with noise (DBSCAN) algorithm to mine the travel patterns of each Smart Card (SC) user in Beijing. For the phenomenon that passengers swipe cards before arriving to avoid the crowd caused by the people of the same destination, the algorithm based on passenger travel frequent items is adopted to guarantee the accuracy of spatial regular patterns. At last, this paper puts forward a model based on density and node importance to gather bus stations. The transportation connection between areas formed by these bus stations can be seen with the help of SC data. We hope that this research will contribute to further studies.

1. Introduction

Traditional studies on passenger travel patterns and passenger segmentation solely focus on passenger physical characteristics or the use of transit user surveys. This classification has little help of knowing passenger travel habits. Therefore, we need another method to study the temporal and spatial regularity. This method must be based on actual data with passenger travel information. SC data meets the needs.

SC data gathered by automated fare collection systems records travel details which are very valuable. However, passenger travel patterns can only be watched through a period time, and a lot of people travel by public transportation in big cities like Beijing daily, which leads to large-scale data and difficult operation. This paper adopts SPARK platform to solve this problem. Several computers are used to build the platform and calculate together.

This paper adopts a systematic approach to mine the travel pattern and search the temporal and spatial regularity using SC data. After the literature review in this section, this paper introduces the SC dataset adopted and the method used to drop invalid data. We consider each item in the dataset a transaction. Then, this paper rebuilds the SC dataset by reconstructing the completed transactions of SC users into a trip, which can recognize SC user transfer behaviour. After the reconstruction, the density-based spatial clustering of application with noise (DBSCAN) algorithm is adopted to mine the travel pattern and obtain the temporal and spatial regularity of each SC user in Beijing. In spatial dimension, this paper designs an algorithm based on passenger travel frequent items to handle the phenomenon that passengers swipe their cards before arriving to avoid the crowd. Then, we put the temporal data and the spatial data together to classify SC users by temporal and spatial features. Finally, this paper puts forward a model based on distance and node importance to gather bus stations into areas without any intersection, and then we assign the SC data into every area to investigate transportation connection between them. In Section 4, this paper sums up some conclusions.

Analysis of SC data has attracted research interest and a lot of researches have been done in the past few years. Catherine Morency et al. (2007) measured transit variability with SC data. They built an object model to understand the relationship between different elements within the transit network, and then used k-means cluster to indicate the spatial variability of passengers [1]. Ka Kee Alfred Chu et al. (2008) detected transfer coincidence based on information of the vehicles and SC data, and then they obtained temporal distribution and cumulative percentage of transfer time [2]. Their group (2010) proposed a methodology to enhance transit trip characterization by adding a multiday dimension to SC transactions. They detected individual, anchor points—precise to an exact address for each SC user. Then, they adopted spatial statistics, spatial analyses with geographic information system, visualizations, and data mining to describe passenger activity space, locations and departure time [3].

In recent years, with the development of the associated modelling methods, solving technology and computing capabilities, the study of SC data has developed rapidly. Jun Liu et al. (2014) presented a traffic monitoring and analysis system for large-scale networks based on Hadoop, an open-source distributed computing platform for big data processing on commodity hardware [4]. Sui Tao et al. (2014) applied a geovisualisation-based method to a large SC database to examine spatial temporal dynamics on BRT systems in Brisbane, Australia. They displayed their analysis result by a thermodynamic chart [5]. Cynthia Chen (2016) introduced how to use big data and small data datasets, concepts, and methods to analyse travel behaviour [6].

With the help of big data, researchers can identify passenger patterns derived by data through some complex models and algorithms. Le Minh Kieu et al. (2015) adopted the DBSCAN algorithm, which can find clusters of arbitrary shapes based on different parameters, to mine the travel patterns based on around 34.8 million transactions made by a million SC users over 15000 transit stops of the bus, city train, and ferry networks. They segmented transit passengers into four identifiable types based on the above research. However, because of the high algorithm complexity, this algorithm takes a long time to cluster convergence when the dataset is very large [7]. Mohamed K. ElMahrsi et al. (2017) proposed two approaches to cluster SC data. The first approach clusters stations based on when their activity occurs; i.e., how trips made at the stations are distributed over time. The second approach makes it possible to identify groups of passengers that have similar boarding times aggregated into weekly profiles [8]. Xiaolei Ma et al. (2017) measured spatial-temporal regularity of individual commuters, including residence, workplace, and departure time, using one-month transit SC data. They divided one day into 48 intervals, which means one interval contains half an hour, to observe temporal regularity [9]. Anne-Sarah Briand et al. (2017) presented a two-level generative model that applied the Gaussian mixture model to regroup passengers based on their temporal habits in their public transportation usage. They observed the year-to-year changes in public transport passenger behaviour [10].

2. Materials and Methods

The Materials and Methods should contain sufficient details so that all procedures can be repeated. It may be divided into headed subsections if several methods are described.

Section 2.1 in this section introduces the dataset used for the case study, as well as the methods for the reconstruction of travel itineraries. This part also introduces some definitions makes a simple statistics analysis. Sections 2.2 and 2.3 analyse passenger travel behaviour in time dimension and spatial dimension. Section 2.4 tries to find the relation between different areas based on station density and node contraction in weighted complex networks.

2.1. Dataset and Reconstruction of Travel Itineraries. The SC data used in this paper come from Beijing, which is one of the largest cities in the world. 6 million SC records are collected by AFC every day in this city. The total dataset contains around 150 million transactions over 7000 transit stops of the bus from October 1, 2015 to October 30, 2015. The dataset includes the following main fields:

(1) CARDID: The unique SC ID, which has been hashed into a unique number to maintain the privacy of the cardholder.

(2) TRADETIME: The time that passengers swipe their cards when they get on the bus.

(3) MARKTIME: The time that passengers swipe their cards when they get off the bus.

(4) LINEID: The transit routes that passengers take.

(5) TRADESTATION: The station that passengers swipe their cards at when they get on the bus.

(6) MARKSTATION: The station that passengers swipe their cards at when they get off the bus.

The SC dataset only contains information of passengers, which can be used when combined with bus operation data. The bus operation data includes the following fields:

(7) XLDM: The same as LINEID.

(8) ZM: The name of bus stops.

(9) ZDXH: The same as TRADESTATION and MARKSTATION.

(10) ZDJD: The longitude of bus stops.

(11) ZDWD: The latitude of bus stops.

In this paper, we call each item collected by AFC a transaction. As we know, one passenger may take buses many times each working day, so there are several transactions for each SC user every day. Sometimes, a bus trip one passenger takes contains two or three transactions. How to construct the travel trip from individual transactions is a fundamental problem before mining the travel patterns.

This paper adopts an algorithm to connect the individual transactions: there are two principles of this algorithm: first, if two transactions can be connected into one trip, the time interval between transactions must be less than 60 mins [11]; then, we sort two transactions by MARKTIME. If two transactions can be connected into one trip, the origin stop of the first transaction must be different from the destination stop of the second.

Here, the first boarding stop and the last alighting stop of a completed trip are defined as the "origin stop" and the "destination stop," respectively. The time interval between the alighting time of a transaction and the boarding time of the next transaction of the same trip is defined as the transferring time. Figure 1 shows the flowchart. The steps of the algorithm are described as follows.

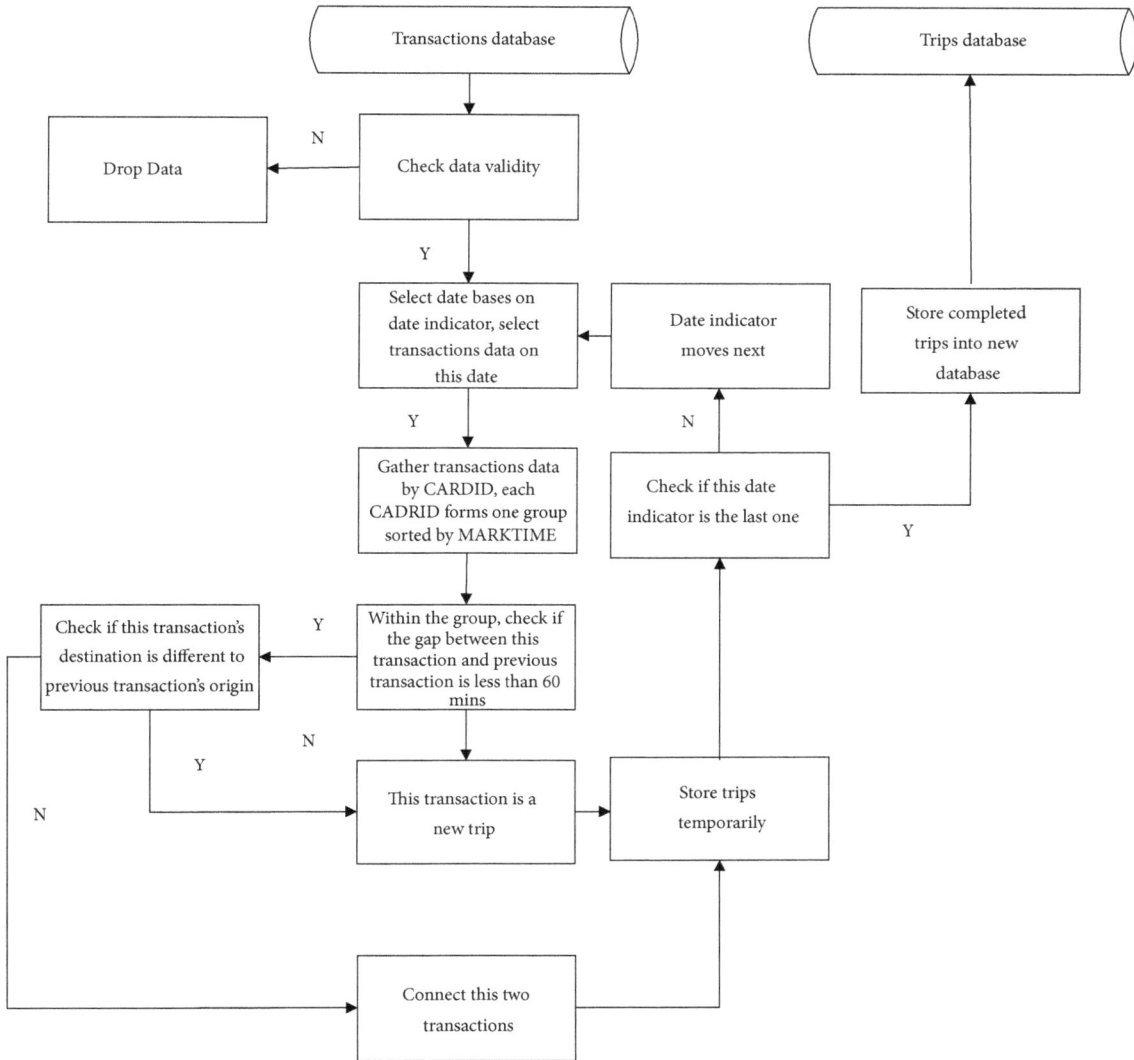

FIGURE 1: Trip reconstruction flowchart.

Step 1. Check data validity. Because of some hardware problems, the data collected by AFC cannot be used, as two items are exactly the same as each other: any data missed in fields "MARKTIME, TRADETIME, LINEID, MARKSTATION, and MARKSTATION" or the MARKTIME equals TRADETIME in one item. By checking data validity, this paper drops around 1% of the transactions data.

Step 2. Set time indicator. Each item in the time indicator represents a date. At first, the time indicator points to the first item and then we select whole data based on the date.

Step 3. All the data selected are classified by "CARDID" to form different groups. Each group represents all the transactions a SC user made on one working day, and the transactions in each group are sorted by "MARKTIME."

Step 4. The time interval between current transaction and previous transaction is calculated. If the time interval is

less than 60 min, the destination of current transaction is compared to the origin of previous transaction. If they are different from each other, these two transactions are connected to one transaction and then we continue to connect two transactions into one trip until two transactions cannot satisfy two principles mentioned above.

Step 5. After calculating all the data of different groups on this date, whether time indicator is in the final position is judged. If not, the time indicator moves to the next and then Steps 2–5 is repeated.

A passenger may take several journeys with the same origin and destination at different time of one day, which has a big influence on the analysis. So, this paper defines that passenger travel times is the number of days which has travel records, namely for each person one travel contains all the different trips recorded on one working day. Based on this principle, the number of travel times a passenger takes cannot be larger than the number of working days within one month.

2.2. Passenger Travel Behaviour Analyses in Temporal Dimension. This part adopts DBSCAN algorithm [12, 13] to mine travel patterns. DBCAN is a clustering algorithm based on density, which has great advantage in the following aspects:

(1) If we consider a passenger travels regularly, he must travel by bus several times within a certain time. The DBSCAN algorithm is proper to this data mining. The number of "a certain time" is the *Eps* in the DBSCAN algorithm, and the number of "several times" is the *minPts* in the DBSCAN algorithm.

(2) As we can see, a passenger may take bus to deal with individual random events. These trips have a lot of differences with regular trip, and we call these trips "noise." This algorithm can find the clusters (regular pattern) and deal with the varying noise effectively.

(3) This algorithm can identify a cluster of any shape and size. It means that we can use this algorithm to obtain various travel regular patterns in consideration of travel frequency.

(4) This algorithm does not require the predetermination of initial cores or the number of clusters. This feature is also essential for travel pattern analysis because the number of patterns from an individual passenger is unknown.

(5) Because of the high complexity of the DBSCAN algorithm, this paper extends this algorithm to a distributed platform, which means that we gather SC data based on CARDID to form a group and then calculate passenger's travel data within each group. After this change, we can use a computer cluster to mine SC data. Each computer in the cluster calculates several group data to increase the speed of calculation.

For a D-dimension dataset containing N points: $X = \{x_1 \ldots x_i \ldots x_n\}$, where $x_i \in R^d$ and DBSCAN algorithm has related definitions as follows:

(1) *Eps* neighbourhood of point x_i: An area with the center x_i and the radius *Eps*.

(2) Density of point x_i: The total number of points within the *Eps* neighbourhood of point x_i. We mark this $density(x_i)$, and the number of $density(x_i)$ is $|N_{Eps}(x_i)|$.

(3) Core point x_i: If the density of point x_i is no less than threshold *minPts*, point x_i is a core point.

The goal of DBSCAN algorithm is to find out the whole core points in dataset X, and then for each core point x, x makes up a cluster together with all the other points whose distance to x is less than *Eps*. This paper choses 20 min as *Eps* in temporal dimension.

For SC data analysis, *minPts* has great meaning in two sides. On one side, the absolute number of the *minPts* must reach a certain value. If the times of a passenger travelled by bus are too small, the regular passenger travel patterns will not be very clear. The number chosen in this paper is 4, which means a passenger must have travel regular record in at least 4 days among 18 working days. On the other side, *minPts* is a relative value, namely, if a passenger travels regularly, the proportion of regular-travel-day number to all the travel times is more than 50%.

2.3. Passenger Travel Behaviour Analyses in Spatial Dimension. This part also adopts DBSCAN algorithm to analyse passenger travel behaviour. Because the density of the bus station in Beijing is large and the frequency of buses is high, there is no need for passengers to choose another station to board or alight. So the *Eps* for passenger travel behaviour analysis in spatial dimension is chosen by 0 m, and the principles to *minPts* is the same as the principles in temporal dimension above. However, in reality, some passengers sometimes choose to swipe their cards in advance in order to ensure the efficiency of alighting, or swipe their cards later to avoid the crowd near SC inductors. For this phenomenon, this part proposes an algorithm based on frequent items to identify the advanced or postponed records.

Step 6. We gather SC data by CARDID for each user to form a 3-dimension set containing N points. The 3 dimensions are LINEDID marked l, MARKSTATION marked m, and TRADESTATION marked t. Then the dataset for each user can be expressed as $X = \{x_1(l_1m_1t_1) \ldots x_i(l_im_it_i) \ldots x_n(l_nm_nt_n)\}$.

Step 7. The occurrences number of each element in the set is calculated, and the elements appearing more than four times are selected as a candidate frequent set, and the other elements are put into an infrequent set. A new dimension, the occurrences number, is added to the candidate frequent set, so the candidate frequent set can be expressed as $X_f = \{x_i(l_im_it_i) \ldots x_j(l_jm_jt_j) \ldots x_k(l_km_kt_k)\}$ and the infrequent set can be expressed as $X_{if} = \{x_u(l_um_ut_u) \ldots x_v(l_vm_vt_v) \ldots x_w(l_wm_wt_w)\}$.

Step 8. The distance between different elements in the candidate frequent set is calculated, which can be expressed as follows: $DistanceL_{ij} = |x_i(l_i) - x_j(l_j)|$, $DistanceM_{ij} = |x_i(m_i) - x_j(m_j)|$, and $|x_i(t_i) - x_j(t_j)|$. If $DistanceL_{ij} = 0$, $DistanceM_{ij} = 1$, $DistanceT_{ij} = 1$, and $c_i > c_j$, we combine x_i and x_j together to form a frequent set.

Step 9. The distance between each element in the infrequent set and each element in the frequent set is calculated. If $DistanceL_{ij} = 0$, $DistanceM_{ij} = 1$, $DistanceT_{ij} = 1$, we change $x_i(l_im_it_ic_i)$ into $x_i(l_im_it_ic_i + 1)$, and drop item x_j from the infrequent set. At least, we put the infrequent set and the frequent set together.

2.4. Bus Station Clustering and Connection Analysis. In a complex bus transit network, we define the number of different lines passing through a bus station as the node importance d, namely, the lager the node importance is, the more lines the bus station connects. Each bus stop has different functions as a node in the network. Some nodes have small node importance but they are very close to other nodes. Some of these nodes can be put together because of the similar roles they play in a certain area, and then the relationship can be seen between different areas or between an area and some nodes.

To gather different nodes, this paper puts forward an algorithm based on density and node importance. The flow of the algorithm is as follows:

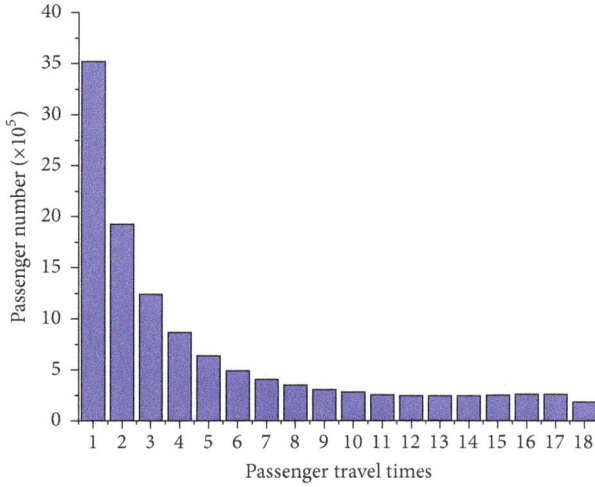

FIGURE 2: The relationship between passenger number and passenger travel times.

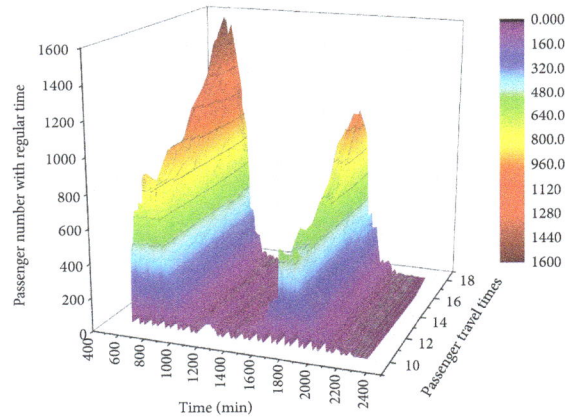

FIGURE 3: The relationship between passenger number with regular time and travel times.

Step 10. This step is similar with DBSCAN. A 3-dimension set containing all bus stops is built, whose dimensions are node importance, longitude and latitude. The dataset can be expressed as $X = \{x_1(d_1 lon_1 lat_1) \ldots x_i(d_i lon_i lat_i) \ldots x_n(d_n lon_n lat_n)\}$. A threshold *Eps* 1000 m is chosen. For bus stop x_i, if the distance between x_j and x_i, which can be calculated by the latitude and longitude, is less than *Eps*, we define x_j as an appendage to x_i. Then, a new dataset containing each bus stop and its appendages is built. The new dataset can be expressed as $X_e = \{x_1(d_1 x_i \ldots x_n) \ldots x_i(d_1 x_1 \ldots x_n) \ldots x_n(d_j x_1 \ldots x_n)\}$.

Step 11. For bus stop x_i and x_j in dataset X_e, if the node importance d_i is larger than d_j, bus stop x_i will absorb the same elements belong to both x_i and x_j. Then each x_i forms an area and each bus stop appears only once in a certain area.

3. Results and Discussion

After calculating by the trip reconstruction algorithm, this paper finds that the total number of passengers travelled by bus on working days in October is 11966945. Around 30% of passengers travelled by bus on only one working day, around 18% of passengers had travel records on over ten working days. The numbers of passengers show little change when the traveling-working day is from 10 to 17. The details show in Figure 2.

The DBSCAN algorithm is used to analyse the travel behaviour for each passenger and identify whether passengers travel regularly. Passenger number with regular travel time or passenger number with regular travel ODs appears below the sum number of passengers whose trips have a certain *Eps* and *Minpts* (two main factors in DBSCAN algorithm). The travel time or ODs of regular passengers is the core point clustered by DBSCAN algorithm for each passenger.

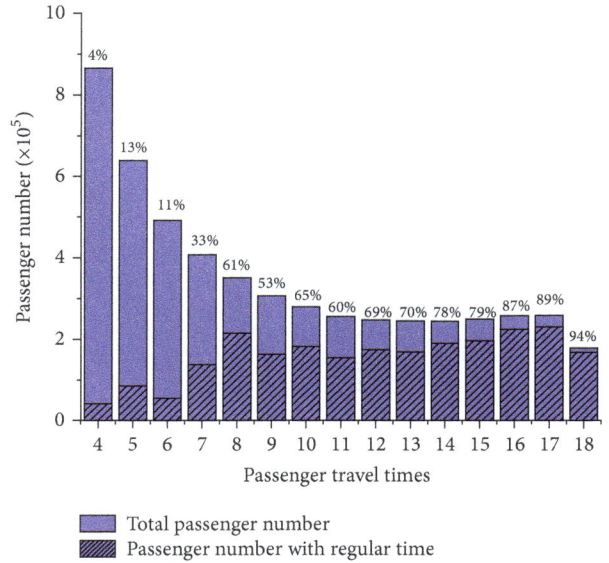

FIGURE 4: Passenger number variation with regular travel time (X axis) for different time (Y axis) and travel times (Z axis).

According to the DBSCAN algorithm introduced above, passenger travel regularity in temporal dimension is analyzed. Passenger number with regular time for different travel times is shown in Figure 3. As passenger travel times increased, passenger number with regular time decreases, but the proportion of passengers with regular time increases.

Figure 4 shows passenger number variation with regular travel time for different time and travel times (from 10 to 18). The morning peak hour begins at 6:30 am and ends at 9:00 am, and the evening peak begins at 17:00 pm and ends at 18:30 pm. Both passenger numbers with regular time in two peak hours are the largest when the passenger travel times is 17. Figure 5 indicates passenger number variation with regular travel time for different time and regular travel times (from 10 to 18). The two peak hours are the same as Figure 4, but passenger numbers with regular time are the largest when the passenger regular travel times is 10.

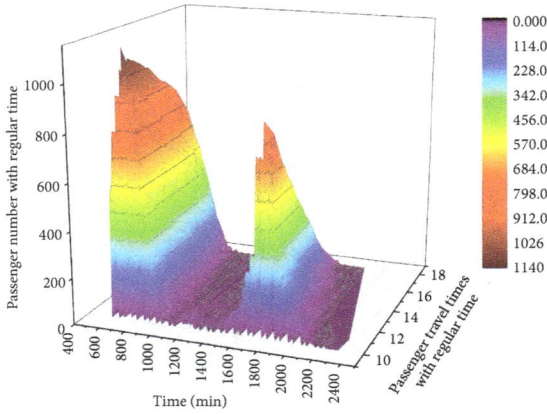

FIGURE 5: Passenger number variation with regular travel time (X axis) for different time (Y axis) and regular travel times (Z axis).

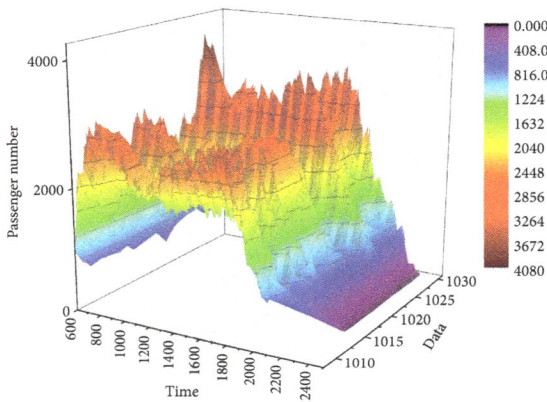

FIGURE 6: Passenger number variation with different time (X axis) and different date (Y axis).

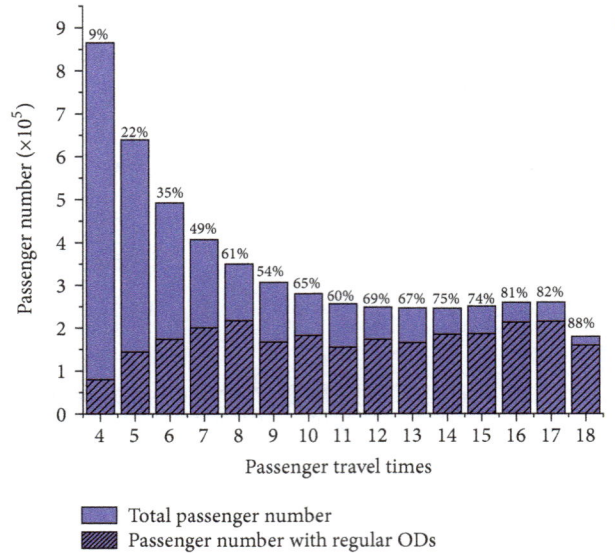

FIGURE 7: Passenger number variation with regular ODs for different travel times.

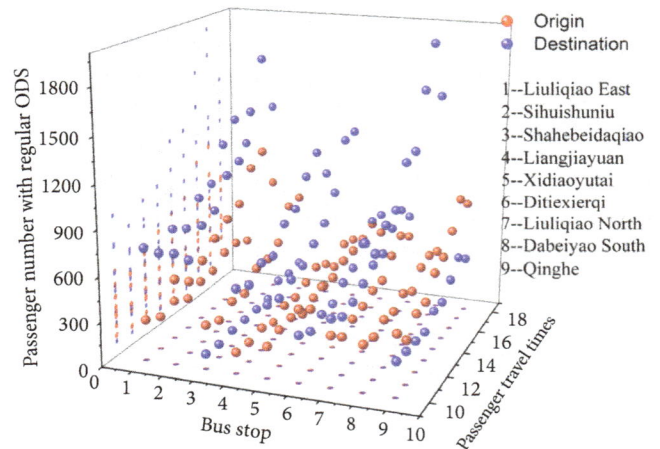

FIGURE 8: Five origin stops and five destination stops (X axis) with the largest passenger number with regular ODs (Z axis) in the morning for different travel times (Y axis).

In the temporal dimension, we can see that the morning peak hour begins at 9:00 for irregular passengers, and the number of passenger during a day varies little with travel time (8:00-20:00). This phenomenon is quite different from this of regular passenger. More details show in Figure 6.

According to the DBSCAN and frequent items algorithm introduced above, passenger number with regular travel ODs is calculated. Passenger number with regular travel ODs for different travel times is shown in Figure 7. As passenger travel times increased, passenger number with regular ODs decreases, but the proportion of passengers with regular ODs increases.

Figure 8 shows five origin stops and five destination stops in the morning with the largest passenger number with regular ODs for different travel times (from 10 to 18). Dabeiyao South stop, located in CBD, is the destination stop with the largest alighting passenger number in the morning. Sihuishuniu stop is the destination stop with the largest boarding passenger number in the morning. It is also an origin stop among the top five origin stops. When passenger travel times are over 10, the peak of passenger travel times is 17 and the peak of passenger travel times with regular ODs is 10.

Figure 9 indicates five origin stops and five destination stops in the evening with the largest passenger number with regular ODs for different travel times (from 10 to 18). The boarding and alighting result in the evening is opposite to that in the morning. The two peaks are the same as that in the morning.

In the spatial dimension, we count the number of irregular passengers at the 5 largest bus stops on ODs per working day. From Figure 10, we can observe that stations1-7 in Figure 10 almost appeared everyday which means that although each passenger traveled irregularly, they had some same destinations. Compared with regular passengers whose target is work area, their destinations are scenic spots, markets, and schools. Station1 (Dongcemen) is close to Tian'anmen Square, and Station2 (Chuangbeixiaoqu) is the

TABLE 1: Detail data for the top five origin stops and destination stops with the most passengers in the morning.

Stop	Lon[°]	Lat[°]	Passenger number (D)	Passenger number (O)	Stop type
①Sihui	116.4903	39.9052	13228	9295	OD
②Xierqi	116.3014	40.0496	5195	6829	O
③Qinghe	116.3417	40.0290	3630	6601	O
④Shahebeidaqiao	116.2626	40.1290	1820	5178	O
⑤Liuliqiao North	116.3040	39.8887	7078	5022	O
⑥Dabeiyao South	116.4552	39.9038	15842	4755	D
⑦Liuliqiao East	116.3114	39.8865	11415	3837	D
⑧Liangjiayuan	116.4641	39.9071	10051	2405	D
⑨Xidiaoyutai	116.2936	39.9226	9122	3883	D

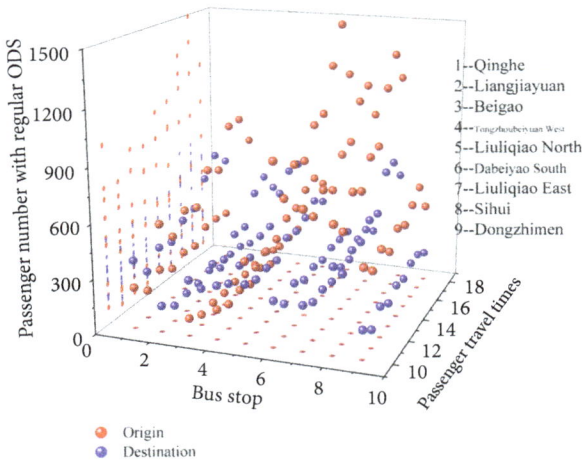

FIGURE 9: Five origin stops and five destination stops (X axis) with the largest passenger number with regular ODs (Z axis) in the evening for different travel times (Y axis).

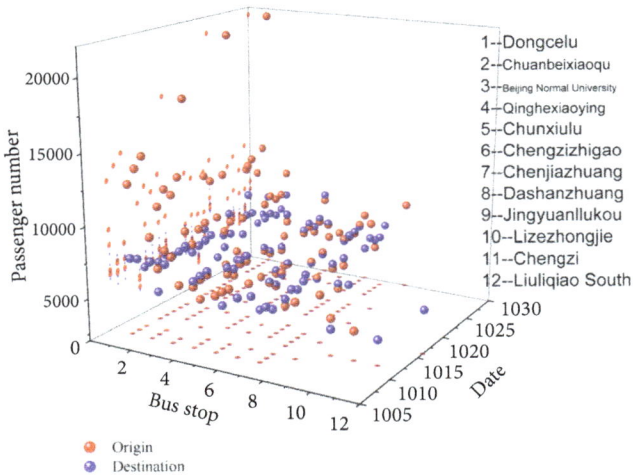

FIGURE 10: Five origin stops and five destination stops (X axis) with the largest passenger number with regular ODs (Z axis) for different date (Y axis).

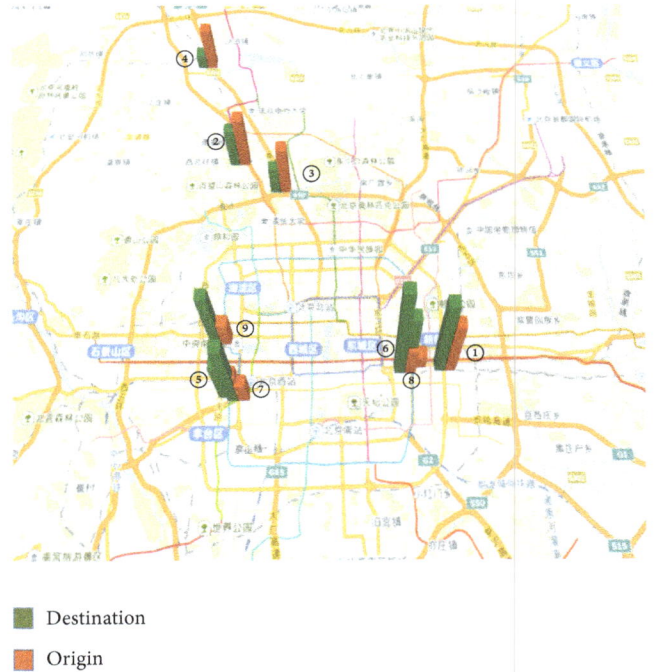

FIGURE 11: The top five origin stops and destination stops with the most passengers in the morning.

main site for the Great Wall. The number of irregular passengers whose destination is Station1 (Dongcemen) changes a lot during different days. This phenomenon is quite different from that of regular passengers.

This paper distributes passengers with regular ODs according to their travel time (morning or evening) and ODs. Then we chose the top five origin stops and destination stops with the most passengers for a detailed analysis. Sihui station belongs to both the top five origin stops and the destination stops. The distribution and detail data of the top five stops in the morning shows in Figure 11 and Table 1.

From Figure 12 and Table 2 we can see that the top 5 destination stops are located in the 3rd ring and three of the top 5 origin stops are located outside the 5th ring. The

TABLE 2: detail data for the top five origin stops and destination stops with the most passengers in the evening.

Stop	Lon[°]	Lat[°]	Passenger number (D)	Passenger number (O)	Stop type
①Dabeiyao South	116.4552	39.9038	3703	14671	O
②Liuliqiao East	116.3114	39.8865	4091	12597	O
③Sihui	116.4903	39.9052	7459	12259	OD
④Liangjiayuan	116.4641	39.9071	1982	9707	O
⑤Dongzhimen	116.4302	39.9408	2729	8239	O
⑥Tongzhoubeiyuan	116.6337	39.9051	5118	2426	D
⑦Qinghe	116.3417	40.0290	5457	3022	D
⑧Liuliqiao North	116.3040	39.8887	5101	6730	D
⑨Beigao	116.5063	40.0101	4589	1121	D

■ Destination

■ Origin

FIGURE 12: The top five origin stops and destination stops with the most passengers in the evening.

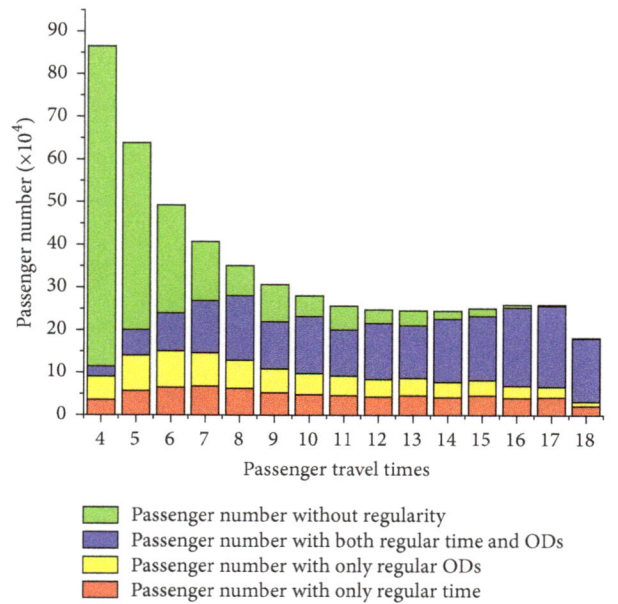

■ Passenger number without regularity
■ Passenger number with both regular time and ODs
■ Passenger number with only regular ODs
■ Passenger number with only regular time

FIGURE 13: The relationship between four type passenger number and passenger travel times.

passenger number with regular ODs in the top 5 destination stops is far bigger than that in the top 5 origin stops, which means the destination of the passengers is concentrated and the origin of the passengers is dispersed. For Sihui and Liuliqiao North, although they are in the top 5 origin stops and their passenger number with regular destinations is bigger than that with regular origins, which means these two stops are main exchange points for passengers in and outside the 3rd ring. Passenger travel behaviour in the evening is quite opposite to that in the morning. The distribution and detail data of the top five stops in the morning show in Figure 12 and Table 2.

According to the analysis in both temporal and spatial dimension, different types of passengers can be obtained. Some passengers travel only with regular time, some of them travel only with regular ODs, and some of them travel regularly in both dimensions, while others travel without regularity. Passenger number for these four types is shown in

Figure 13. We can see that as passenger travel times increased, passenger number without regularity reduced. When the passenger travel times is over 17, only 1% of passengers travelled without regularity.

This paper clusters the bus stations. As a result, more than 2000 areas are identified containing around 7000 bus stops. According to the result of area clustering, this paper analyses the top five origin areas and destination areas with the most passengers and the traffic connection between these areas and other areas in the morning and evening. In the morning, passenger number with regular destinations at the top 5 destination areas is larger than that at the top 5 origin areas, which indicates passengers came from a lot of different areas have several same destinations. The result in the evening is quite opposite to that in the morning. The distribution of the areas shows in Figures 14 and 15, and the detailed data shows in Tables 3 and 4.

TABLE 3: Detail data for the top five origin areas and the top five destination areas in the morning.

Core Stop	OD Type	Passenger number with regular origins	Passenger number with regular destinations
①Liuliqiao North	OD	21902	35248
②Dabeiyao East	OD	20900	57376
③Qinghe	O	19044	11036
④Yanhuang museum	O	17820	21272
⑤Sihui	O	17744	27048
⑥Zhongguancun South	D	11256	36731
⑦Dongsanqi South	D	12478	28126
⑧Sanyuanqiao	D	17035	27860

TABLE 4: Detail data for the top five origin areas and the top five destination areas in the evening.

Core Stop	OD Type	Passenger number with regular origins	Passenger number with regular destinations
①Dabeiyao East	OD	52139	19781
②Liuliqiao North	OD	32129	16167
③Zhongguancun South	O	23828	9896
④Sihui	O	23350	13759
⑤Dongzhimen	O	18466	11463
⑥Tongzhoubeiyuan East	D	13350	17211
⑦Qinghe	D	9946	15493
⑧Tongzhoubeiyuan West	D	4732	13790

- Origin Stops
- Destination Stops
- OD Stops
- Core Stops

FIGURE 14: The top five origin areas and the top five destination areas with the most passengers in the morning.

- Origin Stops
- Destination Stops
- OD Stops
- Core Stops

FIGURE 15: The top five origin area stops and the top five destination area stops with the most passengers in the afternoon.

The paper studies the traffic links between different areas according to the OD data. There are some interesting conclusions based on the study. Among the passengers whose regular destination is CBD area (core stop is Dabeiyao East), 5% and 4.5% of them came from Tongzhoubeiyuan East area and Tongzhoubeiyuan West area, respectively, which are the most closed two areas connecting to CDB. In the evening, 6.6% and 5.4% of the passengers returned to Tongzhoubeiyuan West area and Tongzhoubeiyuan East area, respectively, from CBD area. The distance between CBD and Tongzhoubeiyuan is around 15 km. 24.0% of passengers whose regular origin

TABLE 5: The top five origin and destination areas in the morning and the most closely connected area to them.

Top5 areas	OD Type	Boarding (alighting) passenger number	Closest connecting Area	Alighting (boarding) passenger number	Dist(m)	Prop(%)
Liuliqiao North	O	21902	Yungang	820	15651	3.7
Dabeiyao East	O	20900	Ritanlu	831	1955	4.0
Qinghe	O	19044	Chengfulu South	1200	4431	6.3
Yanhuang Museum	O	17820	Anzhenqiao West	1158	3268	6.5
Sihui	O	17744	Sihui	4256	0	24.0
Dabeiyao East	D	57376	Tongzhou beiyuan East	2810	14954	4.9
Zhongguancun South	D	36731	Beijing Sport University	1992	4391	5.4
Liuliqiao North	D	35248	Gungang	1857	15651	5.3
Dongsanqi South	D	28126	Tiantong beiyuan	2766	2189	9.8
Sanyuanqiao	D	27860	Beigao	1520	7752	5.5

area is Sihui in the morning went to other stops within itself. 70% of passengers whose regular destination is Dongsanqi South in the morning came from several areas which are located within 4 km around Dongsanqi South area. This phenomenon means Dongsanqi South area is a core traffic area gathering a lot of passengers from other areas to take the subway to the city center. More detailed data shows in Table 5.

4. Conclusions

In this paper, four algorithms are used to analyze the temporal and spatial regularity of passengers traveled by bus based on the large scale data of SCs and the traffic relationship between different traffic areas. At first, this paper proposes a trip reconstruction algorithm gathering SC data by CARDID to improve the calculation efficiency using SPARK platform and analyses the times of passengers traveled by bus, that is, the number of days which have travel records in 18 working days. The proportion of passengers with different travel times comes out based on this study. In the temporal dimension, the proportion of passengers who traveled regularly in temporal dimension is obtained and the relationship between this proportion and the times passengers traveled by bus is also described. In the spatial dimension, this paper proposes a data recognition algorithm based on frequent terms to improve the accuracy of SC data and draws some conclusions similar to that in the temporal dimension. According to the temporal and spatial regularities of passengers, passengers are divided into four types: passengers only with regular travel time, passengers only with regular ODs, passengers with both regular travel time and regular ODs, and passengers without regularity. The number of four type of passengers is also obtained. The paper divides the bus area according to the distance between different bus stops and node importance, mainly analyses the passengers with both regular travel time and regular ODs, and determines the traffic connection between different areas.

Conflicts of Interest

The authors declare that they have no conflicts of interest.

Acknowledgments

This research is supported by National Key Technologies Research & Development program (2017YFC0804900).

References

[1] C. Morency, M. Trépanier, and B. Agard, "Measuring transit use variability with smart-card data," *Transport Policy*, vol. 14, no. 3, pp. 193–203, 2007.

[2] K. K. A. Chu and R. Chapleau, "Enriching archived smart card transaction data for transit demand modeling," *Transportation Research Record*, no. 2063, pp. 63–72, 2008.

[3] K. K. A. Chu and R. Chapleau, "Augmenting transit trip characterization and travel behavior comprehension: multi-day location-stamped smart card transactions," *Transportation Research Record*, vol. 2183, pp. 29–40, 2010.

[4] J. Liu, F. Liu, and N. Ansari, "Monitoring and analyzing big traffic data of a large-scale cellular network with Hadoop," *IEEE Network*, vol. 28, no. 4, pp. 32–39, 2014.

[5] S. Tao, J. Corcoran, I. Mateo-Babiano, and D. Rohde, "Exploring Bus Rapid Transit passenger travel behaviour using big data," *Applied Geography*, vol. 53, pp. 90–104, 2014.

[6] C. Chen, J. Ma, Y. Susilo, Y. Liu, and M. Wang, "The promises of big data and small data for travel behavior (aka human mobility) analysis," *Transportation Research Part C: Emerging Technologies*, vol. 68, pp. 285–299, 2016.

[7] L. M. Kieu, A. Bhaskar, and E. Chung, "Passenger segmentation using smart card data," *IEEE Transactions on Intelligent Transportation Systems*, vol. 16, no. 3, pp. 1537–1548, 2015a.

[8] M. K. El Mahrsi, E. Côme, L. Oukhellou, and M. Verleysen, "Clustering Smart Card Data for Urban Mobility Analysis," *IEEE Transactions on Intelligent Transportation Systems*, vol. 18, no. 3, 2016.

[9] X. Ma, C. Liu, H. Wen, Y. Wang, and Y. Wu, "Understanding commuting patterns using transit smart card data," *Journal of Transport Geography*, vol. 58, pp. 135–145, 2017.

[10] A.-S. Briand, E. Côme, M. Trépanier, and L. Oukhellou, "Analyzing year-to-year changes in public transport passenger behaviour using smart card data," *Transportation Research Part C: Emerging Technologies*, vol. 79, pp. 274–289, 2017.

[11] C. Seaborn, J. Attanucci, and N. H. M. Wilson, "Analyzing multimodal public transport journeys in London with smart card fare payment data," *Transportation Research Record*, no. 2121, pp. 55–62, 2009.

[12] L. Duan, L. Xu, F. Guo, J. Lee, and B. Yan, "A local-density based spatial clustering algorithm with noise," *Information Systems*, vol. 32, no. 7, pp. 978–986, 2007.

[13] X. Xu, J. Jäger, and H. P. Kriegel, "A fast parallel clustering algorithm for large spatial databases," *Data Mining and Knowledge Discovery*, vol. 3, no. 3, pp. 263–290, 1999.

12

Evaluating the Impacts of Bus Stop Design and Bus Dwelling on Operations of Multitype Road Users

Jian Zhang[ID],[1] Zhibin Li[ID],[2,3] Fangwei Zhang[ID],[4] Yong Qi,[5] Wenzhu Zhou,[6] Yong Wang[ID],[7] De Zhao,[8] and Wei Wang[3]

[1]Associate Professor in School of Transportation, Southeast University, 2 Si Pai Lou, Nanjing 210096, China
[2]Nanjing University of Science and Technology, Nanjing 210094, China
[3]Professor in School of Transportation, Southeast University, 2 Si Pai Lou, Nanjing 210096, China
[4]Assistant Professor in Shanghai Maritime University, 1550 Haigang Ave, Shanghai 201306, China
[5]Professor in School of Computer Science & Engineering, Nanjing University of Science and Technology, Nanjing 210094, China
[6]Associate Professor in School of Architecture, Southeast University, 2 Si Pai Lou, Nanjing 210096, China
[7]Associate Professor in School of Economics and Management, Chongqing Jiaotong University, 66 Xuefu Road, Nan'an District, Chongqing 400074, China
[8]Postdoctoral Researcher in Department of Civil and Environmental Engineering, National University of Singapore, 1 Engineering Drive 2, E1A 07-03, Singapore 117576

Correspondence should be addressed to Zhibin Li; lizhibin@seu.edu.cn and Fangwei Zhang; fangweizhang@aliyun.com

Academic Editor: Hocine Imine

On urban streets with bus stops, bus arrivals can disrupt traffic flows in the neighboring areas. Different stop designs have distinct influences on the road users. This study aims to evaluate how different types of bus stops affect the operations of vehicles, bicycles, and buses that pass by. Four types of stops that differ in geometric layout are examined. They are termed the shared bike/bus (Type 1), separated shared bike/bus (Type 2), vehicle/bus with inboard bike lane (Type 3), and bus bay with inboard bike lane (Type 4). Data are collected from eight sites in two cities of China. Results of data analysis show that different bus stop designs have quite different impacts on the neighboring traffic flows. More specifically, Type 3 stops create the least bicycle delay but the largest vehicle delay. Type 4 stops have the least impact on bicycle and vehicle operations, but occupy the most road space. Traffic operations are less affected by Type 1 stops than by Type 2 stops. Policy suggestions are discussed regarding the optimal design of bus stops that minimizes the total vehicle delay of all modes.

1. Introduction

In recent years, as urban streets becoming more congested, many cities and countries have considered developing public transit systems within urban areas [1–8]. On urban streets with bus lines, bus stops are usually designed on roadsides to allow travelers to get on/off buses. Bus dwelling could disturb the continuing traffic flow in the neighboring areas posting excessive delays to road users. Evaluating the influence of bus stops on the operations of traffic flow can provide useful information to city planners for the design of bus stops.

Previous studies have evaluated traffic operations near bus stops. Some studies evaluated the traffic flow characteristics in the vicinity of stops [9–13]. For example, Sun and

Elefteriadou [13] analyzed the characteristics of vehicle lane-changing behaviors near the bus stops. Tirachini et al. [11] evaluated the impact of the passenger crowding at the bus stops on the operations and travel time of buses. Some other studies focused on evaluating the conflicts between different road users near the bus stop areas [14–18]. For example, Zhao et al. [16] evaluated the traffic interactions between the motorized and nonmotorized vehicles near a bus stop. However, the above studies did not distinguish the types of bus stop designs.

Until recently, only a few studies have compared the operation features between different types of bus stops [19–22]. For example, Koshy et al. [19] compared the influence of

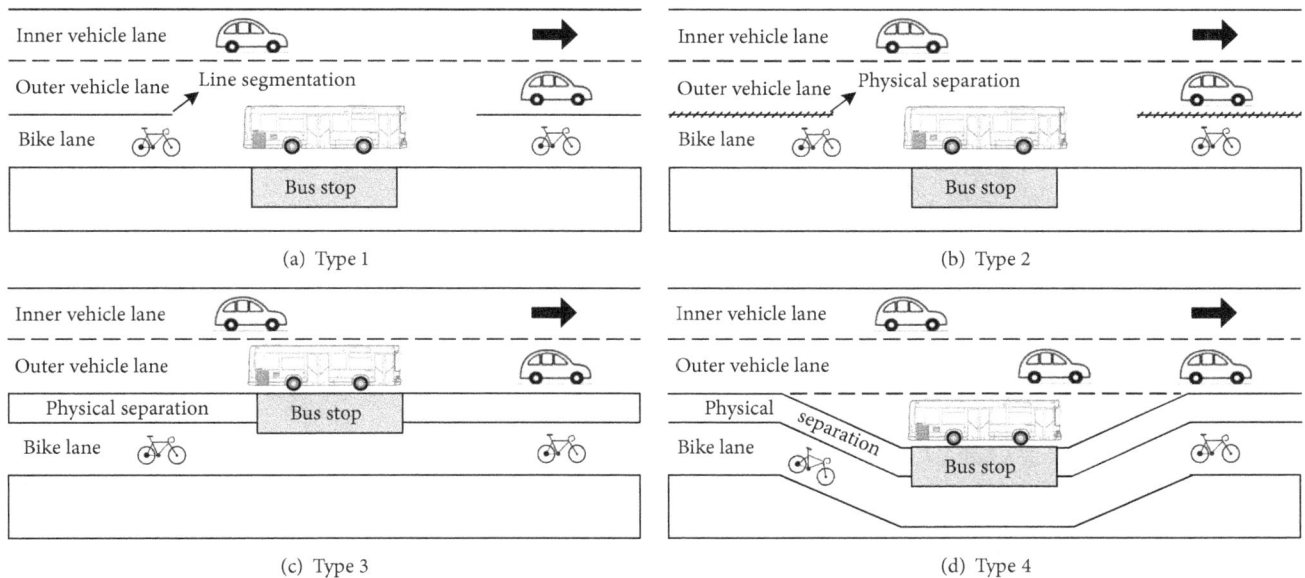

FIGURE 1: Illustration of four types of bus stops [21].

the curbside stops and bus bays on the operation of vehicles. Results showed that the average vehicle speed decreased rapidly at the curbside stops. However, the study was conducted in the simulation environment without validating the findings with actual data. Zhao et al. [20] evaluated the interactions between buses and bicycles at different stops. However, they only analyzed the impacts on bicycle speeds. The study did not evaluate how different bus stop designs impact the operations of vehicles and buses. As a result, it is still not quite clear to policy makers for what type of bus stop should be considered to minimize delay.

The primary objective of the study is to evaluate how different bus stop designs influence the operations of neighboring traffic. Four bus stops that differ in geometric layout were considered. With data collected in the field, the delay of vehicles, bicycles, and buses was evaluated for each stop. In the following section, data collection is presented. In Section 3, methodologies used in the study are introduced. Results of data analysis are shown in Section 4. The policy suggestions are given in Section 5. The paper ends with brief concluding remarks and future work in Section 6.

2. Data Collection

The research team carefully examined the current designs of bus stops on urban streets in several cities of China. Bus stops are usually implemented on the right side of urban streets. Bike lanes are usually provided to accommodate the large cycling demand. After a careful examination, four most common types of bus stops are identified, as illustrated in Figure 1.

Type 1 bus stop contains the shared bike/bus lane (see Figure 1(a)). Bike lane is designed on the right side of the vehicle lanes. Arriving bus occupies the space of bike lane. Bicycles go through the stop from either the right side or the left side of the bus.

Type 2 bus stop contains the separated shared bike/bus lane (see Figure 1(b)). Bike lane is physically separated from the vehicle lanes except for the short section near the bus stop. Arriving bus occupies the space of bike lane. Bicycles go through the stop from either the right side or the left side of the bus.

Type 3 bus stop contains the shared vehicle/bus lane with the separated inboard bike lane (see Figure 1(c)). Bike lane is separated from the vehicle lanes and the bus stop is a curbside design. Arriving bus occupies the space of the outer vehicle lane. Vehicles may shift to the inner lanes to pass the bus or wait after the bus in the outer lane.

Type 4 bus stop contains a bay design with the separated inboard bike lane (see Figure 1(d)). Arriving bus goes into the bay to drop off and pick up passengers. Bus does not occupy the space of vehicle lane or bike lane.

Field investigations were conducted to obtain the traffic flow data in the vicinity of the bus stops. The study sites selected for data collection should satisfy the following requirements: (1) the sites should have typical types of bus stops; (2) there should be no pedestrians in the bus lane and the bike lane; (3) the bus stop should be far away from the upstream and downstream intersections so that there is no spill back traffic; and (4) there are no junctions and taxi/truck loading areas nearby that might disrupt the traffic.

The flow of vehicles and bicycles may affect the delay of road users. For example, if vehicle flow is high, lane changes are more difficult and cars would not be able to avoid the bus by using the inner lane. Besides, with large traffic flow the travel delay may not only be because of the bus dwelling, but also the traffic flow itself. For such considerations, we selected the sites with low traffic demand so that traffic is in free flow condition during investigation. Though bus stops will create more serious problem in congested traffic, in most of the times traffic flow near bus stop is free flowing. Thus, this

TABLE 1: Information of study sties [21].

Number	Street name	Stop type	Bus arrival frequency	Vehicle flow (veh/h)	bicycle flow (bike/h)
Site 1	Hunan Rd, Nanjing	Type 1	1.9 min	653	1023
Site 2	Shanxi Rd, Nanjing	Type 2	1.1 min	726	1333
Site 3	Hongwu Rd, Nanjing	Type 3	1.42 min	707	874
Site 4	Houbiaoying Rd, Nanjing	Type 4	0.8 min	984	612
Site 5	Longsheng Rd, Shanghai	Type 1	1.7 min	1084	1179
Site 6	Renmin Rd, Shanghai	Type 2	1.5 min	953	980
Site 7	Wenchang Rd, Shanghai	Type 3	1.1 min	902	1209
Site 8	Wenhui Rd, Shanghai	Type 4	1.2 min	828	787

study only focused on the general free flow situation leaving the congested situation as a future research task.

Finally, eight streets in the urban areas in Nanjing and Shanghai, China, were selected for data collection. The information of study sites are shown in Table 1. Field data were collected on weekdays with the fine weather in May and June 2014. Video cameras were properly placed on the tall buildings near the investigated sites to capture the overall traffic operations (see Figure 2). Each site contained 1.5 hour video data during nonpeak period from 9:30 to 11:00 AM. The street segment includes three sections which are the upstream section, bus stop section, and downstream section. From the videos we observed that most of bus dwelling maneuvers are within 15 meters near the stop which is used to decide the length of upstream and downstream sections. The total length of each investigated segment is 50 meters.

3. Method

Two methods were considered in this study. The first method was to extract the traffic information from the video camera data. Student's t-test was then used to examine if the difference between groups was statistically significant. The methods are briefly introduced.

3.1. Extraction of Traffic Flow Information. Traffic flow information of bicycles, vehicles, and buses were extracted from the video camera data. As shown in Figure 3(a), the arriving location A and leaving location B are marked in the investigated street section. The distance between the two locations is L=15 meters. During the data processing procedure, the time that each bicycle/vehicle passed the locations A and B was recorded as t_A and t_B. Then the speed of bicycle/vehicle can be calculated as $v=L/(t_A-t_B)$. The type of bike (i.e., electric bike or conventional bike) was recorded by manually check the physical appearances of bikes as well as the motion characteristics of the riders. In addition, the position of bike (i.e., passing from the left side or the right side) was also recorded in the video processing. The position of each vehicle (i.e., inner lane or outer lane) was recorded simultaneously.

With the time information, we can construct the cumulative count curves for the locations A and B separately, as shown in Figure 3(b). The x-axis is the time and the y-axis is the cumulative count of bicycle/vehicle that have passing

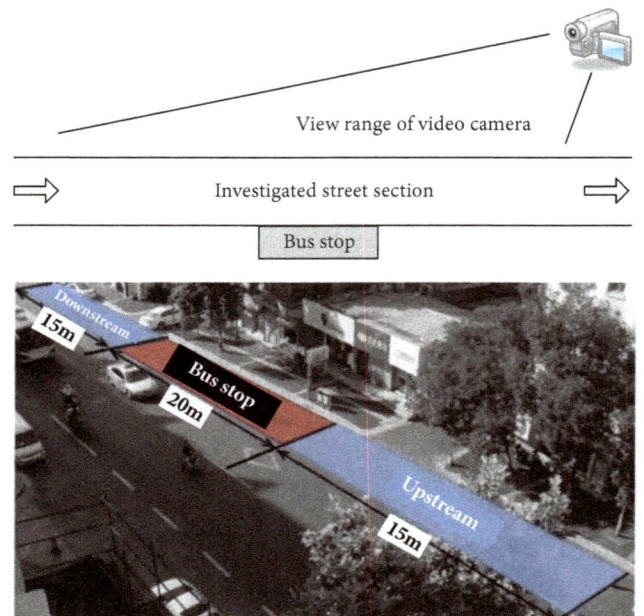

FIGURE 2: Data collection with video camera [21].

the corresponding location before the time point. The slope of the curve within any short period Δt is the number of bicycle/vehicles that passes the location within the period, which can be calculated for the traffic flow. The vertical difference between the two curves at time t can be calculated for the traffic density. The average speed can be calculated for each time period [23].

For each bus, four types of time information were recorded including the time that the bus passed the location A (t_A), the time that the bus completed the full stop (t_{STOP}), the time that the bus started leaving (t_{START}), and the time that the bus passed the location B (t_B). With the time information on locations A and B, the status of bus station, i.e., occupied by bus or no bus, was obtained for each time slice. The operation of buses at different stops was evaluated with the bus stopping and leaving time information.

3.2. Student's t-Test. Student's t-test has been extensively used to identify if the difference between two population

TABLE 2: Statistical tests for speed differences of all bicycles [21].

Speed	With bus at stop			Without bus at stop			ΔV_{Mean}[c] (km/h)	p-value
	Sample size	V_{Mean}[a] (km/h)	V_{Std}[b] (km/h)	Sample size	V_{Mean} (km/h)	V_{Std} (km/h)		
Type 1	38	13.12	2.21	82	15.91	4.83	2.79	<0.001
Type 2	114	14.61	4.48	160	17.37	7.12	2.76	<0.001
Type 3	176	14.56	3.33	158	15.62	5.38	1.06	<0.001
Type 4	28	14.07	1.74	72	15.36	5.06	1.29	<0.005

[a]Mean bicycle speed including electric bicycles and conventional bicycles.
[b]Standard deviation of bicycle speed including electric bicycles and conventional bicycles.
[c]Difference between mean speed with bus at stop and without bus at stop.

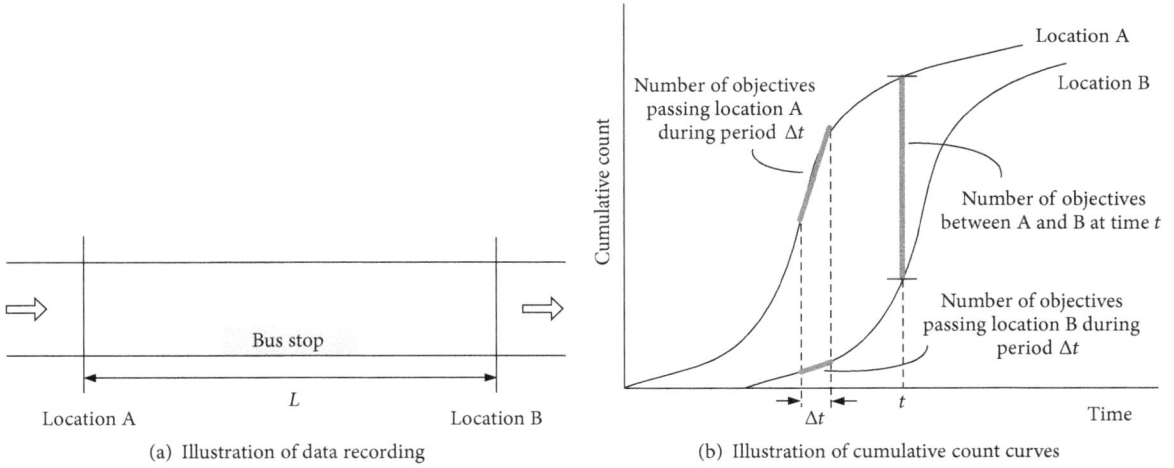

(a) Illustration of data recording

(b) Illustration of cumulative count curves

FIGURE 3: Illustration of data extraction from video camera.

means is statistically significant [24]. In this study, the t-test was conducted to identify if the difference of speeds was statistically significant. Let μ_1 be the mean of average bicycle/vehicle speed when no bus was at the stop station and μ_2 be the mean of average bicycle/vehicle speed when the stop was occupied by bus. The sample standard deviations s_1 and s_2 were obtained for the two groups with the sample size n_1 and n_2, respectively.

The hypothesis states that

$$H_0: \mu_1 = \mu_2 \tag{1}$$

can be rejected if

$$t = \frac{|\mu_1 - \mu_2|}{\sqrt{s_1^2/n_1 + s_2^2/n_2}} \geq t_{a/2} \tag{2}$$

where α is the level of significance and $t_{\alpha/2}$ is the $100(1-\alpha/2)$% percentile of t distribution. The corresponding p-value of the test is given by

$$p = \Pr\left(|t| \geq \frac{|\mu_1 - \mu_2|}{\sqrt{s_1^2/n_1 + s_2^2/n_2}}\right) \tag{3}$$

4. Results of Data Analysis

The impact of bus stop on the bicycle traffic flow was first evaluated. Then the impact on the vehicle traffic was analyzed. We also evaluated the bus operation at different bus stops.

4.1. Impact of Bus Stop on Bicycle Traffic. Speed is an intuitional measure of how traffic flow operates. It reflects the travel time and delay status. Bicycle speed in the situation when bus stop is occupied or there is no bus was calculated. Electric bikes and conventional bikes were analyzed separately. Average bicycle speed at each bus stop is shown in Figure 4. At the Type 1 and Type 2 stops, the bicycle speed when the stop is occupied is obviously lower than that when there is no bus. The reduction of electric bike speed is larger than the conventional bike. At the Type 3 and Type 4 stops, the arrival of bus does not largely impact the speed of bicycles. The electric bicycle speed is reduced slightly.

Student's t-test was conducted to identify if the difference in bicycle speeds was statistically significant. The results are shown in Table 2. All the t-tests are significant at a 95% significance level. It indicates that bus dwelling has a significant impact on the speed of bicycle traffic. Type 1 and Type 2 stop reduce the average speed by 2.76 to 2.79 km/h, because the bus occupies the space of bike lane and blocks the maneuvers of bicyclists. Type 3 and Type 4 stop

(a) Type 1

(b) Type 2

(c) Type 3

(d) Type 4

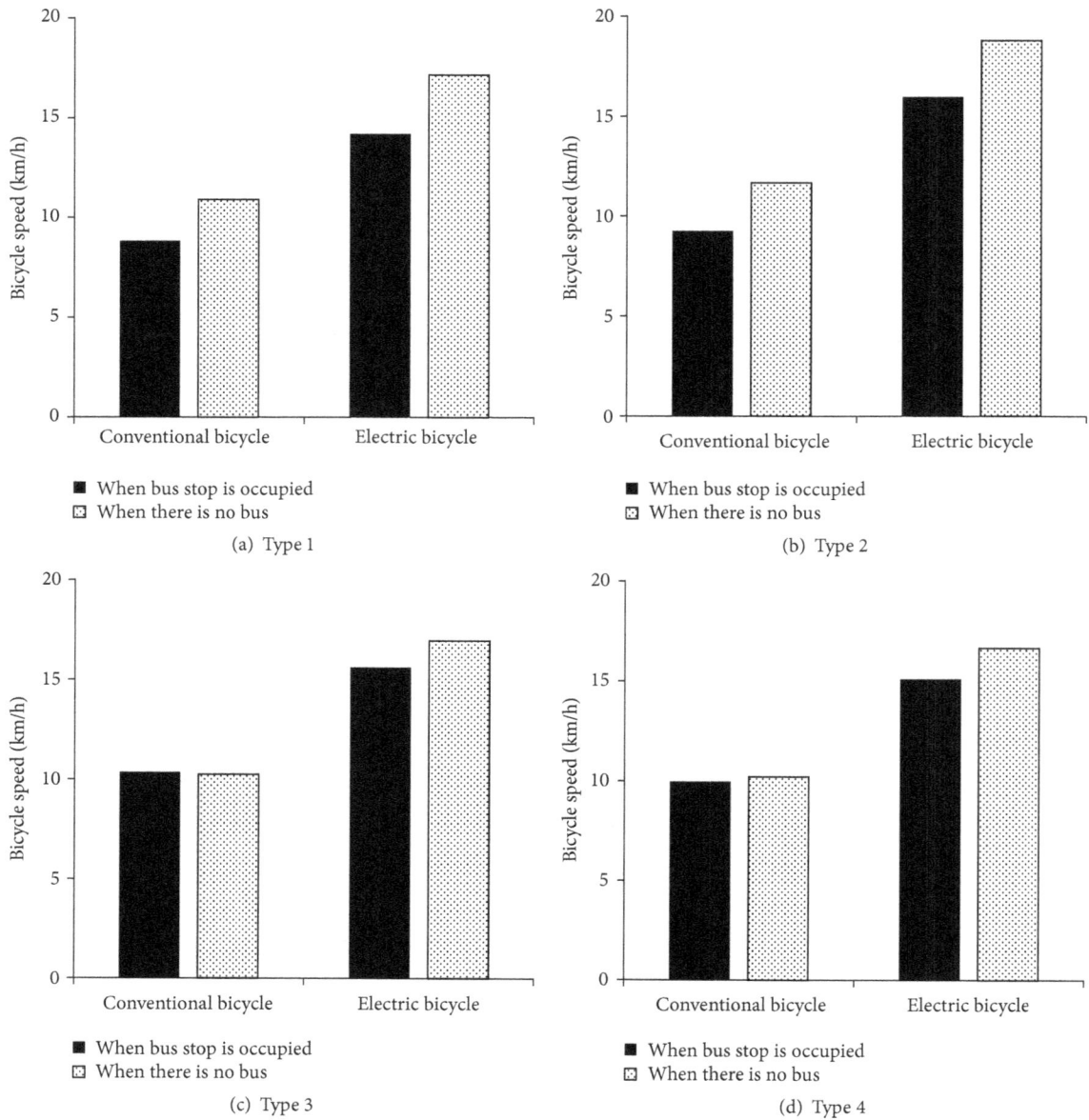

FIGURE 4: Bicycle speeds at different types of bus stops [21].

reduce the average speed only by 1.06 to 1.29 km/h. Because of the physical separation between buses and bicycles, the interactions between them are reduced. The slight reduction of speed is probably due to the passengers' getting on/off behaviors.

4.2. Impact of Bus Stop on Vehicle Traffic. We first validated that vehicle traffic is in free flow condition. The reduction of vehicle speed in the inner lane and outer lane was calculated, as shown in Figure 5. Bus dwelling has a large impact on the vehicle speed in the outer lane, but has a minor impact in the inner lane. The quantitative influences of bus stops were calculated as shown in Table 3. Students' t-test results show that the differences in vehicle speeds are all statistically significant at a 95% significance level.

The results in Table 3 show that Type 3 stop has the largest impact on vehicle speed, followed by the Type 2 stop. The average vehicle speed is reduced by 6.82 and 6.06 km/h, respectively. Type 4 stop only decreases the vehicle speed by 2.19 km/h which is much smaller as compared to other stops. Especially on the outer vehicle lane, the Type 4 bus stop reduces average vehicle speed by 14.3%, which is much lower than that for the Type 1 (28.4%), Type 2 (24.3%), and Type 3 (33.2%). It suggests that the bus bay design in Type 4 stop can remarkably reduce the conflicts between the buses and the vehicle traffic. Type 1 stop has slightly smaller impact on the vehicle speed as compared to Type 2 stop. This would because that driver near the Type 1 stop observes the overall traffic situation better than near the Type 2 stop and would take proactive action to avoid potential travel delay. Type 3 stop

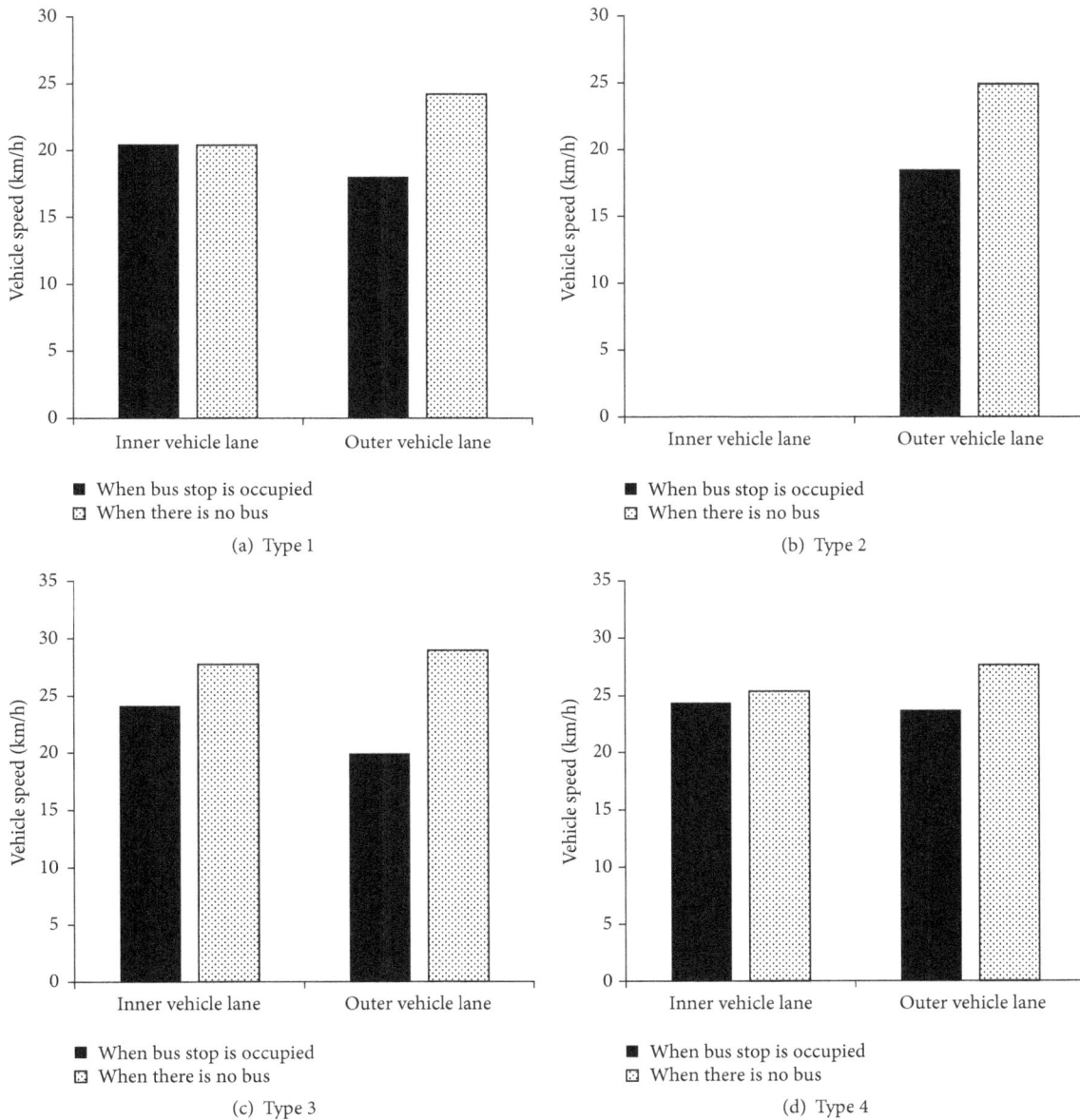

FIGURE 5: Vehicle speeds at different types of bus stop areas [21].

TABLE 3: Statistical tests for speed differences of all vehicles [21].

Speed	With bus at stop			Without bus at stop			ΔV_{Mean}[c] (km/h)	p-value
	Sample size	V_{Mean}[a] (km/h)	V_{Std}[b] (km/h)	Sample size	V_{Mean} (km/h)	V_{Std} (km/h)		
Type 1	26	18.22	3.82	156	22.32	4.94	4.10	<0.001
Type 2	76	18.87	4.66	170	24.93	6.17	6.06	<0.001
Type 3	144	21.43	6.32	178	28.24	7.87	6.82	<0.001
Type 4	102	24.10	7.24	156	26.29	8.29	2.19	<0.001

[a]Mean vehicle speed on the two travel lanes.
[b]Standard deviation of vehicle speed on the two travel lanes.
[c]Difference between mean speed with bus at stop and without bus at stop.

(a) Bus stopping time

(b) Bus leaving time

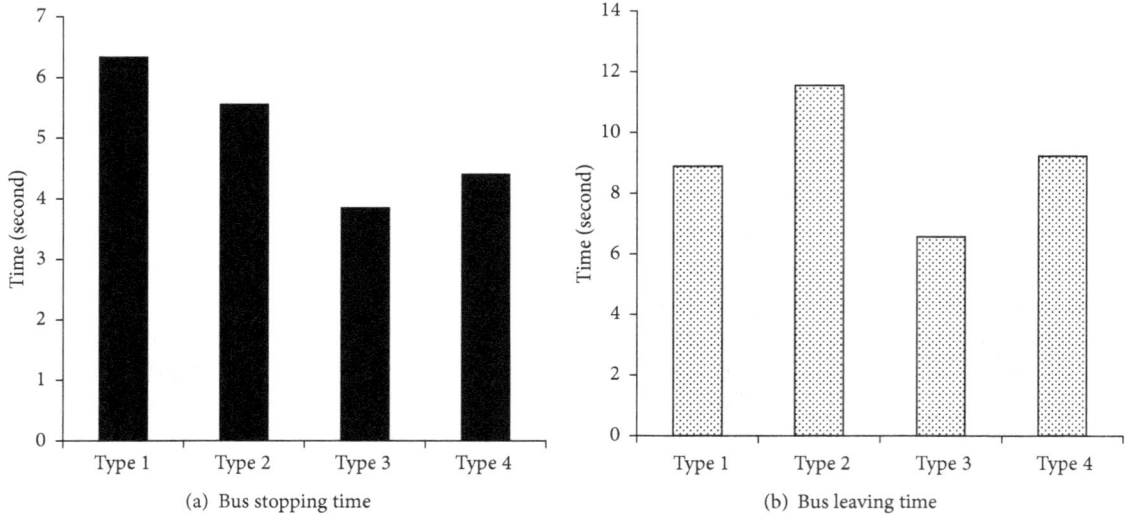

FIGURE 6: Bus operations at different types of bus stop areas [21].

has the largest impact on vehicle speed, because the arrival of bus occupies the vehicle travel lane and blocks the continuing vehicle traffic.

4.3. Impact of Bus Stop on Bus Operation. The study calculated two measures to present the operations of buses. The bus stopping time was calculated as the difference between the time that a bus reaches upstream section (t_A) and the time that the bus fully stops before passengers getting off (t_{STOP}). The measure indicates if a bus can complete the stopping action easily. The bus leaving time was calculated as the difference between the time that a bus starts to run after passengers getting on (t_{START}) and the time that the bus leaves downstream section (t_B). It indicates the easiness of merging into the street mainline. To ensure there is no spillover from stops, the research team excluded the data that contains two or more buses at the same time. The bus stopping time and leaving time at each stop are shown in Figure 6.

The results show that Type 3 stop has the shorted bus stopping and leaving time. It suggests a bus could complete the stopping and merging actions quickly, and the bus operation is less impacted by other road users. A bus also completes the stopping action easily at the Type 4 stop, but has difficulties in merging into the street mainline due to the bay design. Type 1 and Type 2 bus stop have a relatively longer time in the stopping and leaving process than Type 4 stop. This result is a little counterintuitive since people expect that it would be more difficult for a bus to reenter traffic from a taper. The possible reason would be that a bus driver could drive really cautiously to avoid the potential conflicts with bicycles nearby. It takes longer time than avoiding conflicts with vehicles. Type 2 stop has a longer bus leaving time than Type 1 stop, probably because a bus driver needs to avoid the physical separation barrier in the ahead street.

5. Policy Suggestions

Analyses in the above sections suggest that traffic operations are quite different at different types of bus stops, resulting

in different influences on the delay of bicycles, vehicles, and buses. The advantages and disadvantages of the four stops are summarized in Figure 7. Four measurements are considered which are the decrease in bicycle speed, decrease in vehicle speed, bus stopping plus leaving time, and number of occupied lanes. It is quite clear that Type 1 and Type 2 bus stops have a large impact on the bicycle speed and a moderate impact on the vehicle speed. The two stops have a large impact on the bus operation, and require the moderate road space. Type 3 bus stop has the least impact on the bicycle traffic but has the largest impact on the vehicle traffic. Type 4 bus stop has the smallest impact on both the bicycle and the vehicle traffic. It requires the most travel lanes and occupies the largest road resource.

Findings of this study can help city planners to decide which bus stop design should be considered under different situations. More specifically, if city planners would like to reduce the bicycle and vehicle delay, Type 4 stop is recommend for application if urban street has enough space for the bay design. Otherwise, if the road resource is limited and city planners would reduce the vehicle delay, Type 1 stop is recommended. Type 3 stop is recommended if the bus operation is the primary consideration.

A quantitative procedure is proposed to help select bus stops to minimize the total person delay by all modes for the study segment. The utility function u_i for a stop type i is defined as

$$u_i = -\left(D_i^{BIC} + D_i^{VEH} + D_i^{BUS}\right) \cdot (S_i)^{\kappa} \qquad (4)$$

$$D_i^{BIC} = F_i^{E-BIC} \cdot \left(\frac{L}{\overline{V}_i^{E-BIC}} - \frac{L}{\overline{V}_i'^{E-BIC}}\right) + F_i^{C-BIC}$$

$$\cdot \left(\frac{L}{\overline{V}_i^{C-BIC}} - \frac{L}{\overline{V}_i'^{C-BIC}}\right) \qquad (5)$$

(a) Decrease of bicycle speed (km/h)

(b) Decrease of vehicle speed (km/h)

(c) Bus stopping plus leaving time (second)

(d) Number of occupied lanes

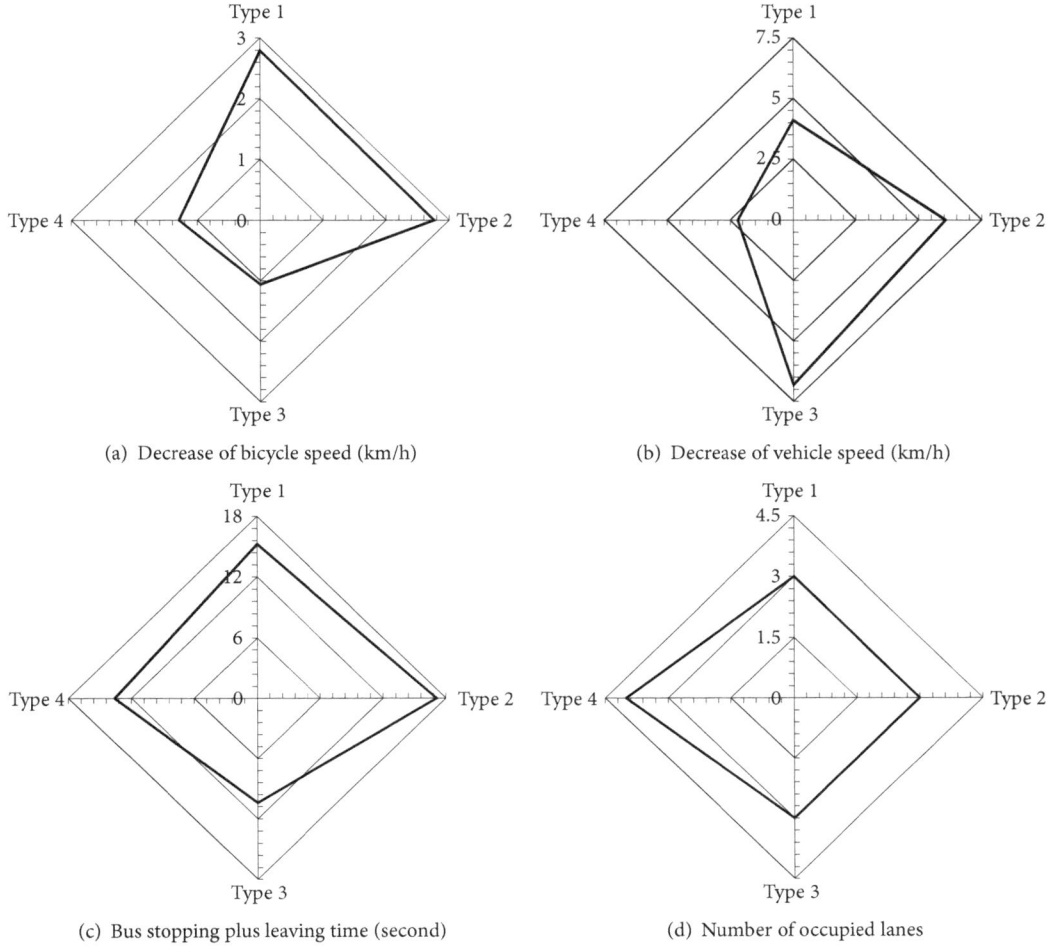

FIGURE 7: Summary of impacts of different bus stops [21].

$$D_i^{VEH} = F_i^{I-VEH} \cdot \left(\frac{L}{\overline{V}_i^{I-VEH}} - \frac{L}{\overline{V}_i'^{I-VEH}} \right) + F_i^{O-VEH}$$

$$\cdot \left(\frac{L}{\overline{V}_i^{O-VEH}} - \frac{L}{\overline{V}_i'^{O-VEH}} \right)$$

(6)

$$D_i^{BUS} = F_i^{BUS} \cdot \overline{P}_i^{BUS} \cdot \left(\overline{T}_i^{STOP} + \overline{T}_i^{LEAV} - \overline{T}_i \right)$$

(7)

where D_i^{BIC} is the bicycle delay at stop i, D_i^{VEH} is the vehicle delay at stop i, D_i^{BUS} is the bus delay at stop i, S_i is the space requirement at stop i, κ is the parameter which decides the importance of road resource, F_i^{E-BIC} and F_i^{C-BIC} are the flow rate of electric and conventional bicycle, $\overline{V}_i^{E-BIC}$ and $\overline{V}_i^{C-BIC}$ are the average bicycle speed when stop is occupied by bus, $\overline{V}_i'^{E-BIC}$ and $\overline{V}_i'^{C-BIC}$ are the average bicycle speed without a bus at stop, L is the street segment length, F_i^{I-VEH} and F_i^{O-VEH} are the vehicle flow rate on the inner and outer lane, $\overline{V}_i^{I-VEH}$ and $\overline{V}_i^{O-VEH}$ are the average vehicle speed when stop is occupied by bus, $\overline{V}_i'^{I-VEH}$ and $\overline{V}_i'^{O-VEH}$ are the average vehicle

speed without a bus at stop, F_i^{BUS} is the bus frequency, $\overline{P}_i^{BUS}$ is the average passenger number per bus, $\overline{T}_i^{STOP}$ is the average bus stopping time, $\overline{T}_i^{LEAV}$ is the average bus leaving time, and $\overline{T}_i$ is the average bus traveling time without dwelling.

In the practical application, the utility value for each bus stop can be calculated using (4) to (7) given the traffic flow information in the street segment of interest. The bus stop with the largest utility value is selected the optimum to minimize the total delay. An example of bus stop selection is given in this section. The geometric and traffic parameters for a street segment in Nanjing are shown in Table 4. The parameter κ is set to be 1. The utility value of each stop is calculated in the table. According to the results, Type 3 stop is considered the best design for the delay minimization. The same procedure can be followed to decide how existing bus stops can be redesigned to improve the traffic operations.

6. Conclusions and Recommendations

This study evaluated the influences of the bus stop designs on the operations of bicycles, vehicles, and buses. Four common stop types were considered. Field investigations

TABLE 4: Procedure for bus stop selection.

Parameters	Value			
Length of street segment (m)	50			
Flow of electric bicycle (bic/h)	600			
Flow of conventional bicycle (bic/h)	400			
Flow of vehicle in inner lane (veh/h)	550			
Flow of vehicle in outer lane (veh/h)	450			
Frequency of bus (bus/h)	12			
Average passenger number per bus (person)	20			
Average bus traveling time without dwelling (sec)	6			
κ	1			
Results	Type 1	Type 2	Type 3	Type 4
Total bicycle delay (h)	0.798	0.734	0.134	0.234
Total vehicle delay (h)	0.319	0.386	0.504	0.181
Total bus delay (h)	0.615	0.741	0.295	0.508
Utility in the delay model	-5.196	-5.583	**-2.798**	-3.690

were conducted to collect traffic information such as flow and speed. Results of data analysis showed that the bus dwelling process had impacts on the operations of different road users. The average bicycle speed was reduced by 1.06 to 2.79 km/h when the bus stop was occupied as compared to the no bus case. The average vehicle speed was reduced by 2.19 to 6.82 km/h. The bus stopping time and leaving time were also evaluated.

The results in the study showed that different bus stop designs had quite different impacts on traffic flow. Type 3 stop had the least impact on bicycle speed and delay, but had the largest impact on vehicle speed and delay. Type 4 stop did not disturb bicycle and vehicle traffic, but occupied the most road space and also limited the number of buses able to stop. The advantages and disadvantages of each stop design were summarized in the study. A quantitative procedure was proposed to help city planners to decide which bus stop to consider for minimizing the total delay of all modes. The same procedure can be followed to decide how to redesign current bus stops to improve the traffic operations.

The study also has several limitations. First, the study only considered the street segments with free flow traffic conditions and low bus volumes. Traffic operations could be very different in the congested traffic situation or with bus spillover from stops. The delay model needs to be modified to accommodate busy traffic conditions. In addition, due to the availability of data, the vehicle and bicycle traffic flow are not controlled when analyzing the bus delay at different stops. More data needs to be collected to estimate the bus delay near stops. Furthermore, the focus of the study is only on delay, but safety is another factor that needs to be considered in the bus stop design. Conflicts between road users and how buses, vehicles, and bicyclists make responses to conflicts could be investigated. Authors recommend that future studies could focus on those issues.

Disclosure

The earlier version of the paper has been presented in the 94st Annual Meeting of Transportation Research Board.

Conflicts of Interest

The authors declare that there are no conflicts of interest for the paper.

Acknowledgments

This research was jointly sponsored by the National Natural Science Foundation of China (61620106002, 51508094, 71871057, and 51608115), the National Key R&D Program in China (2016YFB0100906; 2016YFE0108000), and the Fundamental Research Funds for the Central Universities (2242018K30015, 2242017K40130, and 2242018K41009). The first author's work is also supported by the Zhishan Young Scholar Support Program of Southeast University.

References

[1] X. Ma and Y. Wang, "Development of a data-driven platform for transit performance measures using smart card and GPS data," *Journal of Transportation Engineering*, vol. 140, no. 12, 2014.

[2] M. Hokey, "Public-private partnerships for improving the regional mass transit system: A case study of the Toledo Area Regional Transit Authority," *International Journal of Logistics Systems and Management*, vol. 17, no. 2, pp. 160–179, 2014.

[3] Z. Wang and Y. Chen, "Development of location method for urban public transit networks based on hub-and-spoke network structure," *Transportation Research Record*, no. 2276, pp. 17–25, 2012.

[4] Y. Wang, X. Ma, Z. Li, Y. Liu, M. Xu, and Y. Wang, "Profit distribution in collaborative multiple centers vehicle routing problem," *Journal of Cleaner Production*, vol. 144, pp. 203–219, 2017.

[5] Y. Wang, X. Ma, M. Liu et al., "Cooperation and profit allocation in two-echelon logistics joint distribution network optimization," *Applied Soft Computing*, vol. 56, pp. 143–157, 2017.

[6] Z. Li, W. Wang, C. Yang, and D. R. Ragland, "Bicycle commuting market analysis using attitudinal market segmentation approach," *Transportation Research Part A: Policy and Practice*, vol. 47, no. 4, pp. 56–68, 2013.

[7] Z. Li, W. Wang, C. Yang, and G. Jiang, "Exploring the causal relationship between bicycle choice and trip chain pattern," *Transport Policy*, vol. 29, pp. 170–177, 2013.

[8] Z. Li, W. Wang, C. Yang, and H. Ding, "Bicycle mode share in china: a city-level analysis of long term trends," *Transportation*, vol. 44, no. 4, pp. 1–16, 2017.

[9] W. Gu and M. J. Cassidy, "Maximizing bus discharge flows from multi-berth stops by regulating exit maneuvers," *Transportation Research Part B: Methodological*, vol. 56, pp. 254–264, 2013.

[10] W. Gu, M. J. Cassidy, V. V. Gayah, and Y. Ouyang, "Mitigating negative impacts of near-side bus stops on cars," *Transportation Research Part B: Methodological*, vol. 47, pp. 42–56, 2013.

[11] A. Tirachini, D. A. Hensher, and J. M. Rose, "Crowding in public transport systems: Effects on users, operation and implications for the estimation of demand," *Transportation Research Part A: Policy and Practice*, vol. 53, pp. 36–52, 2013.

[12] M. McCord, R. Mishalani, and X. Hu, "Grouping of bus stops for aggregation of route-level passenger origin-destination flow matrices," *Transportation Research Record: Journal of the Transportation Research Board, No. 2277, Transportation Research Board of National Academics, Washington, D.C., USA*, no. 2277, pp. 38–48, 2012.

[13] D. Sun and L. Elefteriadou, "Research and implementation of lane-changing model based on driver behavior," *Transportation Research Record*, no. 2161, pp. 1–10, 2010.

[14] X. Yang, B. Si, and M. Huan, "Mixed traffic flow modeling near Chinese bus stops and its applications," *Journal of Central South University*, vol. 19, no. 9, pp. 2697–2704, 2012.

[15] X. B. Yang, Z. Y. Gao, B. F. Si, and G. Liang, "Car capacity near bus stops with mixed traffic derived by additive-conflict-flows procedure," *Science China Technological Sciences*, vol. 54, no. 3, pp. 733–740, 2011.

[16] X.-M. Zhao, B. Jia, Z.-Y. Gao, and R. Jiang, "Traffic interactions between motorized vehicles and nonmotorized vehicles near a bus stop," *Journal of Transportation Engineering*, vol. 135, no. 11, pp. 894–906, 2009.

[17] M. Raesaenen, "Functionality of a Bus Stop at Exit or Merging Lanes and Its Impact on Driver Behavior," *Traffic Engineering Control*, vol. 47, no. 1, pp. 29–32, 2006.

[18] S. C. Wong, H. Yang, W. S. A. Yeung, S. L. Cheuk, and M. K. Lo, "Delay at signal-controlled intersection with bus stop upstream," *Journal of Transportation Engineering*, vol. 124, no. 3, pp. 229–234, 1998.

[19] R. Z. Koshy and V. T. Arasan, "Influence of bus stops on flow characteristics of mixed traffic," *Journal of Transportation Engineering*, vol. 131, no. 8, pp. 640–643, 2005.

[20] D. Zhao, W. Y. Wang, Y. Zheng, W. Ji, Wang., and X. Hu, "Evaluation of Interactions Between Buses and Bicycles at Stops," in *Proceedings of the 93rd Annual Meeting of the Transportation Research Board*, Washington, D.C., USA, 2014.

[21] F. W. Zhang, Z. L. Li, D. Zhao, Y. Wang, W. Wang, and J. B. Li. Influences of, "Various Types of Bus Stops on Traffic Operations of Bicycles, Vehicles, and Buses," in *Proceedings of the 94st Annual Meeting of Transportation Research Board*, Washington, D.C., USA, 2015.

[22] B. Wang, S. A. Ordonez Medina, and P. Fourie, "Simulation of Autonomous Transit On Demand for Fleet Size and Deployment Strategy Optimization," *Procedia Computer Science*, vol. 130, pp. 797–802, 2018.

[23] Z. Li, M. Ye, and M. Du, "Some operational features in bicycle traffic flow," *Transportation Research Record: Journal of the Transportation Research Board*, vol. 2520, pp. 18–24, 2015.

[24] S. P. Washington, M. G. Karlaftis, and F. L. Mannering, *Statistical and Econometric Methods for Transportation Data Analysis*, Chapman & Hall, Boca Raton, Fla, USA, 2010.

13

Research on the Impact Scope of Bus Stations Based on the Application of Bus Lanes

Yi Luo [ID] [1,2] **and Dalin Qian** [ID] [1,2]

[1]*School of Traffic and Transportation, Beijing Jiaotong University, Beijing, China*
[2]*MOE Key Laboratory for Urban Transportation Complex Systems Theory and Technology, Beijing Jiaotong University, Beijing, China*

Correspondence should be addressed to Dalin Qian; dlqian@bjtu.edu.cn

Academic Editor: Dongjoo Park

Designation of bus lane means implementing the priority of urban bus system by monopolizing portion of the road resources, which has greater attraction for passengers and can impel the shift of bus passenger flow, thus impacting bus stations. With the aim of mastering the scope and extent of the impact accurately, the article firstly modified the bus station network and the bus transfer network. Furthermore, this paper proposed an algorithm for detecting communities in the improved bus transfer network to mine the transfer relations between any bus routes, and then, on the basis of the improved bus station network, designed a referable bus travel time and put forward an impact model to calculate the absolute impact and relative impact of bus lanes. Finally, the validity of the method was verified according to the actual investigation data. The results show the feasibility and effectiveness of the proposed approach that can obtain the impact of bus lanes on the stations. The research in this paper will be beneficial to the strategy of bus scheduling and also has guiding significance for the evaluation of existing bus lanes or further applications.

1. Introduction

In recent years, the number of motor vehicles owned increased sharply in many cities, which has a negative impact on public transportation. As a result, public transportation vehicles cannot run smoothly and the reliability of transit can hardly be guaranteed. For the lack of road resources and high costs of transport infrastructure, making use of the existing road sections to designate bus lanes has become one of the effective ways to implement public transportation priority. The data have revealed that after the application of bus lanes, compared with the previous mixed traffic, the speed of buses has been greatly increased, parts of them even twice as before [1]. Obviously, the travel condition of buses can be improved and the reliability of the public transport can also be guaranteed through the setup of bus lanes.

Generally speaking, when travelling by urban public transportation, especially commuting, the length of travel time is a prerequisite for passengers to select bus routes. However, the reliability of bus trips has fallen sharply with the incensement of car ownership in recent years; the uncertainty caused by the

crowded mixed traffic environment may even make it more difficult for passengers to determine the actual arrival time. In this case, the reference value of travel time is relatively low. The greatest advantage that bus lane brings to public transportation is to guarantee the reliability of bus trips among some bus stations. In 1994, Abdel-Aty et al. [2] illustrated in a survey about route selection in Los Angeles that fifty-four percent of interviewees regard the reliability of travel time as the most or the second most important factor when they select their primary commute routes. One year later, Abdel-Aty et al. found in a further survey [3] that when the change of travel time is beyond the endurance of passengers, they would rather select a longer but more reliable route. Liu et al. [4] established a mathematical model to study the impact of travel time reliability on passengers' routes selection based on the travel data; they demonstrated that passengers prefer paying more attention to the lowering of variability during trips compared to shortening the travel time. Actually, we also learnt from a filed survey that some passengers are inclined to select bus lane for commuting because of its high reliability, even though that may not shorten their travel time [5]. In addition, according to a traffic survey in Shanghai [6], about seventy-eight percent of interviewees, especially commuters, take the reliability of travel time very seriously; they reckon that the reliability of travel time matters more than the length of travel time. It can be seen that the higher reliability and shorter travel time brought by bus lane will certainly attract some passengers, especially commuters, which can promote the changes in bus trips distribution and have an impact on bus stations. For the transit planners and schedulers, it is a long-term impact, even permanent. Thus, it is very important to understand how many bus stations are impacted and how much the impact is. In this way, we are able to carry out the planning work among the bus stations. On one hand, we can avoid road congestion caused by the shift of passenger flow; on the other hand, we can maximize the utility of bus lanes and improve the passenger transport efficiency of public transit system.

All the time, literatures on the impact of bus lane have been quite substantial. Kiesling and Ridgway (2006) used the cities like San Francisco as examples to discuss what benefits could be derived after the implementation of bus-only lanes [7]. Patankar et al. (2007) developed a microsimulation traffic model to research the impact of BRT dedicated lanes over existing mixed lanes on traffic and commuter mobility [8]; they concluded that the average bus speed is increased and travel time, delay time, and stop time had also significantly decreased. Sakamoto et al. (2007) had identified the effectiveness of the bus priority lane in Shizuoka City, Japan, as a countermeasure for traffic congestion, queue-length and jam-length measurements showed signs of easing of traffic congestion, and travel times of general vehicles were improved [9]. Arasan and Vedagiri (2010) studied the impact of provision of reserved bus lanes on urban roads based on a developed microsimulation [10]. Another research performed by Surprenant-Legault and El-Geneidy evaluated the impact of adding a reserved bus lane on the running times and on-time performance of bus routes; the result showed that reserved lanes had a substantial effect on both service reliability and on-time performance [11]. Tse et al. (2014) studied the safety implications in roads with the setup of bus lanes [12]. Truong et al. (2015) investigated the operational effects of bus lane combinations to establish whether multiple bus lane sections create a multiplier effect in which a series of continuous bus lane sections creates more benefits than several single-lane sections; the results confirm that there is a multiplier effect, so bus travel time benefits and general traffic travel time disbenefits are proportional to the number of links with a bus lane [13]. Bus lanes with intermittent priority (BLIP) were studied by Chiabaut and Barcet (2018); they have demonstrated the strategy of a real-field in Lyon, France, and evaluated the impacts on bus systems performance to show that BLIP can be a promising strategy [14]. The literatures above are mainly divided into two categories: one is the impact on transit operations; the other is the impact on traffic flow. However, less attention has been paid to the bus stations. Actually, the impact on stations can be reflected in the changes of passenger flow in and among corresponding bus stations.

To this end, the paper mainly aims at exploring the impact of bus lane on bus stations. Here, it should be noted that current research suggests the transit network is a typical complex network [15]. Bus stations and relations among them can be studied through the structure of networks. Therefore, it is reasonable to be studied from the perspective of network based on the research content of article. In addition, community structure [16] is an important characteristic of complex network. Nodes in community are tightly connected and connections between communities are sparse. Elements in networks can be clustered by using a rational method of detecting communities. Inspired by the above literatures and ideas, this paper proposes an algorithm to explore the transfer relations through bus transfer network; thereupon we are able to get some sets of bus stations that may be impacted by bus lanes and finally develop an impact model to figure out the accurate impact scope of bus lanes on the stations.

2. Public Transit Networks and Topology

2.1. Methodology. To analyze the properties of public transit networks which are composed of stations and routes, one should start with a proper definition of network topology. Traditionally, the space-L and space-P representation models are widely used [15]. In the space-L network, nodes represent bus, tramway, or underground stations, and an edge connects any two nodes if these two stations are adjacent in at least one common route. Nodes in space-P network have the same meaning with previous topology; however, an edge between two nodes means that there is a direct bus, tramway, or underground route that links them. From the purpose of this paper, on the one hand we need to learn the transfer times between different routes; on the other hand we need to calculate the travel time among stations. Therefore, space-P and space-L are both used for reference.

Each of the networks has its own characteristics. Bus station network not only retains the topology of original network, but also reflects the space and logic relation among stations; based on the bus transfer networks we can learn the transfer relations between different routes. These two networks are shown in Figure 1, in which numbers indicate the stations.

2.2. Network Improvement. In general, sometimes there are no routes between two stations. However, the two stations are so close that people can travel between the two only by walking. As shown in Figure 2, if passengers want to get station 8 or station 10, they may get to station 3 through bus route 1 or get to station 15 through bus route 3 and then walk to station 8 or 10; these traffic routes are rational. However, if the bus station network is built in accordance with the original space-L topology, passengers have to select route 2 to reach destination, which may take a long time.

Considering the above analysis, we set the walking distance equal to 500 meters, with reference to literature [17]. That is to say, any two stations belonging to different bus routes should be linked in the improved bus station network, if the distance between the two is no more than 500 meters. On this basis, the bus transfer network must be adjusted

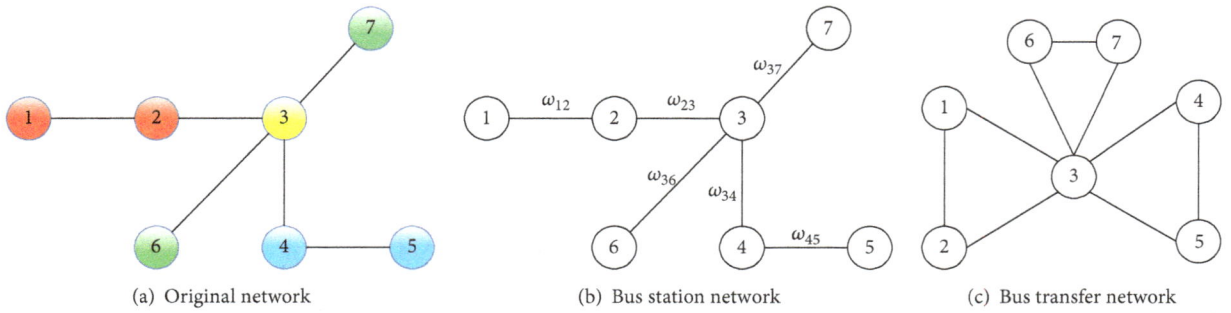

(a) Original network   (b) Bus station network   (c) Bus transfer network

FIGURE 1: (a) An original network is composed of three bus routes. The three routes are displayed in red for route 1, blue for route 2, and green for route 3. Station 3, colored in yellow, is a common station shared by three routes. (b) The corresponding bus station network. The variables floating on the edges are, respectively, RBTT between any two stations which will be explained in the following sections. (c) The corresponding bus transfer network.

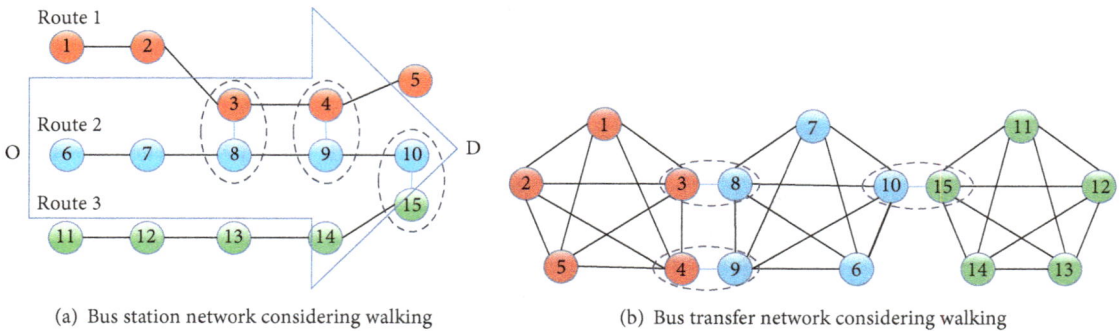

(a) Bus station network considering walking   (b) Bus transfer network considering walking

FIGURE 2: Sketch of bus networks considering walking.

accordingly, as shown in Figure 2(b). It is tantamount to set three new routes between stations 3 and 8, 4 and 9, and 10 and 15; each has only two stations. Obviously, because of these links, passengers can have a new transfer scheme.

Besides, for each route r in Figure 1(b), we set weight coefficients ω_{ij} $\{i, j \in r\}$ to indicate the referable bus travel time (RBTT) between stations i and j during peak hours. This paper holds that the RBTT is related to the Peak bus travel time (PBTT) and the corresponding transit reliability (TR). The PBTT t_{ij} $\{i, j \in r\}$ between any two bus stations i and j denotes the bus travel time in the road section from stations i to j, when the road section has the greatest saturation but not jam. The data of PBTT can be obtained through field survey, Traffic Management Bureau, or traffic flow theory [18]. R_{ij} is the TR of bus travel time from stations i to j; the acquisition method is to observe the number of days d when the bus running time t is equal to or less than t_{ij} within the total working days D of last month and then figure out the results following

$$R_{ij} = \frac{d}{D}\{t \le t_{ij}\}. \tag{1}$$

All historical data in (1) can also be obtained by field survey or Traffic Management Bureau. It is clear that $R_{ij} \in (0, 1]$.

This paper needs to analyze the impact degree of bus lanes through travel time. There are two reasons to use PBTT as

the reference of bus travel time. (1) In the vast majority of urban expressways and arterial roads, the bus travel time is generally approaching or even exceeding PBTT during peak hours. (2) Bus lanes monopolize portion of the road resources to ensure bus priority; thus some of the vehicles are squeezed out to use other road sections, especially around the road sections that include bus lanes. As a result, the saturation of these road sections must be increased. In addition, the purpose of introducing TR is to analyze the probability that the road condition can meet the passengers' demand with recent historical data. The higher the value of TR, the more stable the traffic condition; otherwise passengers are more likely to select other routes. For the bus passengers, it is difficult to estimate the actual travel time; however, judging by historical experience it is feasible. Given the above, we can obtain RBTT as shown in

$$\omega_{ij} = \frac{t_{ij}}{R_{ij}}. \tag{2}$$

For example, if t_{ij} is 30 minutes. We can obtain the RBTT with different TR in 30 workdays according to (2); the results are shown in Figure 3.

As shown in Figure 3, from the view of passengers, RBTT is able to reflect the possible bus travel time in the corresponding road sections. It can be seen that if the TR is lower than 80%, it indicates the road sections being in congested state for at least two days in weekdays and being

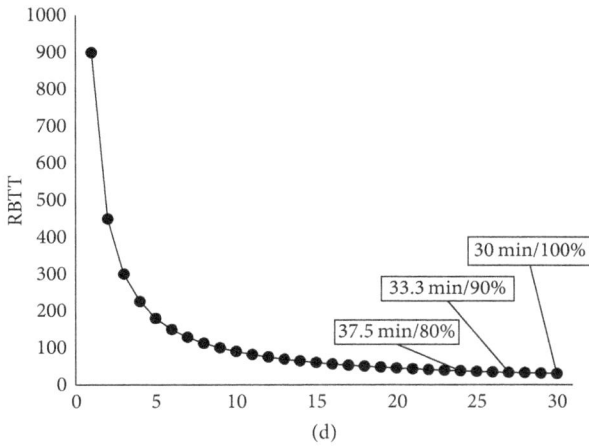

FIGURE 3: Result of RBTT with different TR in 30 workdays.

unable to meet passengers' requirements. The smaller the TR value than 80%, the more likely the commuters to abandon these road sections. This is in line with the reality.

Then, we can obtain the PBTT T_r of route r as shown in

$$T_r = \sum_{i,j \in r} t_{ij}. \tag{3}$$

A bus route r is composed of several relatively independent road sections. The corresponding time offsets will be accumulated in each section when buses pass by and the influence on reliability will be increasing. Therefore, the TR of bus route should be determined by the number of road sections between origin station and arrival station [19]. Assuming that a single bus route r is composed of n stations and $n - 1$ road sections, thus the TR R_r of route r can be obtained as

$$R_r = \sum_{i=1}^{n-1} \sum_{j=2}^{n} a_j R_{ij}. \tag{4}$$

a_j is the reliability coefficient from origin station o to arrival station j as

$$a_j = \frac{j - 1}{n(n - 1)/2}. \tag{5}$$

In conclusion, we can obtain the RBTT W_r of route r as

$$W_r = \frac{T_r}{R_r}. \tag{6}$$

3. Network Analysis and Algorithm

This section will analyze the characteristics of passengers and bus stations which are impacted by bus lanes.

3.1. Network Analysis. There is usually more than one bus route between any two places in cities. As shown in Figure 4, five routes in different road sections can meet the travel demand between OD pair.

Routes 1, 2, and 4 connect with route 3, respectively, as the diagram displays; these are S_1-S_3-S_6, S_3-S_6, and S_7-S_6. Distance between routes 2 and 3 and routes 5 and 3 are nearly the same; however, route 5 has no connection with other routes. Routes 3 and 4 have common bus station S_5. The five routes have the same trip conditions before the setup of bus lanes. Passengers can select the nearest bus stations for trips. This article assumes that the bus lane set is in the road that contains route 3; therefore, route 3 will be the most attractive bus route between OD pair, obviously. Due to the attraction of bus lane, passengers tend to select the routes which are running in bus lanes to replace original routes partially or totally to better meet their travel demands. Generally, they can reach the stations in bus lane by walking or transferring through other bus routes and then continue their trips along the bus lane. However, it should be noted that changes in bus trips are not as flexible as changes in car trips. Not only are bus trips related to the locations of stations, but also they are restricted by the transfer relations between different bus routes. Moreover, bus lanes are usually available during peak hours. For example, most of the valid time of bus lanes in Beijing, Shanghai, and other cities is 7:00 to 9:00 and 17:00 to 20:00. It is clear that the more time the passengers spend to arrive at route 3, the weaker their willingness to select bus lanes.

Further analysis on Figure 4 is as follows.

(1) Passengers near stations S_5 and S_6 are able to select route 3, for they have the best transfer conditions. Passengers near stations S_1, S_3, and S_7 have to transfer one time to reach route 3, which has certain transfer conditions. Routes 2 and 3 have the same distance to route 5, but the passengers who took route 5 before can hardly transfer to route 3 due to transfer restrictions at station S9.

We know that bus transfer is a necessary step for passengers to change routes. However, the time of bus transfer can only passively depend on the arrival time of the previous bus route and departure time of the next one. The reliability of bus transfer is also close to the reliability of the bus running time of the two routes; thus bus transfer has a great uncertainty. The more the times of transfer, the greater the loss of bus travel time.

In addition, we can see from Figure 5 that if passengers only need to transfer one time to reach bus lanes it must happen on bus lanes. Bus lane can guarantee the reliability of travel time, so the bus transfer time will be also guaranteed when passengers transfer from the previous route. Based on this, the acceptability of bus passengers to transfer to bus lanes must be high; moreover, the commuting distance is not very long generally. Based on the characteristics of commuting in peak hours [20] and field research [5], we can see that the times of bus transfer are generally once at most in daily commuting. In view of the above analysis, we will no longer consider the passengers who need to transfer at least two times to reach the stations in bus lane.

(2) The stations S_1, S_3, S_5, S_6, and S_7 are all located in the impact scope of bus lanes, which have a certain probability to transfer. The probability depends on the improvement of travel time after the selection of routes in bus lanes.

In conclusion, we can sum up the characteristics of passengers who are impacted: (1) they reach the bus lanes

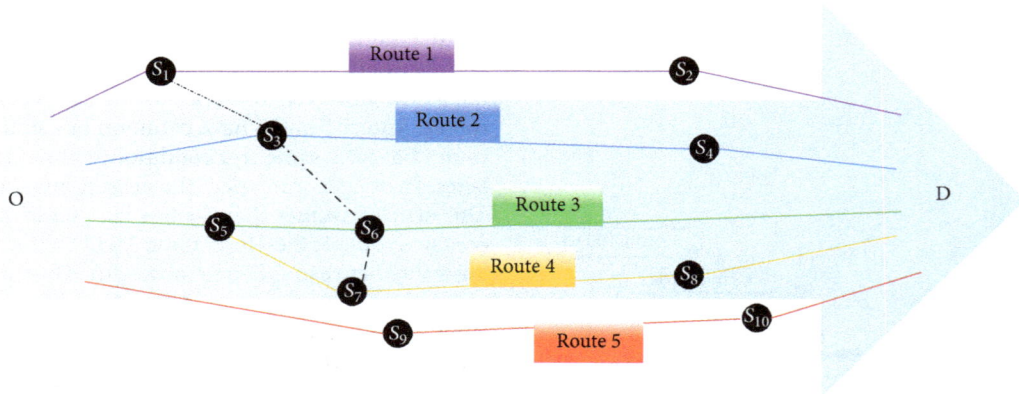

FIGURE 4: Sketch of impact on stations.

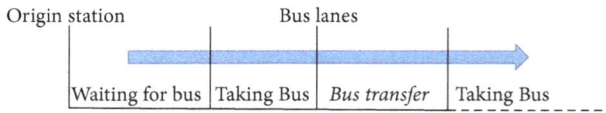

FIGURE 5: Sketch of transfer to bus lanes.

with high convenience; (2) travel time can be acceptable after the selection. Attraction of bus lane is also determined by the two conditions above.

We can divide the acceptable time described in condition (2) above into two types: one type is the travel time being less than before when passengers select routes that include bus lanes. In this case, we call it absolute impact (Abi) on the corresponding bus stations E_a, as follows:

$$E_a = T - A \cdot T_0 \quad \text{if } E_a > 0, \tag{7}$$

where T, T_0 are, respectively, the shortest bus travel time before and after the application of bus lane between OD pairs and A is the transfer factor.

The other type is the travel time being no less than before when passengers select routes that include bus lanes, but it can be accepted by passengers. In this case, we call it relative impact (Rei) on the corresponding bus stations E_r, as follows:

$$E_r = -E_a = A \cdot T_0 - T \quad \text{if } 0 \le E_r \le \eta \cdot T. \tag{8}$$

η is the endurance coefficient, which indicates the increase proportion coefficient of bus travel time that passengers can endure. It can also be obtained by field research. For example, the average travel time in research area is 45 minutes, the survey data revealed that most of the passengers can accept the new and more reliable trip which is no more than 60 minutes; then we can consider setting $\eta = 1/3$. We can believe that there will be a very high transfer probability of passengers under the absolute impact, while they may have a certain probability to transfer under the relative impact [3]. In this way, the traffic managers can get a more accurate determination of the impact.

We define transfer factor $A = 1 + \ln(1+h)$ and h is transfer times [21]; it can be seen that the more the transfer times are,

the greater A is. By substitution into (7)-(8), we can obtain (9)-(10) as follows:

$$E_a = T - [1 + \ln(1 + h)] \cdot T_0 \quad \text{if } E_a > 0 \tag{9}$$

$$E_r = [1 + \ln(1 + h)] \cdot T_0 - T \quad \text{if } 0 \le E_r \le \eta \cdot T. \tag{10}$$

Bus travel time consists of three parts, which are bus waiting time, bus running time, and bus transfer time. This paper assumes that the departure interval of buses is τ in peak hours; that is to say, waiting time in one bus route is τ. Therefore, the more the routes one can use, the shorter the the bus waiting time. Then by substitution into (9)-(10), we can obtain (11)-(12) as follows:

$$E_a = \left(\frac{\tau}{n_0} + W_l \right) - [1 + \ln(1 + h)]$$
$$\cdot \left[\frac{\tau}{n_1} + \left(\frac{\tau}{n_2} + t' \right) \cdot h + t \right] \quad \text{if } E_a > 0 \tag{11}$$

$$E_r = [1 + \ln(1 + h)] \cdot \left[\frac{\tau}{n_1} + \left(\frac{\tau}{n_2} + t' \right) \cdot h + t \right]$$
$$- \left(\frac{\tau}{n_0} + W_l \right) \quad \text{if } 0 \le E_r \le \eta \cdot T, \tag{12}$$

where τ can be obtained through transit planning, such as 2 or 5 minutes. W_l is the RBTT of route l. n_0 and n_1 are, respectively, the number of bus routes available at the origin stations in the trips with or without bus lanes. n_2 is the number of bus routes available at the transfer station; transfer times $h = 0$ or 1. t is the bus running time in the trips with bus lanes; t' represents the average transfer time.

Hence, the impact of bus lanes E can be obtained as follows:

$$E = \begin{cases} E_a & \text{if } E_a > 0 \\ -E_r & \text{if } 0 \le E_r \le \eta \cdot T \\ 0 & \text{else.} \end{cases} \tag{13}$$

E is expressed through bus travel time; the larger E is, the greater the impact is and vice versa. Thus, in this paper, a positive value indicates the absolute impact and a nonpositive value indicates relative impact.

Calculation procedure is described as follows.

(1) Obtain the improved bus transfer network N_P based on the method in Section 2.2 and then detect community structure in N_P to get transfer relations between any two bus stations. The set of stations θ is composed of stations that have zero or one time transfer relations with the stations in bus lanes.

(2) Obtain the improved bus station network N_L based on the method proposed in Section 2.2 and calculate the RBTT among stations.

(3) Designate the OD pairs and based on the Dijkstra or Floyd algorithm calculate the travel time of the shortest path P_{st} and the second-shortest path P_{ss} between the stations in θ and the destination. If bus lanes are not included in path P_{st} or P_{ss}, then set E of the station equal to zero and reselect others.

(4) If the bus lanes are included in path P_{st}, we can calculate the shortest bus travel time T_{st} based on (11). After this, we will change the corresponding weights in N_L to the bus travel time that existed before the application of bus lanes and recalculate the new shortest bus travel time T'_{st} and finally obtain the absolute impact E_a by $T'_{st} - T_{st} = E_a$.

(5) If the bus lanes are included in path P_{ss}, then we can calculate the relative impact E_r through the second-shortest bus travel time T_{ss} in P_{ss} and the shortest bus travel time T_{st}; that is, $T_{ss} - T_{st} = E_r$.

(6) Change back the weights of bus lanes that are modified in Step (4). Calculate and repeat until all stations in θ obtain an impact value E. The values are marked on the corresponding stations in N_L, according to (13). Finally, the accurate scope can be divided into two sets of stations; these are absolute impact (Abi) set θ_a and relative impact (Rei) set θ_r.

3.2. Overlapping Community Detection. According to the calculation procedure in Section 3.1, we first need to explore the transfer relations between bus routes. Three bus routes 1-2-3, 3-4-5, and 3-6-7 are shown in Figure 1(c), respectively. The three routes are three complete subgraphs. Obviously, passengers do not need to transfer when they travel inside a single complete subgraphs. Node 3 is an overlapping node in Figure 1(c); from that we can see that if there are two or more complete subgraphs having at least one overlapping node, passengers need to transfer at most once among the stations which belong to these complete subgraphs. For example, if node 3 is the overlapping node in both 1-2-3 and 3-6-7, passengers need to transfer once if they travel to node 1 or 2 from node 6.

It can be seen that the transfer relations can be explored based on bus transfer networks, so we need an appropriate algorithm to cluster the stations in the bus transfer network. Community can also be referred to as cluster or module, which refers to a set of objects with similar functions or same feature. Thus, community structure detection is a reasonable method for network-mining and element-clustering in complex networks. There are many algorithms to detect communities, according to whether a node belongs to multiple communities, which can be classified into two categories: one

is to detect nonoverlapping communities, such as Spectral Bisection [22], Hierarchical Clustering [23], and GN algorithm [24]; the other is to detect overlapping communities based on Clique Percolation [25], Local Expansion and Optimization [26], or Fuzzy Community [27].

On the basis of the analysis above and the characteristics of bus transfer networks, it can be seen that the algorithm for detecting the overlapping community structure is reasonable. Therefore, based on Clique Percolation and Hierarchical Clustering algorithm, taking Figure 2(b), for example, we proposed the algorithm as follows.

Step 1. Obtain the improved bus transfer network N_P based on the method in Section 2.2; then get the adjacency matrix GP of N_P.

Step 2. Find out all complete subgraphs in N_P as follows:

(1) Select node i ($i = 1, 2, 3, \ldots, N$) as the initial node arbitrarily, where N is the total number of nodes. If node i has no neighbors, set $i = i + 1$ and restart the search.

(2) If node i has a neighbor node j ($j = 1, 2, 3, \ldots, N$; $j \neq i$), then we set i and j together to form a new complete subgraph G_1 which has two nodes. Continue to search new neighbors of node i; if there exist new neighbors that can form a new complete subgraph with G_1, then add them into G_1; otherwise start a new search, by analogy until no node can be added in. The process is shown in Figure 6.

(3) Set $i = i+1$ and continue to search new node, till $i > N$.

Step 3. Based on the Hierarchical Clustering algorithm, each of the complete subgraphs in Step 2 is considered as a new node (community). Then link any two nodes with an edge if the corresponding complete subgraphs have at least one common node. New graph will be formed in this way and repeat operation until all nodes agglomerate into one community. Process is shown in Figure 7. Six complete subgraphs are composed of fifteen nodes, that is, nodes 1 to 5, 6 to 10, and 11 to 15, besides nodes 3 and 8, nodes 10 and 15, and node 4 and 9, respectively. The six complete subgraphs are six communities, which are named "0-transfer-network." That is to say, passengers do not need to transfer when they travel within the six "0-transfer-network" communities, respectively. Similarly, the "1-transfer-network" is composed of two "0-transfer-networks," which represents at least one time to transfer when passengers travel within the "1-transfer network." Finally, the entire network agglomerates into one community as a "2-transfer network." That is to say, the transfer times are at most two when travelling in the whole network. In addition, we need to improve the "0-transfer-network" in the above process. As shown in Figure 7, the complete subgraph which contains only two nodes is processed as one edge. In this way, the three original bus routes have a reasonable transfer relation; otherwise, they would only be three separate routes.

Step 4. Confirm the result of community detection. As everyone knows, there are hundreds of bus routes in cities. This paper detects communities in bus transfer network based on agglomerative algorithm; moreover, we can intuitively

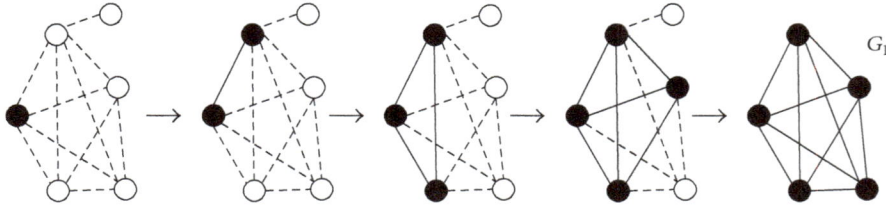

FIGURE 6: Search process of complete subgraph.

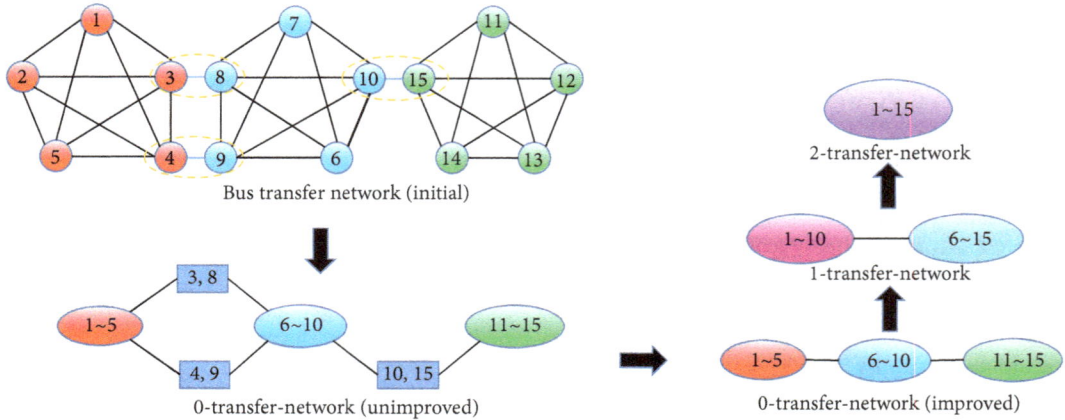

FIGURE 7: Community detection of bus transfer network.

observe transfer relations between any bus routes through tree structure based by modularity Q.

To obtain the best partition of network we should select a level or threshold to cut the tree. It can not only evaluate the quality of detection, but also observe the relations between routes. Newman-Girvan modularity [28] is proposed to evaluate the statistical significance of a given community partition. It is expected that better community partitions will have larger Q-value and vice versa. In reality, the Q-value is often between 0.3 and 0.7. The Newman-Girvan modularity is defined as

$$Q = \frac{1}{2M} \sum_{i,j} \left(a_{ij} - \frac{k_i k_j}{2M} \right) \delta\left(\sigma_i, \sigma_j\right), \qquad (14)$$

where M is the number of edges in network and a_{ij} is element of adjacency matrix of network. k_i and k_j are the degree of nodes i and node j, respectively. δ is equal to one if node i and node j are both in the same community and zero otherwise.

Based on the Newman-Girvan modularity, a huge number of algorithms have been proposed at previous years. However, it can only find out the coarse community structure and cannot detect overlapping communities. Thus, Shen et al. [29] proposed new modularity as shown in

$$Q_c = \frac{1}{2M} \sum_{c \in P} \sum_{uv} \delta_{cu} \delta_{cv} \left(a_{uv} - \frac{k_u k_v}{2M} \right), \qquad (15)$$

where P is universal set of communities. One node may belong to at least two communities in some networks; therefore

we need to improve function δ in (15) and obtain (16) as follows:

$$Q_c = \frac{1}{2M} \sum_{c \in P} \sum_{uv} \alpha_{cu} \alpha_{cv} \left(GP_{uv} - \frac{k_u k_v}{2M} \right), \qquad (16)$$

where

$$\alpha_{cu} = \frac{n_{cu}}{\sum_{c \in P} n_{cu}},$$
$$\alpha_{cv} = \frac{n_{cv}}{\sum_{c \in P} n_{cv}}. \qquad (17)$$

n_{cu} and n_{cv} are the edges of node u and node v in community c, respectively. α needs to be satisfied as follows:

$$\sum_{\forall c \in P} \alpha_{cu} = 1, \quad 0 \leq \alpha_{cu} \leq 1, \ \forall c \in P, \ u \in V. \qquad (18)$$

The physical significance of α is to determine the membership degree based on the link proportion inside and outside a community. Using (16), we calculate the value of modularity in Figure 7 as 0.3984, which shows that the modularity is obvious.

In summary, we can figure out the number of transfer times when passengers travel among the "N-transfer-network" and obtain the relations among any bus routes.

4. Application

In order to meet a huge amount of commuting demand between Tong-Zhou District and downtown Beijing, the Jing-Tong expressway has set bus lanes, which are about 8.6

FIGURE 8: Five square kilometers of Jing-Tong expressway.

kilometers long. During morning peak, the commuters go into town from east side of Jing-Tong expressway. The road network is shown in Figure 8; commuting trips in Tong-Zhou District are mainly accomplished through Chao-Yang North Road, Chao-Yang Road, Jing-Tong expressway, Jian-Guo Road, and so on.

We take the above description, for example, to study the impact of bus lane and verify the results by field investigation.

(1) Determine Bus Transfer Relations in the Research Area. There are 218 bus stations and 35 bus routes of the road network in Figure 9 (data sources: http://beijing.gongjiao.com). We can obtain thirty-five "0-transfer-networks", nineteen "1-transfer-networks," and one "2-transfer network."

(2) Obtain the Set of Stations θ. There are many bus routes running in the bus lanes of Jing-Tong expressway, such as No. 626, No. 668, and No. 322. Refer to the results in step (1); it is concluded that the bus routes such as No. 615, No. 316, and No. 666 need at most one transfer to reach Jing-Tong expressway. This paper takes the OD pair from Tong-Zhou District to Guo-Mao CBD as the research object. In this way, we can get the set of stations θ that may be impacted by bus lanes.

(3) Calculate the Impact Value (E_a and E_r). There are no intermediate bus stations in bus lanes of Jing-Tong expressway; thus the bus stations to enter Jing-Tong expressway are transfer stations in morning peak. We can calculate the results based on the connection between the transfer stations and the stations in θ. In this way, the accurate impact scope of bus lanes θ_a and θ_r is obtained.

(4) Field Research and Results Validation. Take the stations in θ_a and θ_r as center, respectively. Field investigations are carried out in the residential areas which are in the coverage of stations. We collect the information of the residents' routes selections through questionnaire, to investigate whether there exist commuters who change their routes after the application of bus lanes and to verify the transfer proportion under different impact degrees, and then analyze the main impact factors. In this paper, we refer to literatures [17, 30] and set the coverage of bus station as a circular region with radius of 500 meters.

According to θ_a and θ_r, we selected three residential areas for investigation, that is, Liu-Zhuang, Tong-Dian, and Alpha, respectively, as shown in Figure 9(a). Bus stations in Liu-Zhuang and Alpha mainly belonged to the "1-transfer-network," and those in Tong-Dian mainly belonged to the "0-transfer-network." Tong-Dian is the nearest residential area to enter Jing-Tong expressway; Liu-Zhuang and Alpha each have the same distance to the entrance of Jing-Tong expressway; however, the number of bus routes is more in Liu-Zhuang than in Alpha.

We have taken statistics on the shift of passengers during the six months to fully understand the impact of bus lane. 585 questionnaires have been obtained in this survey; the valid number is 563, and the effective rate is 96.24%.

The shift proportion under the absolute influence (Abi) is shown in Figure 9(b). Obviously, from the second month, the transfer proportion is gradually stable, and the maximum shift proportion of Liu-Zhuang is 72%, Tong-Dian is 68%, and Alpha is 54%. Liu-Zhuang and Alpha are relatively far apart from the entrance of Jing-Tong expressway; however, the number of bus routes in Liu-Zhuang is more than Alpha and the travel time can be shorter by at least 20 minutes when commuters select the routes with bus lanes; thus Liu-Zhuang is more significantly impacted. Tong-Dian is the nearest residential area to the entrance of Jing-Tong expressway, so the shift proportion is the highest theoretically. However, the origin station named Ba-Li-Qiao in Tong-Dian is much crowded in morning peak; it is difficult for some commuters to get into the buses and this leads the shift proportion to be lower than Liu-Zhuang. This illustrates the necessity of enhancing public transportation service level. During survey, we also learn that the number of routes in Alpha is small, but some commuters can reach Jing-Tong expressway by bike with the development of bike-sharing in recent years. It illustrates that with more traffic resources that can be used in "1-transfer-network," the shift proportion will be higher.

The shift proportion under the relative impact (Rei) is shown in Figure 9(c). We noticed that, in the first three or four months, the shift proportion gradually increased and then began to stabilize. The reason is that the relative impact value E_r is gradually decreasing. Why? Because part of the road space is exclusive for bus lanes and some of the vehicles are squeezed out to use other road sections during peak hours. As a result, the travel environment of buses in other road sections gets worse and some commuters have to select the routes which include bus lanes. The assumption can be confirmed through the changes in vehicle volume, as shown in Table 1.

In conclusion, after the application of bus lanes, bus stations must be impacted by the shift of passenger flow in bus network. The shift proportion under absolute impact is obviously larger than that under relative impact. However, the shift proportion under relative impact will be changed along with the service level of corresponding road sections.

5. Conclusion

In this paper, a quantitative analysis approach is designed for detecting the accurate impact of bus lane on the stations in transit network. It is not only reasonable, but easy to execute and it has the following advantages.

(1) Through finding the overlapping communities of bus transfer network, we can indicate the possible impact scope

(a)

(b)

(c)

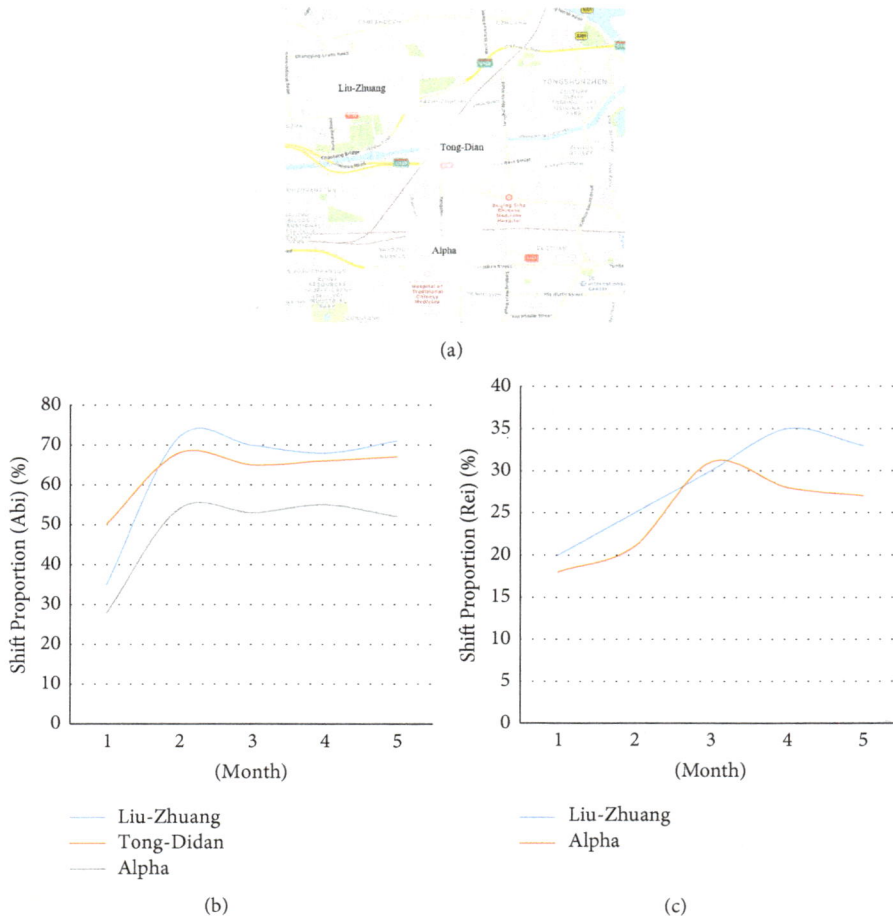

FIGURE 9: Research area and transition probability.

TABLE 1: Changes in vehicle volume after bus lane set in Jing-Tong expressway [5].

| Time period | Vehicle volume | | |
	Before (pcu/h)	After (pcu/h)	Change rate (%)
Morning peak	6488	4242	−34.6
Evening peak	5071	3233	−36.3

after the bus lanes are applied in any road section. In this way, we can try to avoid the expensive public transport investigation.

(2) The analysis shows that the interactions among bus routes are the basis of the impact of bus lanes. On the other hand, improving the level of Public Transport Services and the conditions of the transfer stations in bus lanes are effective measures to deepen the impact.

(3) The impact of bus lane on the stations can be divided into absolute impact and relative impact. Service level of the surrounding road sections may decline after the application of bus lanes and the relative impact could be changed on some stations. This situation can serve as a reference for traffic planners.

(4) From the point of view of networks, based on the status of widely applying bus lanes in many cities, the approach in this paper can also be used to study the impact of multiple bus lanes.

Conflicts of Interest

The authors declare that they have no conflicts of interest.

References

[1] D. Jepson and L. Ferreira, "Assessing travel time impacts of measures to enhance bus operations. Part I: past evidence and study methodology," *Road and Transport Research: a journal of*

Australian and New Zealand research and practice, vol. 8, no. 4, pp. 41–54, 1999.

[2] M. A. Abdel-Aty, R. Kitamura, P. P. Jovanis, and K. M. Vaughn, "Investigation of criteria influencing route choice: initial analysis using revealed and stated preference data," Research Report UCD-ITS-RR-94-12, Institute of Transportation Studies, University of California, Davis, Calif, USA, 1994.

[3] M. A. Abdel-Aty, R. Kitamura, and P. P. Jovanis, "Investigating effect of travel time variability on route choice using repeated-measurement stated preference data," *Transportation Research Record*, no. 1493, pp. 39–45, 1995.

[4] H. X. Liu, W. Recker, and A. Chen, "Uncovering the contribution of travel time reliability to dynamic route choice using real-time loop data," *Transportation Research Part A: Policy and Practice*, vol. 38, no. 6, pp. 435–453, 2004.

[5] L. P. Yang, *The Model of Commute Mode Shift from Private Car to Public Transport*, Beijing Jiaotong University, Beijing, China, 2014.

[6] J. J. Tao and Y. Zhang, "Time Reliability Analysis of Multi-modal Transportation Networks," *Journal of Transportation Systems Engineering and Information Technology*, vol. 15, no. 2, pp. 216–222, 2015.

[7] M. Kiesling and M. Ridgway, "Effective bus-only lanes," *ITE Journal (Institute of Transportation Engineers)*, vol. 76, no. 7, pp. 24–29, 2006.

[8] V. M. Patankar, R. Kumar, and G. Tiwari, "Impacts of bus rapid transit lanes on traffic and commuter mobility," *Journal of Urban Planning and Development*, vol. 133, no. 2, pp. 99–106, 2007.

[9] K. Sakamoto, C. Abhayantha, and H. Kubota, "Effectiveness of bus priority lane as countermeasure for congestion," *Transportation Research Record*, no. 2034, pp. 103–111, 2007.

[10] V. T. Arasan and P. Vedagiri, "Microsimulation study of the effect of exclusive bus lanes on heterogeneous traffic flow," *Journal of Urban Planning and Development*, vol. 136, no. 1, pp. 50–58, 2010.

[11] J. Surprenant-Legault and A. M. El-Geneidy, "Introduction of reserved bus lane: Impact on bus running time and on-time performance," *Transportation Research Record*, no. 2218, pp. 10–18, 2011.

[12] L. Y. Tse, W. T. Hung, and A. Sumalee, "Bus lane safety implications: A case study in Hong Kong," *Transportmetrica A: Transport Science*, vol. 10, no. 2, pp. 140–159, 2014.

[13] L. T. Truong, M. Sarvi, and G. Currie, "Exploring multiplier effects generated by bus lane combinations," *Transportation Research Record*, vol. 2533, pp. 68–77, 2015.

[14] N. Chiabaut and A. Barcet, "Demonstration and evaluation of intermittent bus lane strategy," Research Report 18-00028, Transportation Research Board, Washington DC, USA, 2018.

[15] J. Sienkiewicz and J. A. Hołyst, "Statistical analysis of 22 public transport networks in Poland," *Physical Review E: Statistical, Nonlinear, and Soft Matter Physics*, vol. 72, no. 4, Article ID 046127, 2005.

[16] S. Wasserman and K. Faust, *Social Network Analysis: Methods and Applications*, Cambridge University Press, Cambridge, UK, 1994.

[17] Y. Jiang, P.-C. Zegras, and S. Mehndiratta, "Walk the line: station context, corridor type and bus rapid transit walk access in Jinan, China," *Journal of Transport Geography*, vol. 20, no. 1, pp. 1–14, 2012.

[18] Z. Y. Zhang, *Research on Link Performance Function under the Rules of Bus Priority*, Chongqing Jiaotong University, Chongqing, China, 2015.

[19] W. B. Lu, *Urban Bus Lane Network Optimization Based on Travel Time Reliability*, Southwest Jiaotong University, Chengdu, China, 2012.

[20] Y. L. Chang and Q. Z. Hu, "Optimal line model on urban public traffic line network," *China Journal of Highway and Transport*, vol. 18, no. 1, pp. 95–98, 2005.

[21] W. B. Du, X. L. Zhou, Z. Chen, K. Q. Cai, and X. B. Cao, "Traffic dynamics on coupled spatial networks," *Chaos Solitons & Fractals*, vol. 68, pp. 72–77, 2014.

[22] A. Capocci, V. D. P. Servedio, G. Caldarelli, and F. Colaiori, "Detecting communities in large networks," *Physica A: Statistical Mechanics and its Applications*, vol. 352, no. 2-4, pp. 669–676, 2005.

[23] S. Boccaletti, V. Latora, Y. Moreno, M. Chavez, and D.-U. Hwang, "Complex networks: structure and dynamics," *Physics Reports*, vol. 424, no. 4-5, pp. 175–308, 2006.

[24] M. E. J. Newman, "Detecting community structure in networks," *The European Physical Journal B*, vol. 38, no. 2, pp. 321–330, 2004.

[25] G. Palla, I. Derényi, I. Farkas, and T. Vicsek, "Uncovering the overlapping community structure of complex networks in nature and society," *Nature*, vol. 435, no. 7043, pp. 814–818, 2005.

[26] D. Jin, B. Yang, C. Baquero, D. Liu, D. He, and J. Liu, "A Markov random walk under constraint for discovering overlapping communities in complex networks," *Journal of Statistical Mechanics: Theory and Experiment*, vol. 2011, Article ID P05031, 2011.

[27] V. Nicosia, G. Mangioni, V. Carchiolo, and M. Malgeri, "Extending the definition of modularity to directed graphs with overlapping communities," *Journal of Statistical Mechanics: Theory and Experiment*, vol. 2009, no. 3, Article ID P03024, 2009.

[28] J. Park and M. E. J. Newman, "Origin of degree correlations in the Internet and other networks," *Physical Review E: Statistical, Nonlinear, and Soft Matter Physics*, vol. 68, no. 2, Article ID 026112, p. 026112/7, 2003.

[29] H. Shen, X. Cheng, K. Cai, and M.-B. Hu, "Detect overlapping and hierarchical community structure in networks," *Physica A: Statistical Mechanics and its Applications*, vol. 388, no. 8, pp. 1706–1712, 2009.

[30] R. Cervero, "Walk-and-Ride: Factors Influencing Pedestrian Access to Transit," *Journal of Public Transportation*, vol. 3, no. 4, pp. 1–23, 2001.

14

A Subjective Optimal Strategy for Transit Simulation Models

Agostino Nuzzolo ⓘ **and Antonio Comi** ⓘ

Department of Enterprise Engineering, University of Rome Tor Vergata, Via del Politecnico 1, 00133 Rome, Italy

Correspondence should be addressed to Agostino Nuzzolo; nuzzolo@ing.uniroma2.it

Academic Editor: Monica Menendez

A behavioural modelling framework with a dynamic travel strategy path choice approach is presented for unreliable multiservice transit networks. The modelling framework is especially suitable for dynamic run-oriented simulation models that use subjective strategy-based path choice models. After an analysis of the travel strategy approach in unreliable transit networks with the related hyperpaths, the search for the optimal strategy as a Markov decision problem solution is considered. The new modelling framework is then presented and applied to a real network. The paper concludes with an overview of the benefits of the new behavioural framework and outlines scope for further research.

1. Introduction

Transit network planning requires prediction of bus travel times, on-board loads, and other state variables representing system operations. One way to obtain such variables is to use simulation models [1, 2] which reproduce interactions over time among travellers, transit vehicles, and sometimes also other vehicles sharing the right of way.

In simulation models, a transit supply module is able to support detailed simulation of vehicles serving stops with a given schedule [3], picking up, and dropping off passengers, while monitoring transit vehicles' capacities and speeds. The simulation takes into account when passengers cannot board a vehicle because its capacity limit is already reached. Examples of supply model components are those of the simulators MATsim [3], BUSMEZZO [4], and DYBUS [2]. The simulators perform a within-day dynamic simulation. Each transit vehicle from the departure terminal to that of arrival is followed and, at each bus departure from a stop, the forecasted vehicle travel times, considering the irregularities of the transit services, are updated. Each traveller of a time-dependent origin-destination matrix is followed from origin to final destination, and dynamic routing is applied, taking into account real-time information on current and forecasted states of the transit network. Further, a day-to-day simulation with a traveller learning and forecasting process of service

attributes allows a demand-supply equilibrium condition to be obtained.

While the supply and demand-supply interaction components of transit simulation models are quite well defined in the literature [4], traveller path choice modelling still presents its limits. A case that requires in-depth analysis is that of *multiservice stochastic (unreliable) service networks,* where at some bus stops more than one line is available to reach the destination and some path attributes (e.g., waiting time, on-board time, and on-board occupancy degree) are random variables.

According to the seminal paper by Spiess [5], in the case of multiservice stochastic networks, a stochastic decision approach should be considered, and *optimal travel strategy* modelling should be applied. In stochastic decision theory, an optimal strategy, detailed in Section 2 below, is the behaviour rule that travellers should follow to optimise the expected value of the experienced travel utility.

Two types of travel strategies can be considered from a modelling point of view. One is the *objective* (or *normative*) optimal strategy, which is the behaviour that travellers should follow to optimise the expected value of the experienced travel utility. A different question is the actual strategic behaviour, *subjective* (or *descriptive*) optimal strategy, which travellers adopt, with their cognitive constraints and own perceived path attributes. Drawing on data collected through new ticketing technologies, recent research confirms that, on unreliable transit networks with diversion nodes, subjective travel strategies are sometimes applied [6–11]. These subjective strategies can differ among travellers and very often differ from the objective optimal strategy. Therefore, in transit path choice modelling, a *subjective* optimal strategy should be used, in principle modelling each traveller or at least each traveller category. In practice, such an approach would be very complex, and therefore in the literature a unique optimal strategy is assumed valid for all travellers. Further, in order to determine the applied optimal strategy, until now two main approaches have been followed. In one approach, an *objective optimal strategy* is searched and adopted, such as the optimal strategy reported in Spiess and Florian [12], but in this way neglecting the travellers' cognitive limitations and simplifications. The other, as in BUSMEZZO [4] and DYBUS [2], applies path choice random utility models, and the stochasticity of the services is hidden in the stochasticity of the path choice utilities.

From this analysis of transit path choice modelling applied in simulation, the need arises to adopt in reproducing traveller behaviour not a hypothetical objective optimal strategy, but a subjective strategy-based approach, which is more realistic in relation to the cognitive and computational traveller's capacities and obtained with a stochastic decision approach. This paper proposes such a type of subjective travel strategy approach, defining travellers' utility as combinations of anticipated values through travellers' parameters to estimate, moving from the first investigation performed by Nuzzolo and Comi [13]. The estimation process of such parameters is simplified by the new opportunities offered by big data collecting and processing, which allows effective reverse assignment procedures to be applied [14].

The paper is structured as follows: Section 2 analyses the travel strategy approach in unreliable transit networks and the related routes, while Section 3 considers the search for an optimal travel strategy as a solution to a Markov decision problem. Section 4 presents the proposed behavioural assumption framework and finally Section 5 reports some concluding remarks and future research perspectives.

2. Transit Travel Strategy and Hyperpaths

Let there be an origin-destination pair *od* and an *unreliable* transit service network with *diversion nodes,* that is, nodes where choices are made among different subpaths. Because of transit service stochasticity, rather than relying on a *pretrip* selected single path from origin up to destination, users should adopt a travel strategy *ST* which is [5] a set of coherent behavioural decision rules *(diversion rules)* at diversion nodes, according to random service occurrences (e.g., random arrival times of buses at a stop, random transit vehicle crowding, failure to board, and so on), with the aim of minimising the expected travel cost or maximising the expected travel utility.

Nguyen and Pallottino [15] highlighted the underlying graph structure of Spiess' basic strategy concept, introducing a graph-theoretic framework and the concept of *hyperpath,* which is an acyclic subnetwork, connecting the origin to the destination and including a subset of diversion nodes and a subset of diversion links. At each diversion node, the choice of diversion link depends on the occurrences of transit services and therefore there are certain probabilities for choosing a link among the alternative diversion links [16].

In general, two types of graph representation of a transit service network can be used: *line graph* and *run graph.* While nodes of a *line graph* (see Figure 1) have only spatial coordinates, in a *run graph* the nodes have space-time coordinates *(diachronic graph).* Hence, below we refer to two types of hyperpath representations: *line hyperpaths* and *run hyperpaths.* To each line hyperpath corresponds run hyperpaths with the same spatial nodes, but with different temporal coordinates for each spatial node.

3. Optimal Travel Strategy Search

Although this paper focuses on subjective optimal strategies, objective optimal strategy search methods are first analysed since such methods can suggest efficient search methods for the subjective case as well.

3.1. Objective Optimal Travel Strategies as Solutions to Markov Decision Problems. Path choice in an unreliable service network entails decision making without comprehensive knowledge of possible future evolution of all relevant factors. Hence the outcomes of any decision depend partly on randomness and partly on the agent's decisions. Therefore, in this case a general *theoretical* framework for *objective* optimal strategy search can be found in stochastic decision theory. If path choice is considered as decision making in a Markov decision process (MDPs), the Markov decision problem (MDPm) approach can be considered, as, for example, reported in Nuzzolo and Comi [17] and as summarised below for the reader's convenience.

A Markov decision process (MDPs; [18]) can be defined by the quintuple $(T; SS^\tau; A_s^\tau; p^\tau[s'/s, a]; r^\tau[s, a])$, where

(i) T is a *set of stages* τ at which the decision maker observes the state of the system and may make decisions,

(ii) SS is the *state space*, where SS^τ refers to the possible system states for a specific time τ,

(iii) A_s^τ is the set of possible actions that can be taken after observing state s at time τ,

(iv) $p^\tau[s'/s, a]$ are the *transition probabilities*, determining how the system will move to the next state. In particular, $p^\tau[s'/s, a]$ defines the transition to state s' belonging to $SS^{\tau+1}$ at time $\tau + 1$ and, as a Markov process, only depends on state s and chosen action a at time τ. P is the set of transition probabilities,

(v) $r^\tau[s, a]$ is the *reward function*, which determines the consequence for the decision maker's choice of action a while in state s, and R is the reward set. In our cases, the value of the reward depends on the next state of

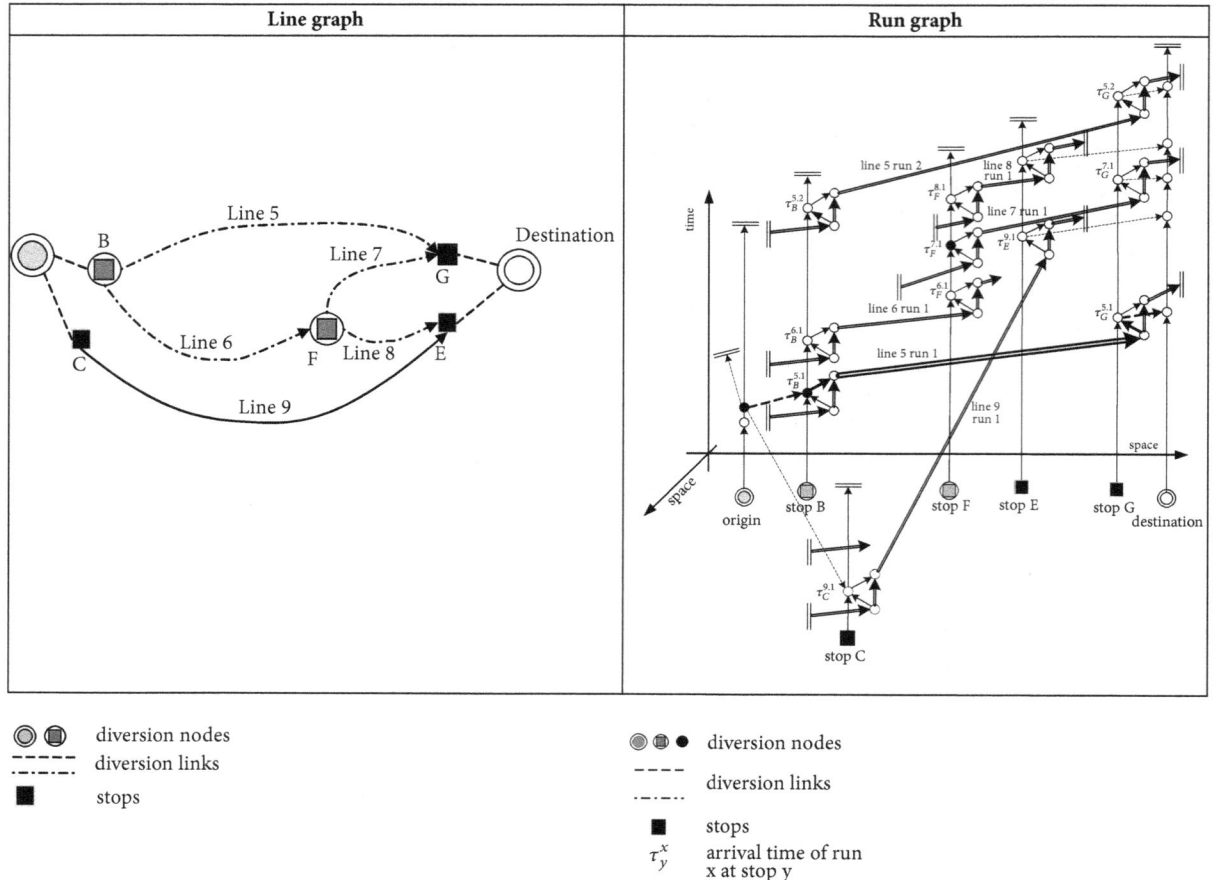

FIGURE 1: Example of line and diachronic graphs.

the system, effectively becoming an *expected reward*, expressed as

$$r^\tau[s,a] = \sum_{s' \in SS^{\tau+1}} r^\tau[s,a,s'] \cdot p^\tau\left[\frac{s'}{s},a\right] \quad (1)$$

where $r^\tau[s,a,s']$ is the relative reward when the system is next in state s'.

An MDPs with a specified optimality criterion (hence forming a sextuple) is called a *Markov decision problem MDPm*. Policies π are essentially functions that regulate, for each state, which actions to perform. The solution of an MDPm provides the decision maker with an *optimal policy* π^* that associates to states SS actions A optimising a predefined objective function.

3.2. Objective Optimal Travel Strategy as an Optimal Policy of MDPm. Given a *run* service network, the optimal travel strategy ST^* can be seen as the optimal policy π^* of a finite and discrete *MDPm*, considering that

(i) the set T is the set of times (τ) when the traveller is at a diversion node s and a diversion link has to be chosen;

(ii) the state space set SS is the set of diversion nodes among which travellers can move;

(iii) an action a is a set of diversion links among which travellers can choose with a given diversion rule and the action set A_s^τ is the set of actions a;

(iv) the change in the time of traveller location within the diversion node set consists in a Markov process;

(v) the transition probabilities $p^\tau[s'/s,a]$ are the probabilities of going from a diversion node (s) to each of the following diversion nodes (s') if action a is applied;

(vi) the reward function $r^\tau[s;a]$ is the expected utility of applying action a at diversion node s;

(vii) the optimal policy π^* gives the best sequence of actions, considering the expected utility up to destination.

To represent an MDPm, a *state-action tree* can be used. At every diversion node, each action can be represented with a set of outgoing links to the next diversion nodes. In Figure 2, in relation to the diachronic graph, the decision tree is reported. For example, at diversion node F three different actions are possible: (1) using run 7.1 (action a_7) and hence stop G; (2) using run 8.1 (action a_8), and hence stop E; (3) using both run 7.1 and run 8.1, with the diversion rule of comparing the expected utility of boarding the first arriving

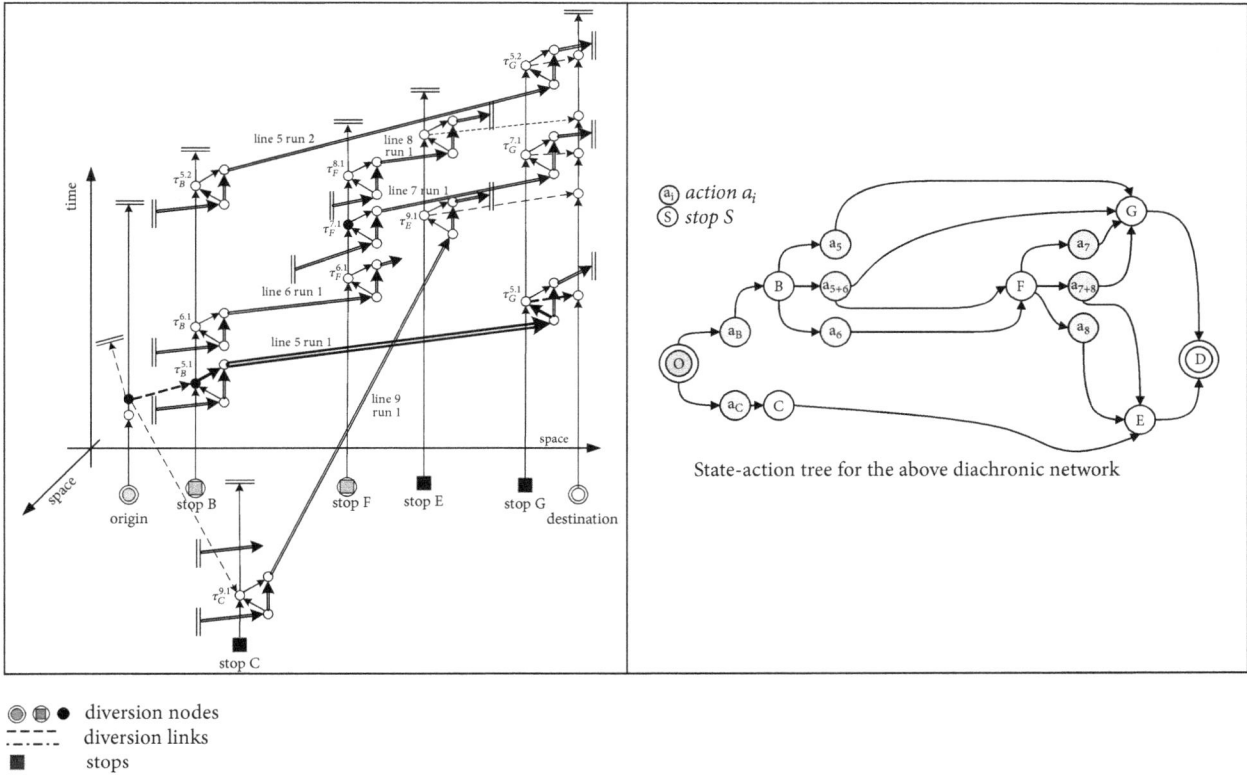

FIGURE 2: Example of run hyperpath with a diversion choice at origin.

run and the expected utility of the next run and then choosing the best (action a_{7+8}). With regard to transition probabilities, consider the case of diversion node F in Figure 2. If action a_{7+8} is applied, the probability of moving onto node G is equal to the probability of using line 7, and the probability of moving onto node E is equal to the probability of using line 8. If action a_7 is applied, the probability of going onto node G is equal to 1. The same holds for action a_8 and node E.

3.3. Objective Optimal Strategy Search Methods.

As explored above, the search for an *objective* optimal travel strategy in a transit network is equivalent to the solution of a Markov decision problem, MDPm. This solution, when the transition probabilities and the expected rewards are known or computable, can be found through exact linear or dynamic programming algorithms. In particular, efficient network algorithms based on the Bellman equation [19] can be used, as in Nguyen and Pallottino [15]. For example, in the case of the optimal strategy reported in Spiess and Florian [12], hypotheses of random arrivals of buses and users at stops and limited information on services allow the transition probabilities to be computed analytically, although such hypotheses are often not congruent with the case studies in question. A more recent example is the dynamic routing of Gentile [20]. Note that in this case operating conditions are assumed for the transit system in several cases very different from the real ones. In order to take into account the specific case study conditions without knowledge of transition

probabilities, some authors use MDPm approximate solution approaches, such as enforcement learning methods (see, for example, the simulator MILATRANS, in [21]). However, this approach requires processes of exploration and exploitation with excessive computation times to reproduce each event. Other authors, in order to consider the actual conditions of the case study, use an adaptive routing problem in a stochastic time-dependent transit network, in which the link travel times are discrete random variables with known probability distributions [22]. Nuzzolo and Comi [11, 17] indicate a way to estimate the transition probabilities and the expected rewards for intelligent transit networks and thus apply an exact *objective* optimal strategy search method.

3.4. Subjective Optimal Strategy Search.

In order to find the *subjective* optimal strategy given the actual conditions of the case study, some authors assume diversion rules which are too complex in relation to travellers' cognitive capacity. For example, a comparison of optimal subhyperpaths is applied by Nuzzolo et al. [2] in the simulator DYBUSRT.

4. The Proposed Behavioural Framework

In this paper, an approach is proposed which applies path choice behavioural modelling based on a dynamic subjective travel strategy and defined in the framework of a Markov decision problem. The proposed model, an advanced version of that presented in Comi and Nuzzolo [13], allows for

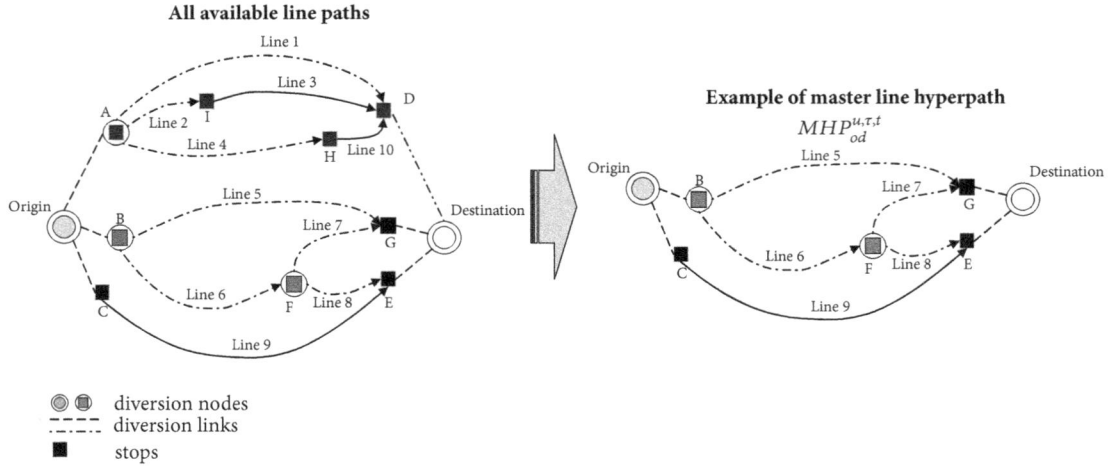

FIGURE 3: Example of a master line hyperpath.

service occurrences and information provided to travellers and considers some travellers' cognitive limitations and simplifications. In the following subsections, the proposed behavioural framework is presented and examined in the MDPm perspective. Further, some application examples are reported.

Traveller behavioural assumptions are defined in the context of

(i) an unreliable or stochastic and within-day dynamic transit service network with diversion nodes;

(ii) transit users who often travel on the origin-destination (O-D) pair (*frequent users*) and are equipped with advanced mobile route planners with real-time individual predictive information, supplying a set of suitable lines and relative path attributes (i.e., travel time components) from current position to destination;

(iii) subjective optimal strategy-based travel behaviour.

4.1. Traveller Behavioural Hypotheses

4.1.1. Master Hyperpaths. Given an O-D pair *od* and its set of available paths at time τ, traveller *u*, as a frequent user on O-D pair *od*, and with the support of an advanced transit trip planner, is assumed to consider a subset of line paths feasible for the traveller. That is, paths that satisfy some logical and behavioural constraints, sometime called a mental map (see, e.g., [21]) and here called *master line hyperpath* $MHP_{od}^{u,\tau}$ (Figure 3). Due to the randomness of transit services, travellers do not refer exactly to time τ but to a time slice $\Delta\tau$ (e.g., $\Delta\tau = \tau \pm 5$ min.), even if, for simplicity, we continue to use τ below.

As a master line hyperpath *MHP* can depend on time slice τ and day *t*, due to within-day and day-to-day dynamicity of the transit service, it is indicated as $MHP_{od}^{u,\tau,t}$. A master line hyperpath can be dynamically upgraded at each diversion node with respect to the service state at time τ of day *t*. For

example, information on disrupted lines allows such lines to be eliminated.

4.1.2. Experienced Path Utility. Given a *line service graph*, a travel strategy *ST* is defined through a *line hyperpath HP* from origin *o* to destination *d*, with a set of diversion nodes and a *diversion rule* dr_i, for each diversion node *i*, which determines the diversion link choice behaviour at that node. Hence a strategy *ST* will be indicated as *ST*[*HP*; **dr**], with **dr** the set of diversion rules dr_i. Note that on a service network, several strategies and therefore several relative hyperpaths can be used. Given a diversion rule and an objective function *Of*, the strategy $ST^*[HP^*, \boldsymbol{dr}]$ which optimises this function is the *optimal strategy conditional upon the diversion rule* **dr** *and the objective function Of*, with HP^* the relative optimal hyperpath.

As a result of random service occurrences and traveller's choices according to a diversion rule**dr**, each feasible path *k* from the origin to the destination has a certain probability of use and its *experienced path utility* $TU_k^{u,\tau,t}$ is a random variable.

Therefore, it can be assumed that travellers consider the average *ATU* of long-period experienced values of all random *TU* relative to all paths of strategy *ST*. Thus, the *subjective optimal strategy* ST^* is the strategy with maximum average experienced utility ATU^* perceived by the traveller.

4.1.3. Dynamic Travel Choices and Diversion Rule. We assume that a traveller *u*, in order to optimise his/her travel utility, applies the following dynamic travel behaviour: "*Given a master line hyperpath, at each diversion node an optimal diversion link is chosen (with the diversion rule reported below) and the relative path is used up to the next diversion node, where a new optimal diversion link is chosen and used.*"

The proposed diversion rule dr_i is composed as follows: given a master line hyperpath $MHP_{od}^{u,\tau,t}$, at diversion node *i* and time τ of day *t*, traveller *u* considers all the diversion links *il*, associates to each of them an *anticipated utility*, defined below, and chooses the diversion link il^* with maximum anticipated utility.

4.1.4. Diversion Link Anticipated Utility. Given a diversion node i and a diversion link il, the anticipated utility AU_{ij} is obtained by summing:

(i) the anticipated utility AU_k of the subpath from diversion node i up to the next diversion node w, including the diversion link il;

(ii) the *nodal* anticipated utility AU_w of the diversion node w up to the next nodes w'.

For example, the anticipated utility of link B-F of Figure 2 is given by the anticipated utility of subpath B-F plus the nodal anticipated utility of node F, which in turn is a function of the anticipated utility of subpaths F-E-D and F-G-D.

4.1.5. Anticipated Utility of Subpaths (AU_k). Given subpath k up to the next diversion node w, the *anticipated utility* AU_k at time τ of day t is a linear function of the vector $AX_k^{u,\tau,t}$ of its attributes AX, *anticipated* by traveller u at time τ of day t:

$$AU_k^{u,\tau,t} = \sum_j \eta_j^u \cdot AX_{k,j}^{u,\tau,t} \tag{2}$$

with η_j^u parameters of the utility function. In turn, the attributes AX anticipated by travellers are functions of path attributes forecasted (if any) by travellers and those forecasted by the information system:

$$AX_{k,j}^{u,\tau,t} = \xi^u \cdot FX_{k,j}^{\tau,t} + \left(1 - \xi^u\right) \cdot PX_{k,j}^{u,\tau,t}, \tag{3}$$

where

(i) $AX_{k,j}^{u,\tau,t}$ is the j-th *anticipated* attribute value at time τ of day t;

(ii) $FX_{k,j}^{\tau,t}$ is the j-th attribute value *forecasted by the information system*;

(iii) $PX_{k,j}^{u,\tau,t}$ is the value (if any) of j-th attribute forecasted by traveller u at time τ of day t (traveller forecasting process);

(iv) $\xi^u \in [0, 1]$ is the *weight* given by traveller u to the information provided, dependent on the traveller's compliance with the information system.

4.1.6. Traveller Forecasted Attributes of a Path. Assuming that travellers use an exponential smoothing forecasting method [23], the values $PX_{k,j}^{u,\tau,t}$ of the j-th attributes forecasted by traveller u at time τ of day t are assumed as

$$PX_{k,j}^{u,\tau,t} = \upsilon^u \cdot TX_{k,j}^{u,\tau,t-1} + \left(1 - \upsilon^u\right) \cdot PX_{k,j}^{u,\tau,t-1} \tag{4}$$

where

(i) $TX_{k,j}^{u,\tau,t-1}$ is the value of the j-th attribute experienced by traveller u at time τ of day t-1;

(ii) $PX_{k,j}^{u,\tau,t-1}$ is the value of the j-th attribute forecasted by traveller u, at time τ of day t-1;

(iii) $\upsilon^u \in [0, 1]$ is the weight given to attributes experienced on day t-1, depending on the memory process of traveller u.

4.1.7. Nodal Anticipated Utility of Next Diversion Node w (AU_w). The nodal anticipated utility AU_w at time τ of day t, of the diversion node w with G_w subpaths k' up to their next diversion nodes w', is obtained by travellers as a function of the anticipated utilities of these subpaths k':

$$AU_w^{u,\tau,t} = \sum_{k' \in G_w} p^{u,\tau,t}\left[k'\right] \cdot AU_{k'}^{u,\tau,t} \tag{5}$$

where $p^{u,\tau,t}[k']$ is the *perceived share* of using path k' at time τ in the past days and $AU_{k'}^{u,\tau,t}$ is the anticipated utility of subpath k' at time τ of day t. It is assumed that the values of shares $p^{u,\tau,t}[k']$ perceived by traveller u at time τ of day t are given by

$$p^{u,\tau,t}\left[k'\right] = \frac{\sum_{t'=1}^{t-1} \alpha_{k'}^{u,\tau,t'}}{\sum_{j \in G_w} \sum_{t'=1}^{t-1} \alpha_j^{u,\tau,t'}} \tag{6}$$

where

(i) $p^{u,\tau,t}[k']$ is, at time τ of day t, the perceived share of using path k';

(ii) $\alpha_{k'}^{u,\tau,t'}$ is the weight given by the traveller to path k' in relation to day t';

$$\alpha_{k'}^{\tau,t'} = \begin{cases} 0 & \text{if path } k' \text{ was not used at day } t' \\[2mm] \dfrac{\phi^u}{(t-t')} & \text{if path } k' \text{ was used at day } t', \end{cases} \tag{7}$$

with ϕ^u the parameter of the traveller's memory process.

In the learning process, travellers search for the optimal weights ξ which maximise the average experienced utility (ATU), as simulated in the application test of Section 4.2 below.

4.1.8. Example of Diversion Choices. As an example of a diversion choice, consider the choice at origin O of the first boarding stop in the master hyperpath of Figures 2 and 3. Traveller u is assumed:

(i) to identify, within the master line hyperpath, the set of diversion links with the root on O, in our case the links O-B and O-C;

(ii) to associate an anticipated utility AU_{ol} to each diversion link ol, in our case:

(a) for link O-C as the anticipated utility of path O-C-E-D;

(b) for link O-B as the sum of the anticipated utility of link O-B and nodal anticipated utility of node B, which considers the anticipated utilities of path B-G-D and the anticipated utility of path B-F plus the nodal utility of diversion node F.

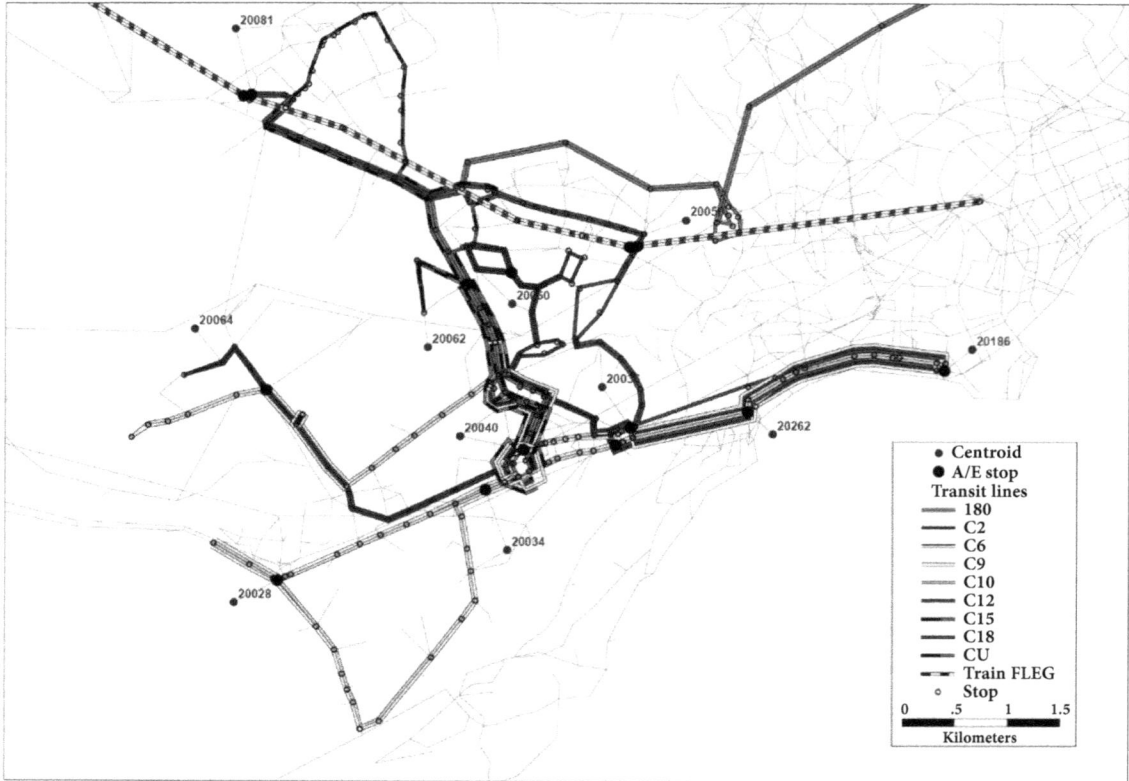

FIGURE 4: The application network [2].

(iii) to use the diversion link ol^* with the maximum anticipated utility AU^*.

Subsequently, at time τ', when traveller u is at the (first boarding or interchanging) stop i and a run r of a line belonging to the run master hyperpath arrives (as depicted in Figure 2), s(he) is assumed:

(i) to consider the diversion link il_r to board run r and the diversion link il_w to wait;

(ii) to associate an *anticipated* utility to each of the two above diversion links;

(iii) to compare the anticipated utilities of these diversion links;

(iv) to board run r if the anticipated utility associated with the link incorporating run r is greater than the maximum anticipated utility associated with waiting link il_w;

(v) if the traveller does not board run r, the process is reapplied when the next run arrives.

4.1.9. Model Parameter Estimation. The application of the presented model requires the knowledge of the following parameters:

(i) η_j are the parameters of the anticipated utility function AU_k,

(ii) $\xi^u \in [0, 1]$ is the weight given by travellers to the information provided,

(iii) $\nu^u \in [0, 1]$ is the weight given to attributes experienced on day t-1,

(iv) ϕ^u is the parameter of traveller's memory process of the perceived share of using path k'.

Parameters η_j can be obtained, for example, with standard stated-preference surveys and aggregate random utility model calibration. Parameters ν, ξ, and ϕ can be obtained applying a reverse assignment procedure [14], minimising the distance between measured alighting and boarding (or onboard) counts and those obtained through the model [24, 25].

4.2. An Application to a Real Network. An application of the proposed path choice modelling, with a unique subjective optimal strategy and the same parameters ν, ξ, and ϕ for all travellers, within the assignment model in DYBUSRT [2], was carried out for the same network as in the authors' other studies in the field of run-oriented transit assignment. The aim of the application was to assess how different values of parameters ξ and hence different combinations of the forecasted utilities of the traveller and information system affect expected values of average experienced utility ATU.

The service network (Figure 4) was obtained from the real service structure of the Fuorigrotta district in Naples (Italy), whose bus running time variation coefficients were appropriately modified for the purpose of the simulation. The

TABLE 1: Application results.

Running time cv	ξ	Average value of TU*	Running time cv	ξ	Average value of TU*
	0.00	-1343.1		0.00	-504.1
	0.25	-1058.0		**0.25**	**-410.9**
0.20	0.50	-992.2	0.75	0.50	-420.6
	0.75	**-713.7**		0.75	-521.8
	1.00	-978.8		1.00	-491.7
	0.00	-874.5		0.00	-477.9
	0.25	-704.3		**0.25**	**-376.7**
0.50	**0.50**	**-649.4**	1.00	0.50	-407.5
	0.75	-938.3		0.75	-495.5
	1.00	-973.8		1.00	-547.7

*The average experienced utility ATU is computed for 60 simulation days.

study area consists of 11 traffic zones served by 11 transit lines, which supply 245 runs (with an average of 6 runs per hour on each transit line) from 7:00am to 10:00am on a typical workday (day t).

As regards the master line hyperpath, according to the literature on choice set formation and as reviewed by Bovy [26], the master set of path alternatives was generated from the set of all available paths and then considering logical constraints to avoid loops, successive boarding of the same run or the use of opposite lines, and behavioural constraints to eliminate unrealistic alternatives in terms of maximum values of attributes, such as number of transfers, transfer time, access and egress times, and schedule delay. Combining the residual paths, a master line hyperpath from each origin o to each destination d was generated. Level-of-service attributes composing path utilities were calculated by using a diachronic graph, whose service subgraph consists of about 10,400 nodes and 20,100 links. The experienced path utility function is the same as that reported by Nuzzolo et al. [2].

The results entail the reproduction of an initial transient of about 60 days to set up the traveller's prior knowledge of path attributes and to reach an equilibrium state, followed by 30 replications of each simulation period, aiming to obtain statistically significant estimates of state variable expected values (i.e., confidence interval method with specified precision at 95%). Anticipated attributes are estimated assuming parameter ν equal to 0.3 [27, 28], while ϕ was hypothesized equal to 1.

The assignment algorithm is coded in C++ and data are managed with a Postgres 9.1 DBMS. As the programming code is optimised to use the latest technologies in the field of multicore CPU processing, simulation times strictly depend on the CPU architecture (i.e., number of cores and processors) and on the operating system. Referring to the above-mentioned three-hour morning period of a workday (i.e., 7:00am - 10:00am), simulation takes 35 seconds on a computer with an Intel Core 2 Duo 3.33GHz, 8Gb RAM, running on Mac-OSX. This time is reduced to 12 seconds if we use a computer equipped with two Intel Core i7 293 GHz, 16Gb RAM, running on MS-Windows 7.

Four different coefficient variations of bus running times were used to consider different levels of service unreliability.

The results (see Table 1) indicate that the weights used for combining the utilities in question strongly influence the average experienced utility and that the weights to use in order to minimise the experienced travel disutility strongly depend on the unreliability of the transit system. As expected, with increasing transit service unreliability and hence with increasing forecasting failures, the best overall performances are obtained with the use of a low ξ parameter, to give much more weight to personal than to system forecasted attribute values.

4.3. The Proposed Behavioural Framework from an MDPm Perspective. If the behavioural framework with the proposed diversion rule is applied, the subjective optimal strategy found at a diversion node can be considered as an approximate solution of a MDPm, where

(i) the master hyperpath is found by considering quite simple logical and behavioural constraints (see Figure 3);

(ii) the perceived shares of use of subpaths k' at time τ in previous days, from the diversion node w up to the next diversion node, are proxies for transition probabilities (see (6));

(iii) the anticipated utilities are proxies for expected rewards (see (2)). Indeed, the anticipated utilities are functions of the anticipated path attributes, given by a combination of the values forecasted by the information system and the values forecasted by travellers. The information-system forecasted attribute values, if obtained through statistical forecasting methods, are estimates of expected values. The traveller's forecasted attribute values are obtained through exponential smoothing methods, hence proxies of expected values. Thus the anticipated utilities can be assumed to be proxies of expected utilities;

(iv) the traveller, at each diversion node, considers as an action only that of choosing among all available diversion links. Referring to the example depicted in Figure 2, the state-action trees are simplified, as reported in Figure 5, where at node B the only possible action is a_{5+6} while at node F it is action a_{7+8}.

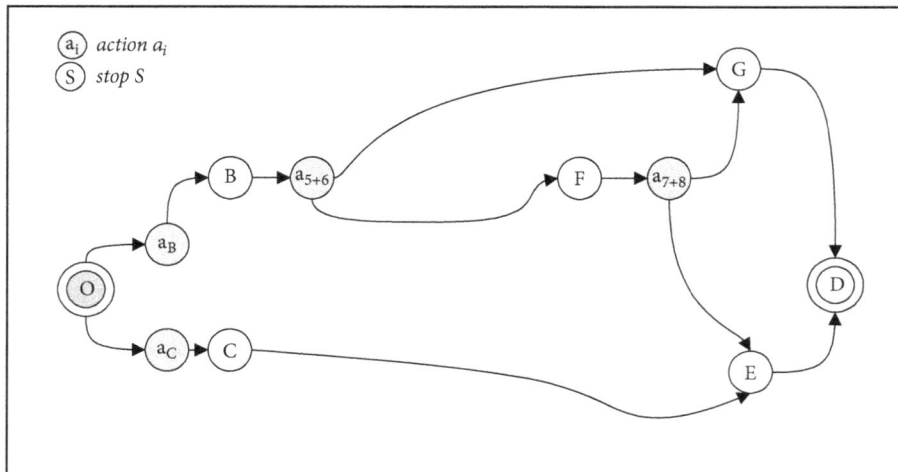

FIGURE 5: Example of reduced action tree of Figure 2.

5. Conclusions and Research Perspectives

This paper sought to overcome some limits of transit path choice modelling, especially that concerning the use of an objective optimal travel strategy for multiservice stochastic networks, instead of subjective strategies. A path choice model was therefore developed by using a dynamic subjective travel strategy. Further, the model was defined in the framework of a Markov decision problem. The optimal subjective strategy can be considered as the solution of a simplified MDPm with approximate transition probabilities and approximate expected rewards. It takes into account service occurrences and the information provided to travellers and applies a diversion rule that considers some of the travellers' cognitive limitations and simplifications.

Even if the proposed modelling framework requires several model parameters, the new opportunities resulting from the availability of a large quantity of data obtained from automated data collecting allow model parameter estimation and upgrading to be more easily achieved, for example, by using the reverse assignment method recalled in the paper. This same data availability helps to obtain new models of travel strategy generation for different categories of users, to be used as subjective travel strategies in assignment models. Therefore, the next steps in this research will be the setup and testing of an overall procedure, including inverse assignment parameter estimation, on the test network. In the near future, through a greater deployment of bidirectional communication between travellers and information centres, a suitable quantity of data will be available, making it possible, at least in theory, to calibrate not only individual model parameters, but also specific subjective strategy-based transit path choice models.

Further research should explore master line hyperpath modelling and the development of travel strategies within theories other than that of expected utility. In addition, the introduction of stochastic path choice models which take into account user perception errors and analyst modelling errors is another possible modelling improvement.

Conflicts of Interest

The authors declare that they have no conflicts of interest.

References

[1] O. Cats, H. N. Koutsopoulos, W. Burghout, and T. Toledo, "Effect of real-time transit information on dynamic path choice of passengers," *Transportation Research Record*, no. 2217, pp. 46–54, 2011.

[2] A. Nuzzolo, U. Crisalli, A. Comi, and L. Rosati, "A Mesoscopic Transit Assignment Model Including Real-Time Predictive Information on Crowding," in *Journal of Intelligent Transportation Systems*, vol. 20, no. 4, Taylor & Francis, 2016.

[3] M. Rieser, "Modeling Public Transport with MATSim," in *The Multi-Agent Transport Simulation MATSim*, A. Horni, Nagel. K., and K. W. Axhausen, Eds., License: CC-BY 4.0, pp. 105–110, Ubiquity Press, London, UK, 2016.

[4] O. Cats, "Dynamic Modelling of Transit Operations and Passenger Decisions [Ph.D. thesis]," KTH, School of Architecture and the Built Environment (ABE), Transport Science, Traffic and Logistics, Stockholm, Sweden, 2011.

[5] H. Spiess, *On optimal route choice strategies in transit networks*, Centre de Recherche sur les Transports, Université de Montréal, Montreal, Canada, 1983.

[6] J.-D. Schmöcker, H. Shimamoto, and F. Kurauchi, "Generation and calibration of transit hyperpaths," *Transportation Research Part C: Emerging Technologies*, vol. 36, pp. 406–418, 2013.

[7] A. Fonzone, J. D. Schm, F. Kurauchi et al., "Strategy Choice in Transit Networks," in *Proceedings of the Eastern Asia Society for Transportation Studies*, vol. 9, 2013.

[8] M. N. Hassan, T. H. Rashidi, S. T. Waller, N. Nassir, and M. Hickman, "Modeling transit user stop choice behavior: Do travelers strategize?" *Journal of Public Transportation*, vol. 19, no. 3, pp. 98–116, 2016.

[9] F. Kurauchi and J. D. Schmöcker, Eds., *Public Transport Planning with Smart Card Data*, CRC Press, Taylor & Francis Group, 2017.

[10] M. Hickman, "Transit Origin-Destination Estimation," in *Public Transport Planning with Smart Card Data*, F. Kurauchi and J. D. Schmöcker, Eds., CRC Press, Taylor & Francis Group, 2017.

[11] A. Comi and A. Nuzzolo, "A dynamic strategy-based path choice modelling for real-time transit simulation," in *Modelling Intelligent Multi-Modal Transit Systems*, A. Nuzzolo and W. H. K. Lam, Eds., pp. 152–173, CRC Press, Taylor & Francis Group, Boca Raton, Fla, USA, 2017.

[12] H. Spiess and M. Florian, "Optimal strategies: a new assignment model for transit networks," *Transportation Research Part B: Methodological*, vol. 23, no. 2, pp. 83–102, 1989.

[13] A. Nuzzolo and A. Comi, "Transit travel strategy as solution of a Markov decision problem: Theory and applications," in *Proceedings of the 5th IEEE International Conference on Models and Technologies for Intelligent Transportation Systems, MT-ITS 2017*, pp. 850–855, Italy, June 2017.

[14] A. Nuzzolo and W. Lam, "Real-time reverse dynamic assignment for multiservice transit systems," in *Modelling Intelligent Multi-modal Transit Systems*, A. Nuzzolo and W. H. K. Lam, Eds., CRC Press, 2017.

[15] S. Nguyen and S. Pallottino, "Equilibrium traffic assignment for large scale transit networks," *European Journal of Operational Research*, vol. 37, no. 2, pp. 176–186, 1988.

[16] A. Khani, M. Hickman, and H. Noh, "Trip-based path algorithms using the transit network hierarchy," *Networks and Spatial Economics*, vol. 15, no. 3, pp. 635–653, 2015.

[17] A. Nuzzolo and A. Comi, "Dynamic optimal travel strategies in intelligent stochastic transit networks," in *Transportmetrica A: Transport Science*, Taylor & Francis, 2018.

[18] M. L. Puterman, *Markov Decision Processes: Discrete Stochastic Dynamic Programming*, vol. 414, John Wiley & Sons, 2009.

[19] R. Bellman, "A Markovian decision process," Technical report, DTIC Document, 1957.

[20] G. Gentile, "Time-dependent Shortest Hyperpaths for Dynamic Routing on Transit Networks," in *Modelling Intelligenti Multi-Modal Transit Systems*, A. Nuzzolo and W. H. K. Lam, Eds., CRC Press, Taylor & Francis Group, Boca Raton, Fla, USA, 2017.

[21] M. Wahba, *MIcrosimulation Learning-based Approach to TRansit ASsignment [Ph.D. thesis]*, Graduate Department of Civil Engineering, University of Toronto, Toronto, Canada, 2008.

[22] T. Rambha, S. D. Boyles, and S. T. Waller, "Adaptive transit routing in stochastic time-dependent networks," *Transportation Science*, vol. 50, no. 3, pp. 1043–1059, 2016.

[23] W. A. Woodward, H. L. Gray, and A. C. Elliott, *Applied time series analysis with R*, CRC Press, Boca Raton, Fla, USA, 2nd edition, 2017.

[24] A. Nuzzolo, U. Crisalli, L. Rosati, and A. Ibeas, "STOP: A Short term Transit Occupancy Prediction tool for APTIS and real time transit management systems," in *Proceedings of the 2013 16th International IEEE Conference on Intelligent Transportation Systems: Intelligent Transportation Systems for All Modes, ITSC 2013*, pp. 1894–1899, Netherlands, October 2013.

[25] A. Nuzzolo and A. Comi, "Advanced public transport and intelligent transport systems: new modelling challenges," in *Transportmetrica A: Transport Science*, vol. 12, no. 8, pp. 674–699, 2016.

[26] P. H. L. Bovy, "On Modelling Route Choice Sets in Transportation Networks: A Synthesis," in *Transport Reviews: A Transnational Transdisciplinary Journal*, vol. 29, no. 1, pp. 43–68, Taylor & Francis, 2009.

[27] E. Cascetta, *Transportation Systems Analysis: Models and Applications*, Springer, 2009.

[28] G. E. Cantarella, "Day-to-day dynamic models for Intelligent Transportation Systems design and appraisal," *Transportation Research Part C: Emerging Technologies*, vol. 29, pp. 117–130, 2013.

15

Exploring the Intermodal Relationship between Taxi and Subway in Beijing, China

Shixiong Jiang⑩,[1] Wei Guan ⑩,[1] Zhengbing He,[2] and Liu Yang[1]

[1]*MOE Key Laboratory of Urban Transportation Complex System Theory and Technology, Beijing Jiaotong University, Beijing 100044, China*
[2]*College of Metropolitan Transportation, Beijing University of Technology, Beijing, China*

Correspondence should be addressed to Wei Guan; weig@bjtu.edu.cn

Academic Editor: Dongjoo Park

Taxi is an indispensable mode in the urban public transportation. Although many studies have explored the travel patterns of taxi trips, few have combined taxi and subway to reveal their intermodal relationship. To bridge the gap, this study utilized taxi's trajectory data to investigate its relationship with subway. Considering the multifaceted relationship between taxi and subway in operation, taxi trips are categorized into three types, namely, subway-competing, subway-extending, and subway-complementing taxi trips. The characteristics of each type of taxi trips reflect the specialties and their interactions with subway. The origin/destination distributions of taxi and subway trips are compared and analyzed. Furthermore, the supply and demand of taxi within the buffer zone of each subway station are analyzed to reflect the difficulty of hailing taxis. The negative binomial regression models are used to explore the relationship between taxi trips and subway ridership. The results show that there is a significantly positive correlation between taxi trips and subway ridership.

1. Introduction

Taxi is a flexible on-demand public transportation, which provides passengers door-to-door services without the requirement of private car ownership. In addition, taxi has the potential to satisfy the travel demand unmet by other modes of public transportation. However, the multimodal relationships between taxi and public transit, especially subway, have not attained enough attention. Understanding the relationship can assist in providing more satisfying transportation services and encouraging mode shift from automobile to subway.

A body of studies have explored the characteristics of taxi trips. Using 20 million trajectories with fine granularity collected from more than 10 thousand taxis in Beijing, the taxis' traveling displacements in urban areas were found to follow an exponential distribution [1]. Moreover, Wang et al. [2] indicated that the displacement distributions of taxi trips follow exponential laws in two displacement ranges, while the trip duration and interevent time distributions can be approximated by log-normal distributions. In addition,

there are some researches conducted from different aspects. Based on a preference travel survey, it was indicated that high population density in a user's residence area is related to longer journey durations [3]. The historical data were used to predict the number of vacant taxis in the given area and period with the prior probability distributions [4] and to predict demand distributions with respect to contexts of time, weather, and location [5]. Kamga et al. [6] found that taxi supply exhibits variations due to the decisions of taxi drivers, which are driven by both ridership levels and trip characteristics.

The pattern of subway ridership is also investigated by many researchers. Lin and Shin [7] found that daily ridership is positively affected by the floor-space area of the station areas, negatively affected by the percentage of four-way intersections, and insignificantly affected by mixed land use. Regarding travel impedance, Choi et al. [8] showed that subway travel time has the most influence on subway ridership, which represents an intrinsic property of trip length distribution. In London subway, Roth et al. [9] observed that intraurban movement patterns are strongly heterogeneous in terms of volume, and there is a polycentric structure composed of large flows organized around a limited number of activity centers. Similarly, Xu et al. [10] found a hierarchical urban polycentric structure composed of large concentrated flows at urban activity centers in Beijing Subway. Jiang et al. [11] observed that the displacements of subway trips follow the gamma distribution. The subway station ridership can be attributed to many factors, such as land use [8, 12], weather [13], and access mode [8, 14]. Jun et al. [15] found that the population and employment density, land use mix diversity, and intermodal connectivity all have a positive impact on subway ridership, but differ in their spatial ranges. Besides, the accessibility impacts the first/last mile transport connectivity to/from the major public transit lines [16]. Walking, cycling, and bus are widely adopted as access modes for the subway stations [17–19].

Due to the congestion and pollution caused by automobile, the government is promoting the modal shift from private cars to public transit. Using data from Xi'an, the metro transit's influences on the mobility instrument ownership [20] and private car driving [21] are examined. The cross-sectional analysis showed that metro is negatively associated with auto ownership whereas it has a positive association with bike ownership [20]. In addition, moving into metro neighborhoods is negatively associated with change in driving [21]. Huang et al. [21] concluded that metro development and the design of station-area neighborhoods have the potential to reduce driving, mitigate its impact on environment, and slow the growth of traffic congestion. Besides, the combination of taxi trips and subway ridership data can provide useful information to identify the underserved areas by public transport [22]. Examining the spatial relationship of taxi trips' origins/destinations and subway stations, Wang and Ross [23] categorized taxi trips into three types: transit-competing, transit-complementing, and transit-extending ones, to explore the inner interaction. Moreover, the geographically weighted regression was implemented to find that the urban form has a large impact on urban taxi ridership [24]. Medium income level was found to reduce the number of taxi trips at particular places and the accessibility to subways was positively associated with the taxi ridership. A high correlation between public transit ridership and taxi trips was explained by the direct demand for taxi service from major transit stations [25].

Except for the urban transport, the intercity traffic is an indispensable part for large cities. Air transport provides services in fixed terminals that require other modes, such as subway, taxi, and private car, to access them. The choice of access modes is influenced by many factors, including access time, access distance, demographic, and cost [26–28]. The quality of access mode has a significant impact on the service level of air travels, and taxi is an important mode to access airports [29]. The emerging taxi Global Positioning System (GPS) trip datasets provide the opportunity to extract travel patterns for a particular region [30]. The airport travel modal share for taxis ranges between 6% and 35% for major airports in the US, the UK, and Japan [29–31]. Spatial variables were

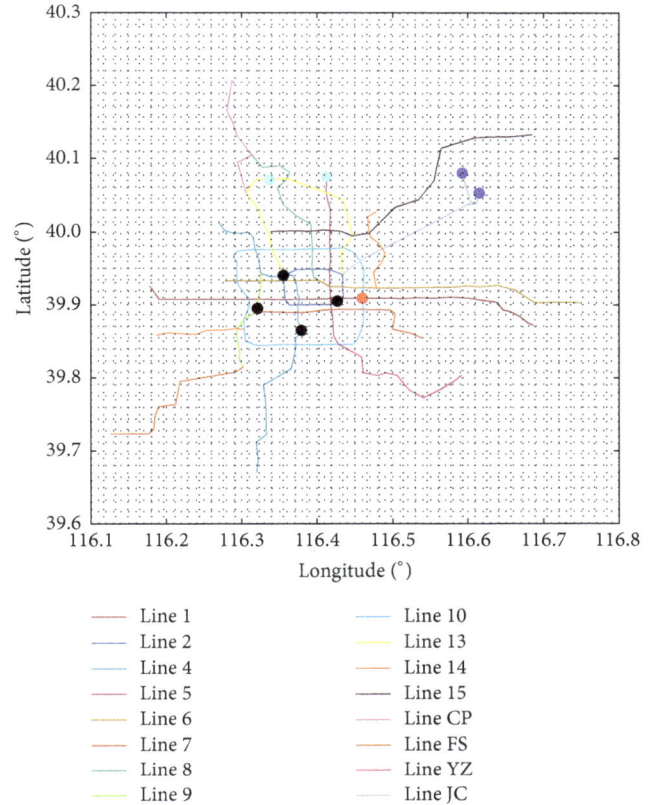

FIGURE 1: *Beijing Subway map and some important points of interest.* The blue point denotes the airport terminal, the black point is the railway station, the red point denotes the center business district, and the cyan point is the large residential area.

found to have the highest impact on the taxi drivers' airport pick-up decisions, followed by temporal and environmental variables [30]. In general, passengers will accept longer travel time and distance for the access travel than those for the egress travel [32].

Although many studies have investigated the travel patterns of different modes, the intermodal relationship has been rarely considered. To fill the gap, this paper aims to explore the relationship between taxi and subway. According to its relationship with subway, the taxi trips are categorized as subway-competing, subway-extending, and subway-complementing taxi trips. Besides, the arrival and departure taxi trips within the buffer zone of subway stations are analyzed, which are compared with the subway ridership. Furthermore, the negative binomial regression models are utilized to explore the relationship between taxi trips and subway ridership.

2. Study Area and Data Description

The study area is the urban part of Beijing, China (see Figure 1). There were about 67.5 thousand taxis in Beijing, which completed 0.67 billion trips (7.20% of all trips) in 2014 [33]. As for the subway system, there were 328 stations and

527 tracks in operation, with an average daily ridership of 9.28 million in 2014 [33].

To explore the relationship between taxi and subway, four datasets are collected, including the taxi GPS data, the subway ridership, the locations of subway stations and airports, and the points of interest (POIs) dataset. The GPS data generated by more than 46,000 taxis [34, 35] during January 12th–18th, 2015, were attained [33], which contained more than 2.9 million taxi trips. The interval of the GPS data is at around 1 minute. Each record contains the location (longitude and latitude), instantaneous velocity, operational status, and so on. The objective time period was chosen since there were no major holidays or unusual weather events. A ridership dataset was available from the Automatic Fare Collection (AFC) system in Beijing Subway during December 30th-31st, 2014. It is a 268 × 268 origin-destination (OD) matrix, which reflects the flows between any pair of subway stations. The two datasets were collected during the common winter days without special events, although they are not in the same time period because of their different sources. In addition, the locations (longitude and latitude) of airports (Terminal 2 (T2) and Terminal 3 (T3) in Beijing Capital International Airport and Beijing Nanyuan Airport), subway stations, and the POIs dataset are collected from a map website [36] to explore how taxi trips interact with these fixed service points.

The connection between taxi and bus is not taken into consideration in this study. There are around 1020 bus routes in Beijing by 2016. Bus stops are widely spread around the city. Taking bus stops into the classification would have classified most of taxi trips as 'bus-competing' ones. Although the majority of taxi trips could be completed by bus for the large extent of the bus system, multiple transfers and long out-of-vehicle time of bus travel make it inferior to taxi [23].

3. Methodology

To explore the relationship between taxi and subway, the taxi trips are categorized into three types, denoted as subway-competing, subway-extending, and subway-complementing taxi trips [23]. And the three types are defined as follows:

(1) Subway-competing taxi trips refer to the trips which can be replaced by taking subway and acceptable walking.

(2) Subway-extending taxi trips provide connections from/to subway stations.

(3) Subway-complementing taxi trips satisfy the travels which cannot be served by subway due to the fixed routes and operation times.

Although different subway stations own various operation times, it is assumed that 5:30–23:00 is the operation time for all subway stations to simplify the problem [37].

3.1. Identify Subway-Competing Taxi Trips. Transit-competing taxi trips are trips which can be served by the subway system [23]. Subway competes with taxi only when both origin and destination of a trip are within the catchment area of subway stations because passengers can only enter

and exit the subway system in stations. The walking distance threshold is widely set as 800 m to access subway stations [38, 39]. Taking the road structure into consideration, the tolerant walking displacement is set as 600 m. In this study, a taxi trip is classified as subway-competing type when its pick-up and drop-off locations have subway stations within 600 m. Additionally, the taxi trip should happen during the subway operation time. The logical process to identify the three types of taxi trips is presented in Figure 2.

3.2. Identify Subway-Extending Taxi Trips. Transit-extending taxi trips provide connection from/to subway [23]. It requires identifying taxi trips that are most likely to connect with subway stations. For such trips, taxis would originate or end close to the entrances/exits of subway stations to transfer conveniently, while another trip end is beyond the catchment area of any subway station. With regard to the size of subway stations, the maximal displacement between the entrance/exit and the station center is set as 300 m. Based on the specific situations, two subtypes of subway-extending taxi trips are defined. The first subtype is the absolute subway-extending taxi trips that provide service between the subway stations and outskirts without subway service. The second subtype can be partly served by subway, and the other part also needs taxis to complete. The rules to select the subtype of subway-extending taxi trips are as follows: (1) only one end of a trip has a subway station within 300 m, and the other end of the trip does not have any subway station within 2 km [23]; (2) the trip happens during the subway's operation time.

The requirements for the second subtype of subway-extending taxi trips are different. The rules are as follows: (1) only one end of a trip has a subway station within 600 m, and the other end of the trip does not have any subway station within 2 km [23]; (2) the trip happens during the subway's operation time; (3) the displacement between the assumed entering and exiting subway stations is more than 2 km.

3.3. Identify Subway-Complementing Taxi Trips. Transit-complementing taxi trips serve the travel demand that cannot be satisfied or connected by the subway system in space or time [23]. After excluding the previous two types of taxi trips, the remained taxi trips belong to this type.

3.4. Identify Arrival and Departure Trips around Stations. As subway provides high quality travel service, subway stations' adjacent areas receive great development, which generates a large number of travel demands. To explore the taxi's supply and demand around subway stations, the area within 2 km from a subway station is regarded as its buffer zone. The buffer zone's originating/ending taxi trips reflect the corresponding demand/supply. The gap between the supply and demand shows the difficulty of hailing a taxi around a subway station.

In addition, taxi trips related to T3 are identified. In general, most taxis enter T3 to pick up and drop off passengers. With regard to its size, the threshold is set as 400 m to identify the related trips. If the pick-up point of a taxi trip is within the catchment area of T3, the taxi trip would be marked as

TABLE 1: Summary of descriptive statistics of variables adopted in the negative binomial regression models.

Variable	Mean	SD	Min	Max
Subway ridership	76459.5	65428.9	1731	381867
Automobile services	44.8	65.9	0	475
Life facilities	188.4	124.8	5	655
Recreation facilities	53.5	35.1	2	213
Medical facilities	48.4	30.7	1	153
Accommodations	34.2	27.3	1	180
Scenic spots	14.3	27.2	0	237
Government agencies	87.6	68.9	2	378
Transport hubs	121.1	70.1	11	367
Finance facilities	55.7	46.2	0	311
Companies	88.1	73.7	1	402

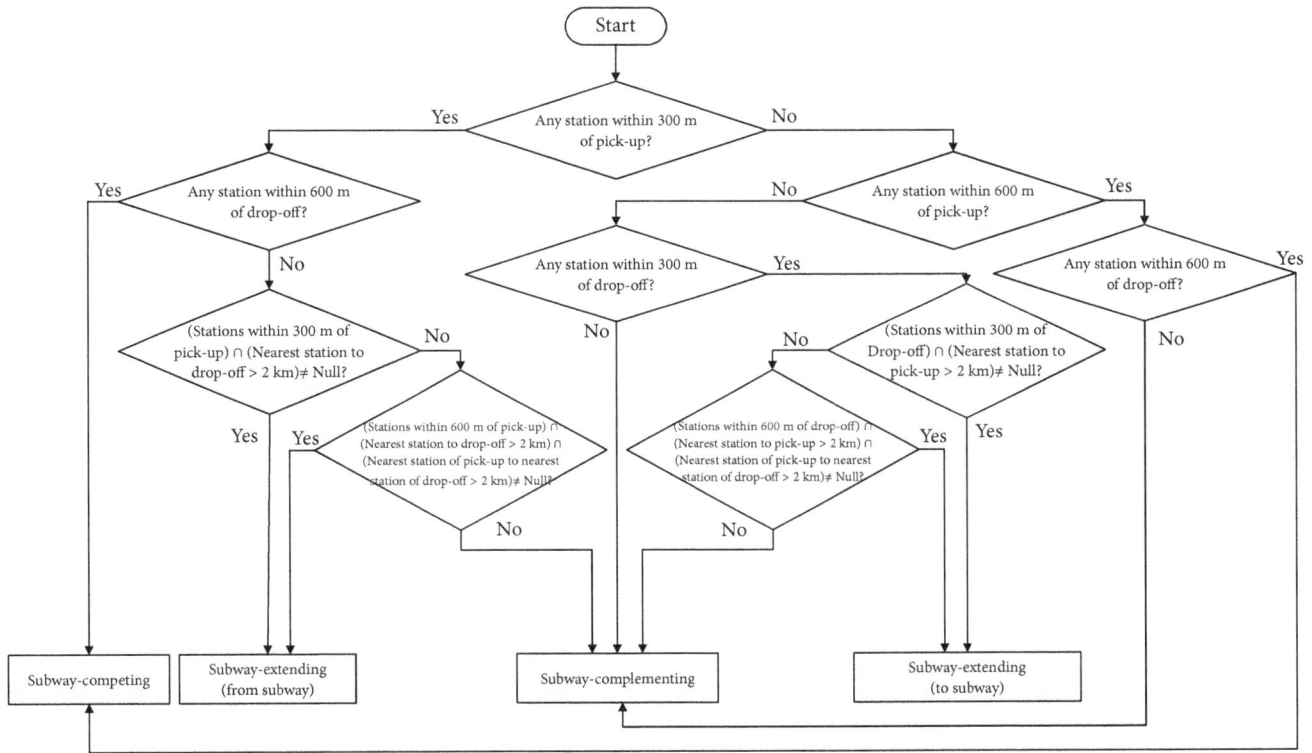

FIGURE 2: Logical flow chart to select subway-competing, subway-extending, and subway-complementing taxi trips.

a trip from T3. Similarly, a taxi trip that drops off within the catchment area of T3 is marked as a trip to T3.

3.5. Negative Binomial Regression Model. To investigate the effect of contributing factors on the taxi ridership and to assess the effect of subway ridership in particular, the negative binomial regression models were fitted [40]. The taxi trips in the subway station buffers, subway-competing taxi trips, and subway-extending taxi trips are explored, respectively. The independent variables used in this study include subway ridership, automobile services, life facilities, recreation facilities, accommodations, scenic spots, government agencies, transport hubs, finance facilities, and companies (see Table 1).

4. Results

4.1. Characteristics of Taxi Trips. Among the 2,918,143 taxi trips, there are 19.32% for subway-competing, 4.58% for subway-extending, and 76.10% for subway-complementing taxi trips (see Table 2), which indicates that many travels cannot be satisfied by the subway system. The responding proportions are quite different from those in New York City [23], in which the transit-competing taxi trips took up 58.50% of all taxi trips.

4.1.1. Travel Distance. The travel distance distributions of the three types of taxi trips are different (see Figure 3). Overall, the average travel distance is 9.59 km for all taxi trips, and

TABLE 2: Characteristics of three types of taxi trips.

Characteristic	Subway-competing	Subway-extending	Subway-complementing
Average travel distance (km)	9.61	14.26	9.30
Average travel time (min)	22.23	26.99	20.33
Count	563,851	133,546	2,220,746
Percentage (%)	19.32	4.58	76.10

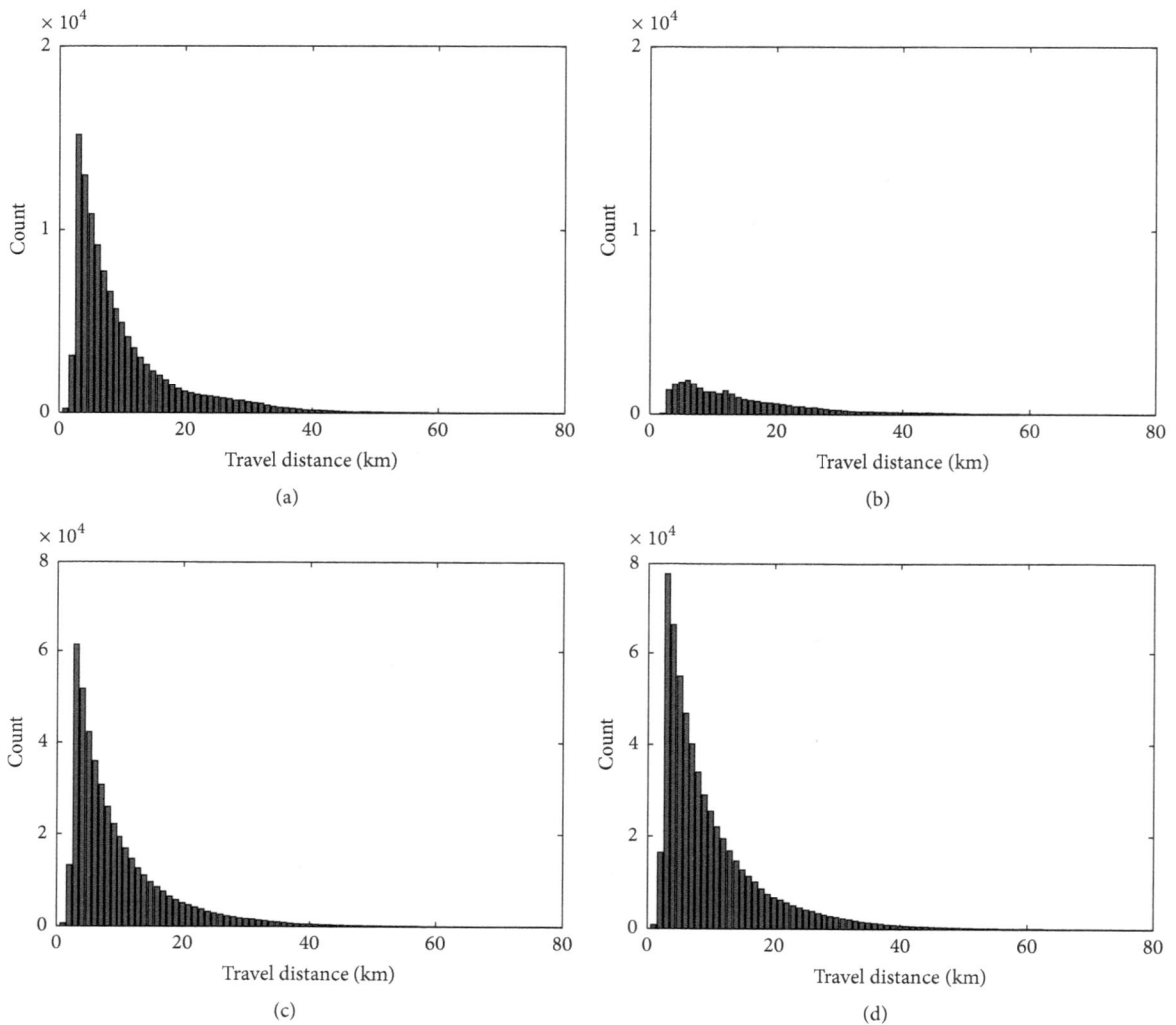

FIGURE 3: *The travel distance distributions of taxi trips.* (a) Subway-competing taxi trips. (b) Subway-extending taxi trips. (c) Subway-complementing taxi trips. (d) All taxi trips.

89.6% of taxi trips are less than 20 km. For the subway-competing and subway-complementing taxi trips, 67.8% and 68.3% of trips are within 10 km, respectively. Besides, a 20-km travel distance covers 88.9% of subway-competing taxi trips and 90.5% of subway-complementing ones. However, the travel distance is apparently longer for the subway-extending taxi trips. There are only 45.9% of subway-extending taxi trips within 10 km and 77.8% of those within 20 km, which implies that they are more likely to relate to peripheries.

4.1.2. Spatial Distribution. The pick-ups and drop-offs spatial distributions of the three types of taxi trips are shown in Figures 4, 5, and 6, respectively. The numbers of pick-ups and drop-offs are mapped at the grid cell level (0.01° *latitude* × 0.01° *longitude*). Subway-competing taxi trips are concentrated on the urban area, while subway-extending taxi trips tend to spread in the peripheries. Subway-complementing taxi trips are widely spread in spatial scale.

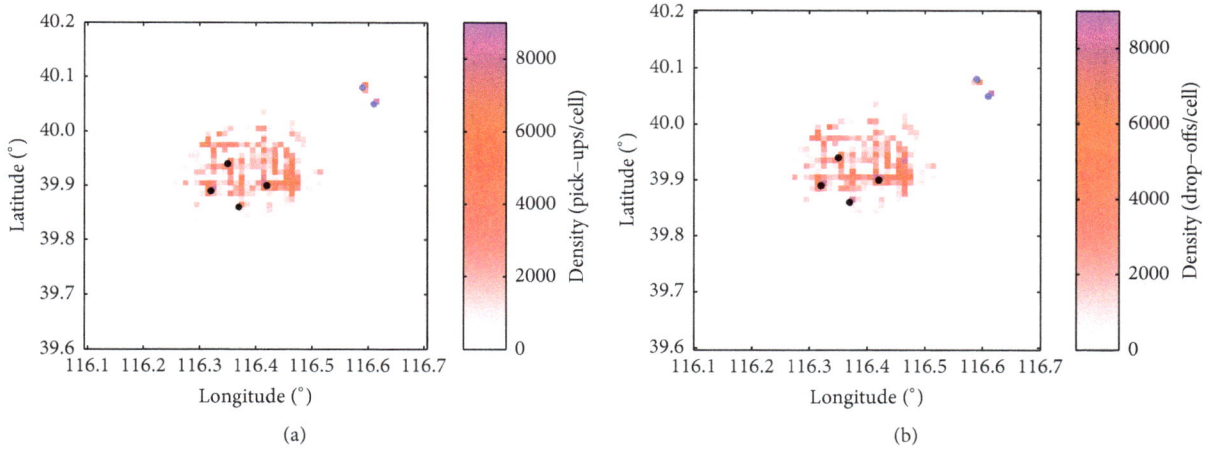

FIGURE 4: *The spatial distributions of subway-competing taxi trips.* (a) Pick-ups. (b) Drop-offs.

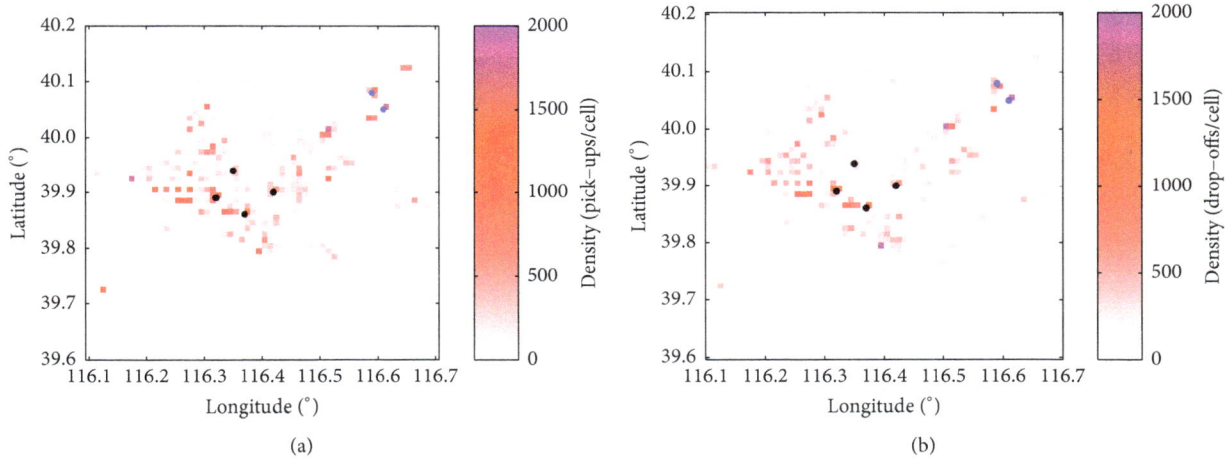

FIGURE 5: *The spatial distributions of subway-extending taxi trips.* (a) Pick-ups. (b) Drop-offs.

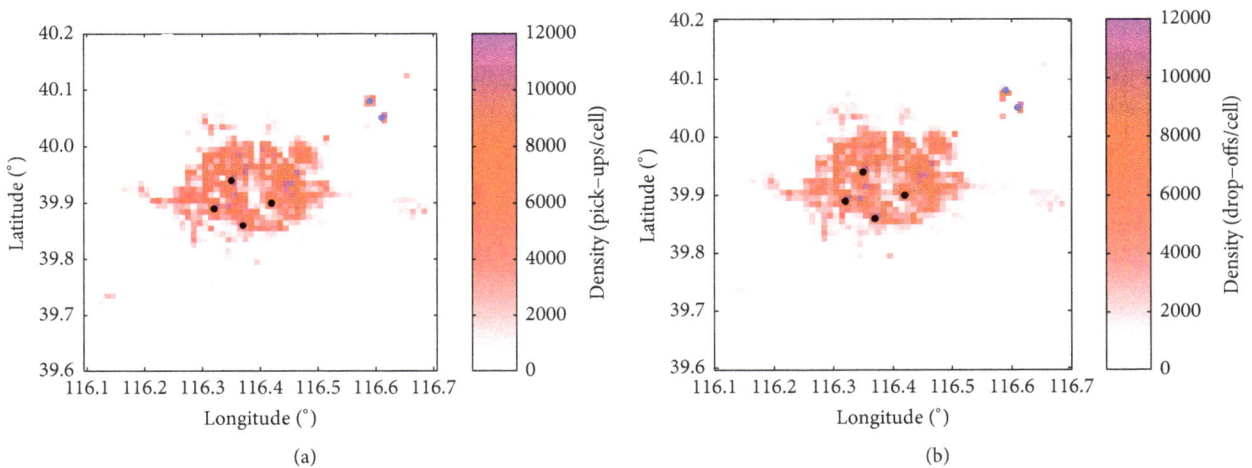

FIGURE 6: *The spatial distributions of subway-complementing taxi trips.* (a) Pick-ups. (b) Drop-offs.

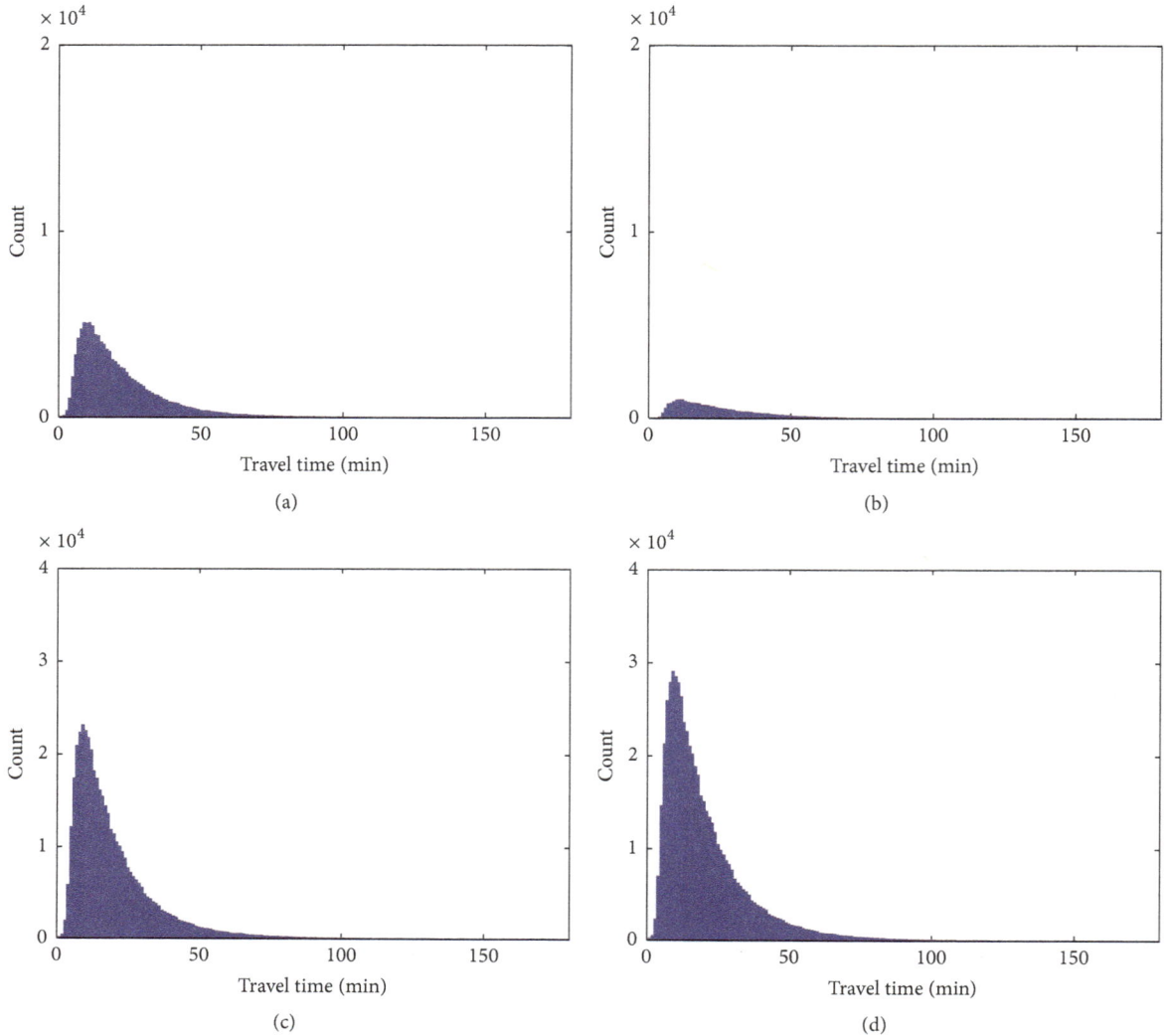

FIGURE 7: *The travel time distributions of taxi trips.* (a) Subway-competing taxi trips. (b) Subway-extending taxi trips. (c) Subway-complementing taxi trips. (d) All taxi trips.

4.1.3. Travel Time. The travel time is also an important indicator for both passengers and taxi drivers. The distributions of travel times are plotted in Figure 7. Overall, the travel time averages at 21.00 min and 96.7% of taxi trips are less than 60 min. In detail, there are 78.7%, 69.3%, and 82.6% of trips within 30 min for subway-competing, subway-expending, and subway-complementing taxi trips, respectively. However, the three types have similar percentages around 95% within 60 min.

Combining the travel distance and travel time, the average speed can be computed. The average speed of subway-extending taxi trips reaches 33.49 km/h, which is significantly higher than the other two types. The average speeds are only 26.86 km/h and 29.11 km/h for subway-competing and subway-complementing taxi trips, respectively.

4.1.4. Departure Time. The departure time distributions of taxi trips are presented in Figure 8. Generally, over 91.0%

of taxi trips happen from 8:00 to 24:00, with two peaks at around 10:00 and 14:00, and the bottom is at 4:00 in the early morning. Because subway does not operate all day, subway-competing and subway-extending taxi trips are limited in the operation time. The subway-competing taxi trips reach the peak at about 13:00.

The average travel distance and average travel time by time of day are plotted in Figure 9. The three types of taxi trips show a similar trend, while the average travel distance of subway-extending taxi trips is significantly higher than other two types. The average distance has a peak at around 6:00 and a bottom at the midnight. However, the average travel time does not show higher value. The average travel distance is normal in the morning and evening peak hours, while the average travel time shows apparent peaks during peak hours, which might be attributed to the traffic congestion.

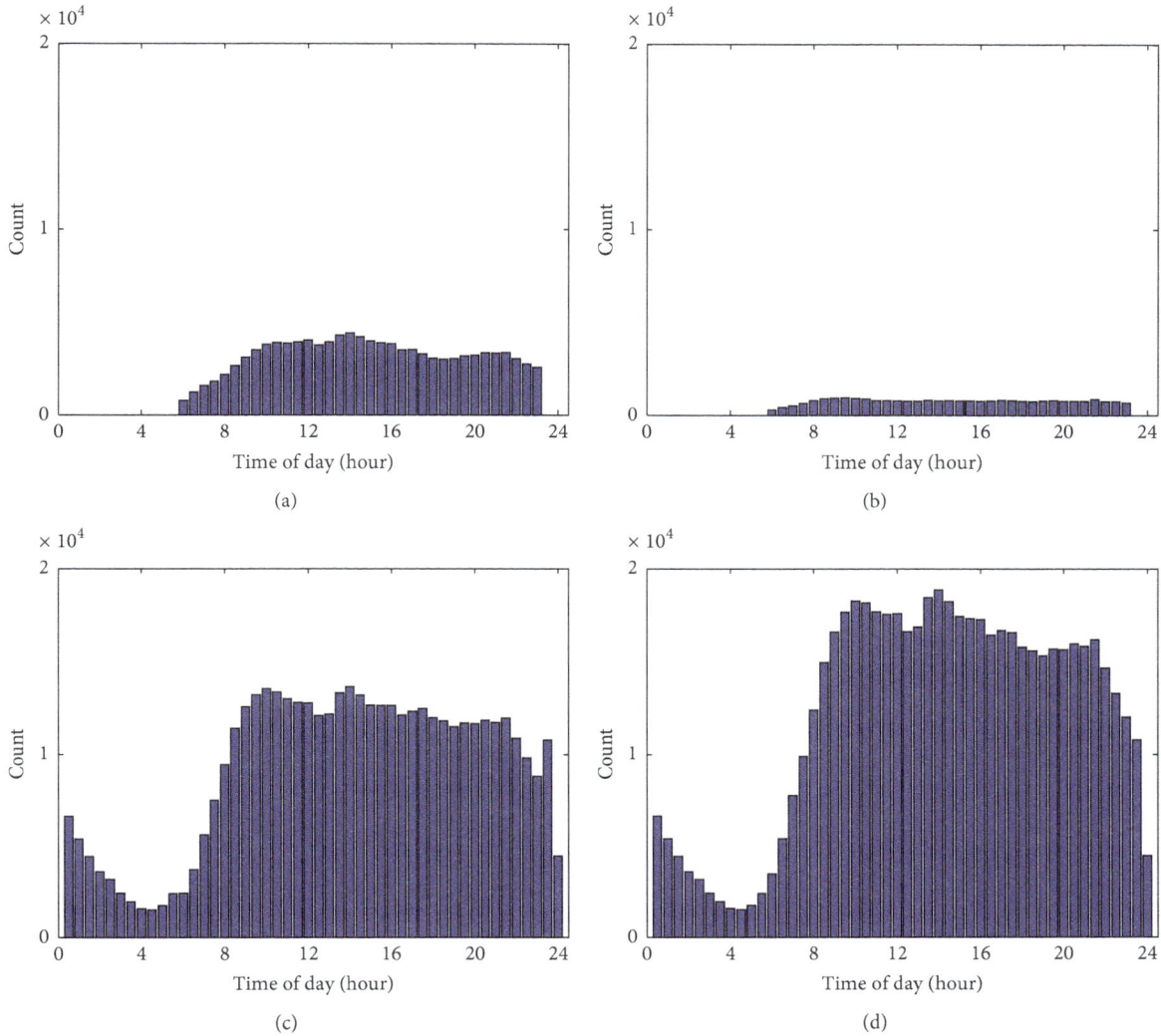

FIGURE 8: *The departure time distributions of taxi trips.* (a) Subway-competing taxi trips. (b) Subway-extending taxi trips. (c) Subway-complementing taxi trips. (d) All taxi trips.

4.2. Spatial Distribution of Arrival and Departure Trips

4.2.1. Subway Ridership Spatial Distribution.
The spatial distributions of subway ridership are presented in Figure 10. There are some stations with apparently higher ridership, with most being in the downtown area and some in the peripheral areas. According to Figure 10, the arrival and departure trips show a similar spatial distribution, with the Spearman correlation of 0.990 ($p < 0.01$), because the subway trips are more likely to be commuting ones, which would be round trips in a day. To explore the difference between the arrival and departure ridership, the gap between arrival and departure is plotted in Figure 11. Most subway stations with a large gap (more arrival than departure) are near the railway stations and airports. On the contrary, many subway stations in residential areas have less arrivals than departures.

4.2.2. Spatial Distribution of Taxi Trips in Buffer Zones of Subway Stations.
The arrival and departure trips of taxi in the buffer zones are plotted in Figure 12. When a taxi arrives its destination, it will be vacant, which means the supply of taxi. Similarly, the originating taxi trip implies the demand for taxi. Generally, the arrival and departure of taxi trips also show similar spatial distribution, with the Spearman correlation of 0.988 ($p < 0.01$). The high pick-up and drop-off zones are located in the urban areas, and they are more dispersed than the subway trip distribution. Besides, there are also some exceptions. For example, T2 and T3 also show a high value of pick-ups and drop-offs.

The buffer zones of subway stations have developed rapidly and there are a large amount of travel demands. The gap between supply and demand of each buffer zone can represent the difficulty of hailing taxis (see Figure 13). The over-demand areas are mainly located in the north central part of the city. On the other hand, there are also some over-supply areas spreading around the suburbs. Particularly, the buffer zones of airports and railway stations are the most over-supply areas.

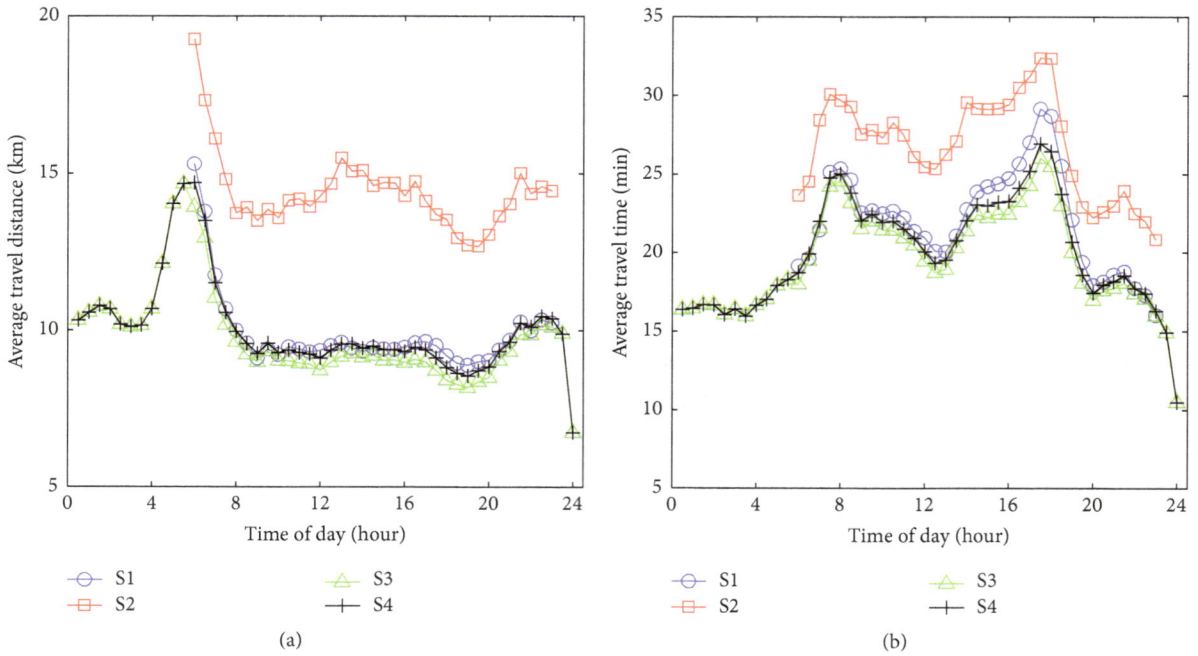

FIGURE 9: *The average travel distance and average travel time by time of day.* (a) The average travel distance. (b) The average travel time. (S1: subway-competing taxi trips; S2: subway-extending taxi trips; S3: subway-complementing taxi trips; S4: all taxi trips.)

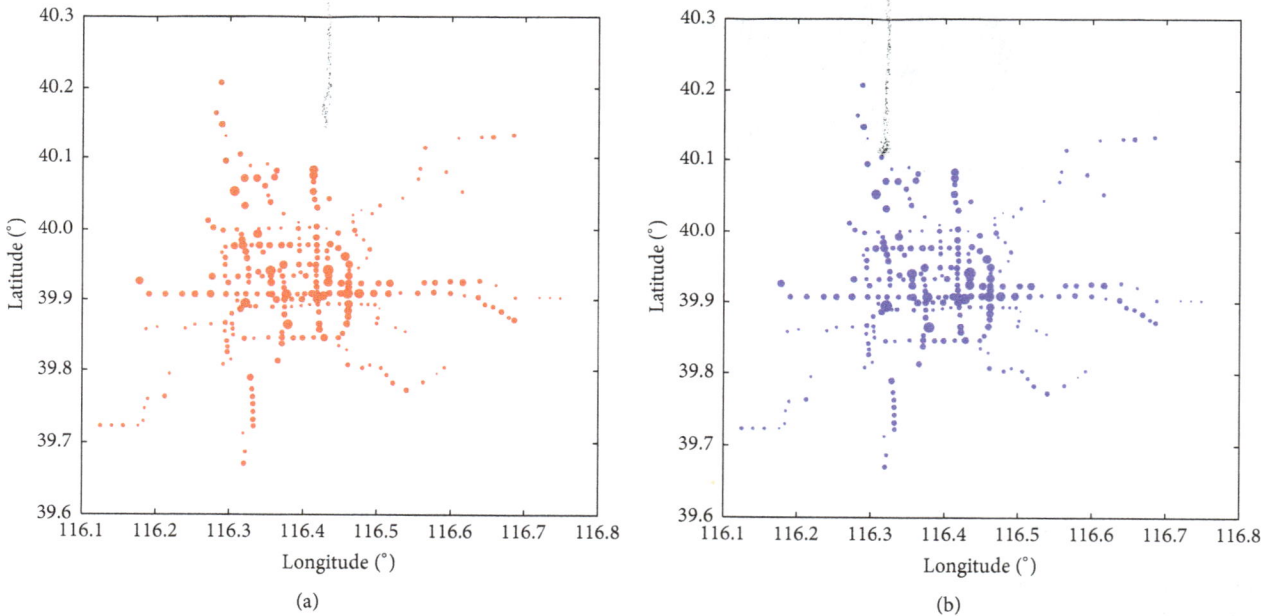

FIGURE 10: *The arrival and departure of subway trips.* (a) Arrival of subway trips. (b) Departure of subway trips.

According to Figures 10 and 12, the ridership spatial distributions are different between subway and taxi. The subway system has some stations with extremely high ridership, while the taxi trips are more dispersed. It is interesting to find that there are much more arrival trips than departure ones by subway and taxi in both railway stations and airports. Besides, it is necessary to consider the ridership patterns of subway and taxi in time of day.

4.2.3. Subway and Taxi Trips in T3. For large cities, there are a great quantity of travelers adopting subway and taxi to access airports. Understanding the travel patterns of these trips can help to provide more appropriate access/egress services. Taxi trips to airports and those from airports are identified according to the pick-up and drop-off locations. The arrival and departure of taxi and subway trips related to T3 are presented in Figure 14. There is an apparent peak at 7:00 in the

TABLE 3: Characteristics of taxi trips connecting to T3.

Characteristic	From T3	To T3
Average travel distance (km)	29.15	26.83
Average travel time (min)	41.40	35.01
Count	22,903	25,706

TABLE 4: Results of the negative binomial regression models for subway-competing taxi trips and taxi trips in buffer zones of subway stations.

Parameter	Model 1: subway-competing taxi trips			Model 2: taxi trips in buffer zones		
	Coefficient	Standard error	p-value	Coefficient	Standard error	p-value
Intercept	7.02**	0.185	0.000	8.47**	0.135	0.000
Subway ridership	5.39e-06**	1.04e-06	0.000	2.84e-06**	6.54e-07	0.000
Automobile services	-4.17e-03**	8.77e-04	0.000	-1.37e-03*	5.49e-04	0.013
Life facilities	-6.55e-03**	9.43e-04	0.000	-3.47e-03**	6.28e-04	0.000
Recreation facilities	7.58e-03**	2.23e-03	0.001	5.83e-03**	1.55e-03	0.000
Accommodations	7.28e-03*	2.99e-03	0.015			
Medical facilities				5.96e-03**	2.03e-03	0.003
Government agencies	4.80e-03**	1.40e-03	0.001	2.53e-03**	9.21e-04	0.006
Transport hubs	3.20e-03+	1.65e-03	0.052	4.47e-03**	1.19e-03	0.000
Finance facilities	1.37e-03**	3.21e-03	0.000	6.01e-03**	1.53e-03	0.000
Companies	-2.66e-03*	1.31e-03	0.043			

Note. **: $p < 0.01$, *: $p < 0.05$, +: $p < 0.10$.

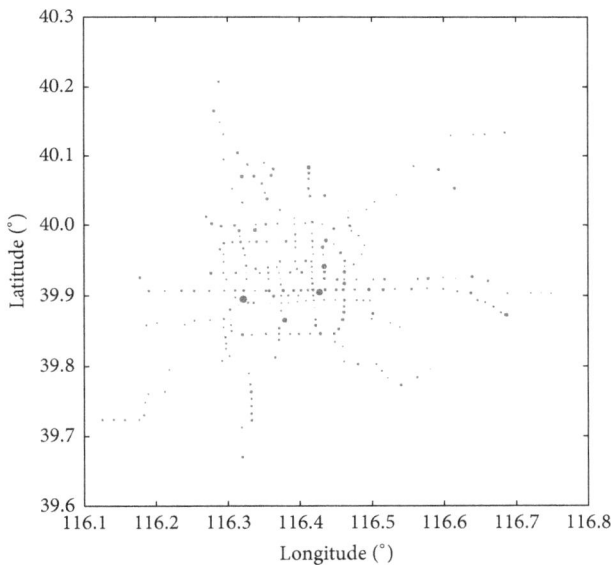

FIGURE 11: *The gap between arrival and departure subway trips. The blue circles denote that the arrival trips are more than departure trips. The red circles denote that the arrival trips are less than departure trips.*

morning for taxi trips to T3. There are a stable amount of taxi trips originating from airports during 10:00-24:00 and some during 0:00-2:00, while the departure trips are much lower in other time periods of a day, but for taxi trips to airports, it is few in the midnight. From 8:00 to 16:00, the number of taxi trips is stable, then it decreases slowly. As shown in Figure 14, it is unbalanced between the supply of taxi and the demand

for taxi at T3. From 4:00 to 16:00, there is more supply than demand, while it is opposite in other time periods. It is found that it is easier for taxi drivers to pick up passengers at T3 after 16:00.

Comparing the taxi and subway trips at T3, the taxi and subway trips from T3 have a similar overall trend, low volumes in the morning and high volumes in the afternoon. However, the taxi trips to T3 see the peak at 6:00, while the subway trips to T3 reach the peak at 19:00. To conclude, the taxi and subway trips from T3 have similar patterns, while those trips to T3 have different patterns.

In all, there are more taxi trips to airports than those from T3 (see Table 3). As the airports are rather far from the downtown and there are few taxi demands in adjacent areas, about 2,800 taxis would go back to the downtown without any passenger in one day. The average travel distance and average travel time of taxi trips to T3 are slightly less than those of taxi trips from T3.

4.3. Relationship between Taxi Trips and Subway Ridership. The variables used in the negative binomial regression models were selected carefully to avoid collinearity, and different models are selected according to the Akaike information criterion (AIC) and log-likelihood values [22, 23]. Table 4 shows the results of the negative binomial regression models for estimating the number of subway-competing taxi trips and taxi trips in the buffer zones of subway stations. The results show that subway ridership, recreation facilities, government agencies, transport hubs, and finance facilities are positively associated with taxi trips in both models. A high correlation between subway ridership and taxi trips is consistent with the previous study [25] and can be explained by the direct

(a)

(b)

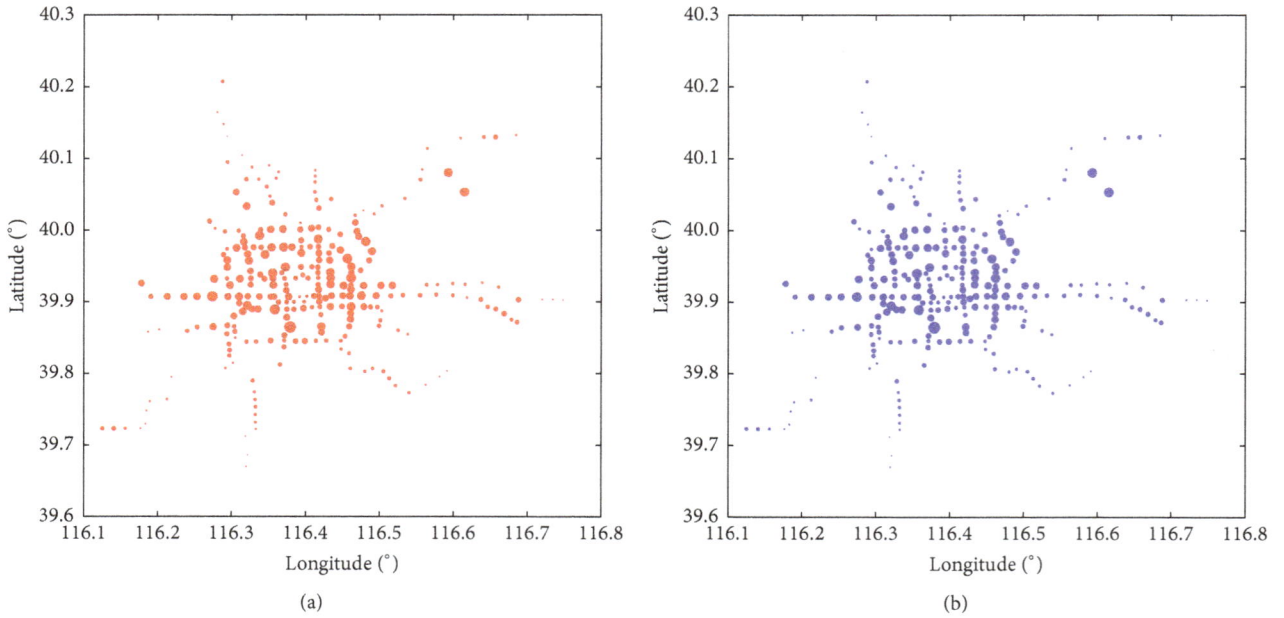

FIGURE 12: *The arrival and departure of taxi trips in buffer zones of subway stations.* (a) Arrival of taxi trips. (b) Departure of taxi trips.

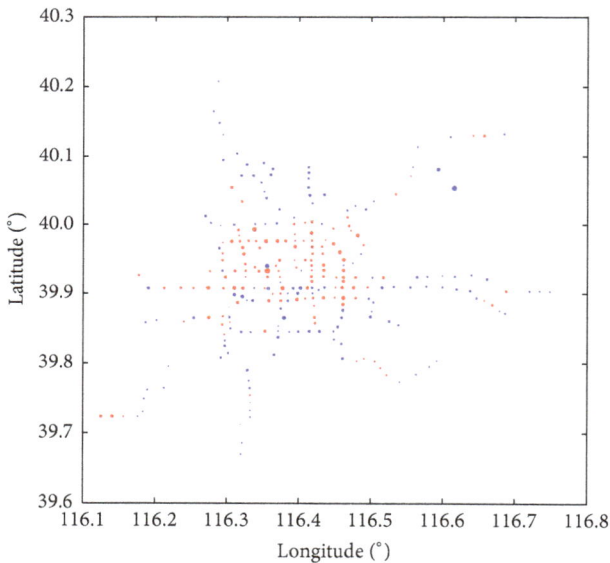

FIGURE 13: *The gap between arrival and departure taxi trips in buffer zones of subway stations.* The blue circles denote that the arrival trips are more than departure trips. The red circles denote that the arrival trips are less than departure trips.

FIGURE 14: *The taxi and subway trips from/to T3 by time of day.*

demand for taxi service from subway stations [22, 25]. The coefficients of subway ridership, recreation facilities, and government agencies are larger in model 1 than those in model 2. However, automobile services and life facilities variables have negative coefficients in model 2 and larger negative coefficients in model 1. Besides, accommodations variable has a positive coefficient in model 1, which is more related to visitors from other cities. It suggests that

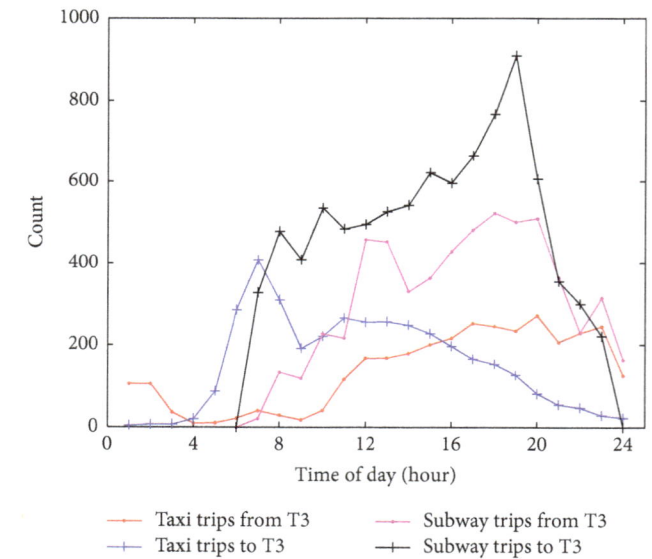

visitors prefer taxi than subway. The companies variable has a negative coefficient in model 1, which is likely to relate to daily commuters who tend to take subway. Furthermore, medical facilities variable is positively associated with taxi trips in station buffers in model 2, which indicates that passengers who are not well in health prefer to take a taxi to get better services.

In addition, the subway-extending taxi trips are also explored by the negative binomial regression model. Three variables associated with the subway-extending taxi trips are presented in Table 5. Subway ridership and transport hubs

TABLE 5: Results of the negative binomial regression model for subway-extending taxi trips.

| Parameter | Model 3: subway-extending taxi trips | | |
	Coefficient	Standard error	p-value
Intercept	4.58**	0.258	0.000
Subway ridership	3.14e-06*	1.33e-06	0.018
Scenic spots	-6.19e-03**	2.34e-03	0.008
Transport hubs	6.42e-03**	1.56e-03	0.000

Note. **: $p < 0.01$, *: $p < 0.05$, +: $p < 0.10$.

variables have positive coefficients, and scenic spots variable has a negative coefficient.

5. Discussion and Conclusions

Taxi service has been an important component of public transportation for a long time, which is the most flexible mode of public transportation as it can provide door-to-door services. The subway system has developed rapidly and provides fast mass transit services in Beijing. As a result, there exist the competition and corporation between taxi and subway at the same time.

This study investigated the relationship between taxi and subway, with the taxi GPS data, subway ridership data, locations of subway stations, and POIs data. Based on the relationship between taxi and subway, the taxi trips are categorized into three types, subway-competing, subway-extending, and subway-complementing ones. The results showed that the subway-complementing taxi trips take the majority of all trips, which indicates that many trips cannot be satisfied by the subway system alone and the subway network requires further enhancement. In detail, the average travel distances of subway-extending taxi trips are apparently longer than the other two types, which is different from the finding of Wang and Ross [23]. Besides, the average speed of subway-extending taxi trip is significantly higher than the other types. As for spatial distributions, the subway-competing taxi trips are serving the urban area with high subway availability while the subway-extending taxi trips are more likely to serve the peripheral areas. There are different functions of taxi trips and the demand for each function would change with the development of the subway system.

Subway has received rapid development for its sustainability. The efficiency and attractiveness of subway system rely not only on the operation of the subway network but also on the performance of access modes. As an access mode, taxi can increase the access distance for the subway station than walk and bicycle. However, the cost is the major constraint of taxi as an access mode. Fortunately, the emerging of ridesharing helps to decrease the fare as well as the difficulty of hailing a taxi. In addition, the bike-sharing offers a cheap and convenient opportunity to access subway stations, which makes subway more competitive when compared with private cars [41].

The spatial distributions of origin and destination are significantly correlated for both subway and taxi trips. However, the two modes show great differences in spatial distribution. According to the supply and demand of taxis within the buffer zones of subway stations, it was found that there are over-demand of taxis in the urban area and over-supply in the peripheries. Another interesting finding is that both the subway and the taxi have more arrival trips than departure trips in the regions of railway stations and airports.

The results of different negative binomial regression models for different situations indicate that subway ridership plays an important role in explaining differences in taxi trips. In addition, there are several variables positively associated with taxi trips in model 1 and model 2, including recreation facilities, government agencies, and transport hubs. However, automobile services and life facilities have significantly negative correlations with taxi trips. In model 1 and model 2, most of independent variables are similar and they have particular variables, respectively. Accommodations variable has a positive correlation with subway-competing taxi trips in model 1, while companies variable has a negative correlation. In model 2, medical facilities variable is positively correlated with taxi trips in buffer zones of subway stations.

Understanding the relationship between taxi and subway can help to integrate different modes to satisfy passengers' travel demand better. In the future works, comparison between the travel times by taxi and subway with the same origin and destination will be made to identify their respective advantaged travels. However, there are some limitations in this study. The sociodemographic characteristics of passengers are not taken into consideration, which might affect travel demands. The classification method lacks validation, and the distance thresholds utilized in this study require further calibration with more actual data.

Conflicts of Interest

The authors declare that there are no conflicts of interest regarding the publication of this paper.

Acknowledgments

The authors would like to acknowledge the financial support for this study provided by the National Natural Science Foundation of China (nos. 71621001, 91746201) and Fundamental Research Funds for Central Universities (no. 2017YJS112).

References

[1] X. Liang, X. Zheng, W. Lv, T. Zhu, and K. Xu, "The scaling of human mobility by taxis is exponential," *Physica A: Statistical*

Mechanics and its Applications, vol. 391, no. 5, pp. 2135–2144, 2012.

[2] W. Wang, L. Pan, N. Yuan, S. Zhang, and D. Liu, "A comparative analysis of intra-city human mobility by taxi," *Physica A: Statistical Mechanics and its Applications*, vol. 420, pp. 134–147, 2015.

[3] Z. Christoforou, C. Milioti, P. Dionysia, and M. G. Karlaftis, "Investigation of taxi travel time characteristics," in *Proceedings of the Transportation Research Board 90th Annual Meeting*, 2011.

[4] S. Phithakkitnukoon, M. Veloso, C. Bento, A. Biderman, and C. Ratti, "Taxi-aware map: identifying and predicting vacant taxis in the city," in *Proceedings of the International Joint Conference on Ambient Intelligence*, pp. 86–95, 2010.

[5] H.-W. Chang, Y.-C. Tai, and Y.-J. J. Hsu, "Context-aware taxi demand hotspots prediction," *International Journal of Business Intelligence and Data Mining*, vol. 5, no. 1, pp. 3–18, 2010.

[6] C. Kamga, M. A. Yazici, and A. Singhal, "Hailing in the rain: Temporal and weather-related variations in taxi ridership and taxi demand-supply equilibrium," in *Proceedings of the Transportation Research Board Annual Meeting*, 2013.

[7] J.-J. Lin and T.-Y. Shin, "Does transit-oriented development affect metro ridership?: Evidence from Taipei, Taiwan," *Transportation Research Record Journal of the Transportation Research Board*, vol. 2063, no. 2063, pp. 149–158, 2008.

[8] J. Choi, Y. J. Lee, T. Kim, and K. Sohn, "An analysis of Metro ridership at the station-to-station level in Seoul," *Transportation*, vol. 39, no. 3, pp. 705–722, 2012.

[9] C. Roth, S. M. Kang, M. Batty, and M. Barthélemy, "Structure of urban movements: polycentric activity and entangled hierarchical flows," *PLoS ONE*, vol. 6, no. 1, Article ID e15923, 2011.

[10] Q. Xu, B. Mao, and Y. Bai, "Network structure of subway passenger flows," *Journal of Statistical Mechanics Theory & Experiment*, vol. 2016, no. 3, Article ID 033404, 2016.

[11] S. Jiang, W. Guan, W. Zhang, X. Chen, and L. Yang, "Human mobility in space from three modes of public transportation," *Physica A Statistical Mechanics Its Applications*, vol. 483, 2017.

[12] J. Zhao, W. Deng, Y. Song, and Y. Zhu, "What influences Metro station ridership in China? Insights from Nanjing," *Cities*, vol. 35, pp. 114–124, 2013.

[13] J. Li, X. Li, D. Chen, and L. Godding, "Assessment of metro ridership fluctuation caused by weather conditions in Asian context: Using archived weather and ridership data in Nanjing," *Journal of Transport Geography*, vol. 66, 2018.

[14] J. Zhao, W. Deng, Y. Song, and Y. Zhu, "Analysis of Metro ridership at station level and station-to-station level in Nanjing: An approach based on direct demand models," *Transportation*, vol. 41, no. 1, pp. 133–155, 2014.

[15] M.-J. Jun, K. Choi, J.-E. Jeong, K.-H. Kwon, and H.-J. Kim, "Land use characteristics of subway catchment areas and their influence on subway ridership in Seoul," *Journal of Transport Geography*, vol. 48, pp. 30–40, 2015.

[16] S. Chandra, M. E. Bari, P. C. Devarasetty, and S. Vadali, "Accessibility evaluations of feeder transit services," *Transportation Research Part A: Policy and Practice*, vol. 52, pp. 47–63, 2013.

[17] G. Sun, J. Zacharias, B. Ma, and N. M. Oreskovic, "How do metro stations integrate with walking environments? Results from walking access within three types of built environment in Beijing," *Cities*, vol. 56, pp. 91–98, 2016.

[18] Y.-H. Cheng and Y.-C. Lin, "Expanding the effect of metro station service coverage by incorporating a public bicycle sharing system," *International Journal of Sustainable Transportation*, pp. 1–12, 2018.

[19] J. Xiong, Z. He, W. Guan, and B. Ran, "Optimal timetable development for community shuttle network with metro stations," *Transportation Research Part C: Emerging Technologies*, vol. 60, pp. 540–565, 2015.

[20] X. Huang, X. Cao, J. Yin, and X. Cao, "Effects of metro transit on the ownership of mobility instruments in Xi'an, China," *Transportation Research Part D: Transport and Environment*, vol. 52, pp. 495–505, 2017.

[21] X. Huang, X. Cao, J. Yin, and X. Cao, "Can metro transit reduce driving? Evidence from Xi'an, China," *Transport Policy*, 2018.

[22] H. H. Hochmair, "Spatiotemporal Pattern Analysis of Taxi Trips in New York City," *Transportation Research Record*, vol. 2542, pp. 45–56, 2016.

[23] F. Wang and C. L. Ross, "New potential for multimodal connection: exploring the relationship between taxi and transit in New York City (NYC)," *Transportation*, pp. 1–22, 2017.

[24] X. Qian and S. V. Ukkusuri, "Spatial variation of the urban taxi ridership using GPS data," *Applied Geography*, vol. 59, pp. 31–42, 2015.

[25] L. Kattan, A. De Barros, and S. C. Wirasinghe, "Analysis of work trips made by taxi in Canadian cities," *Journal of Advanced Transportation*, vol. 44, no. 1, pp. 11–18, 2010.

[26] Y.-C. Chang, "Factors affecting airport access mode choice for elderly air passengers," *Transportation Research Part E: Logistics and Transportation Review*, vol. 57, pp. 105–112, 2013.

[27] G. Akar, "Ground access to airports, case study: Port Columbus International Airport," *Journal of Air Transport Management*, vol. 30, pp. 25–31, 2013.

[28] I. Gokasar and G. Gunay, "Mode choice behavior modeling of ground access to airports: A case study in Istanbul, Turkey," *Journal of Air Transport Management*, vol. 59, pp. 1–7, 2017.

[29] Z. Ma, M. Urbanek, M. A. Pardo, J. Y. Chow, and X. Lai, "Spatial welfare effects of shared taxi operating policies for first mile airport access," *International Journal of Transportation Science and Technology*, vol. 6, no. 4, pp. 301–315, 2017.

[30] M. Anil Yazici, C. Kamga, and A. Singhal, "Modeling taxi drivers' decisions for improving airport ground access: John F. Kennedy airport case," *Transportation Research Part A: Policy and Practice*, vol. 91, pp. 48–60, 2016.

[31] M. L. Tam, M. L. Tam, and W. H. K. Lam, "Analysis of airport access mode choice: a case study in Hong Kong," *Journal of the Eastern Asia Society for Transportation Studies*, vol. 6, pp. 708–723, 2005.

[32] M. Givoni and P. Rietveld, "The access journey to the railway station and its role in passengers' satisfaction with rail travel," *Transport Policy*, vol. 14, no. 5, pp. 357–365, 2007.

[33] BJTRC, *Beijing Transport Annual Report (2015)*, 2017, http://www.bjtrc.org.cn/.

[34] Z. He, L. Zheng, P. Chen, and W. Guan, "Mapping to cells: A simple method to extract traffic dynamics from probe vehicle data," *Computer-Aided Civil and Infrastructure Engineering*, vol. 32, no. 3, pp. 252–267, 2017.

[35] Z. He and L. Zheng, "Visualizing Traffic Dynamics Based on Floating Car Data," *Journal of Transportation Engineering, Part A: Systems*, vol. 143, no. 5, 2017.

[36] Amap, *The coordinates picker system in Amap*, 2017, http://lbs.amap.com/console/show/picker.

[37] S. Beijing, *The adjacent station distance of Beijing Subway*, 2017, http://www.bjsubway.com/station/zjgls/.

[38] D. B. Hess, "Access to public transit and its influence on ridership for older adults in two U.S. cities," *Journal of Transport and Land Use*, vol. 2, no. 1, pp. 3–27, 2009.

[39] A. M. El-Geneidy, P. R. Tétreault, and J. Surprenant-Legault, "Pedestrian access to transit: Identifying redundancies and gaps using a variable service area analysis," in *Proceedings of the Transportation Research Board Meeting*, 2010.

[40] H. Eftekhar, L. Creemers, and M. Cools, "Effect of Traveler's Nationality on Daily Travel Time Expenditure Using Zero-Inflated Negative Binomial Regression Models: Results from Belgian National Household travel Survey," *Transportation Research Record*, vol. 2565, pp. 65–77, 2016.

[41] S. Jiang, W. Guan, Z. Wang, and l. Yang, "Improving the accessibility of metro system by using bike-metro integration," *Working paper*, 2018.

16

Improving Bus Operations through Integrated Dynamic Holding Control and Schedule Optimization

Shuozhi Liu ⓘ,[1] **Xia Luo ⓘ,**[1] **and Peter J. Jin**[2]

[1]*School of Transportation and Logistics, Southwest Jiaotong University, 111 Second Ring Road Beiyiduan, Chengdu 610031, China*
[2]*Department of Civil and Environmental Engineering, Rutgers, the State University of New Jersey, CoRE 613, 96 Frelinghuysen Road, Piscataway, NJ 08854-8018, USA*

Correspondence should be addressed to Xia Luo; xia.luo@263.net

Academic Editor: Antonino Vitetta

Bus bunching can lead to unreliable bus services if not controlled properly. Passengers will suffer from the uncertainty of travel time and the excessive waiting time. Existing dynamic holding strategies to address bus bunching have two major limitations. First, existing models often rely on large slack time to ensure the validity of the underlying model. Such large slack time can significantly reduce the bus operation efficiency by increasing the overall route travel times. Second, the existing holding strategies rarely consider the impact on the schedule planning. Undesirable results such as bus overloading issues arise when the bus fleet size is limited. This paper explores analytically the relationship between the slack time and the effect of holding control. The optimal slack time determined based on the derived relationship is found to be ten times smaller than in previous models based on numerical simulation results. An optimization model is developed with passenger-orient objective function in terms of travel cost and constraints such as fleet size limit, layover time at terminals, and other schedule planning factors. The optimal choice of control stops, control parameters, and slack time can be achieved by solving the optimization. The proposed model is validated with a case study established based on field data collected from Chengdu, China. The numerical simulation uses the field passenger demand, bus average travel time, travel time variance of road segments, and signal timings. Results show that the proposed model significantly reduce passengers average travel time compared with existing methods.

1. Introduction

Maintaining the reliability and efficiency of bus services is critical to ensure their competitiveness and popularity over the use of private cars. The ideal situation is that all buses can keep up to the schedule and travel evenly along the bus route at a predetermined headway (headway is defined as the time gap between two consecutive buses). In reality, however, due to the existence of various perturbations, it is difficult for buses to keep their headways to the designated value. Eventually, some buses bunch up together and start to travel in pairs. This phenomenon is referred to as bus bunching. Fundamentally, the bus bunching phenomenon is triggered by external randomness and amplified by bus system's volatile nature. In a stochastic traffic environment, buses travel in different speeds and cater to different number of waiting passengers because of external randomness (e.g., changing

traffic conditions, different driving behaviors, and fluctuating passenger demands), which causes buses' headways to deviate from the target value. Subsequently, due to the volatility of bus system, any headway deviation tends to increase over time and result in bunching phenomenon at last. This volatile nature is firstly explained by Newell [1]: due to the fact the number of waiting passengers at a stop is proportional to bus headway, a delayed bus with larger headway has to dwell at the stop for a longer time to collect passengers, resulting in a further vehicle delay. Meanwhile, the following bus has fewer passenger to serve and travels relatively faster. Consequently, the process of headway deviation will continuously accelerate under positive feedback loop and lead to bunching phenomenon at last.

Bus bunching is unexpected because it increases passengers' waiting time, leads to crowding conditions in slow buses, and makes passengers' travel time unpredictable. There are mainly two reasons for the increment of waiting time. First, as illustrated by Welding [2], the average waiting time of all passengers increases with the increment of headway deviation, which is named as ordinary waiting time (OWT) in this paper. Second, due to bus capacity limit, some slow buses may become overloaded at some stops, and parts of the waiting passengers will suffer a much longer waiting time to wait for the next bus. We call it as extra waiting time (EWT). Various bus control strategies are proposed to mitigate bus bunching, such as transit signal priority, bus speed regulation, stop-skipping strategy, and etc., of which, holding control strategy is supposed to be one of the most basic and effective one.

Traditionally, transit agencies insert slack times into bus schedules, and require all buses to depart on time at control stops (Barnett [3], Rossetti and Turitto [4]). However, the slack times required are usually very huge and will significantly reduce the efficiency of bus operation. Nowadays, with the rapid development of automated vehicle location (AVL), automated passenger counter (APC) and smartcard payment in transit system, agencies are able to monitor passenger demands, traffic conditions, and passenger loads more efficiently, and then control bus holdings adaptively in real time (Hanaoka [5]). Basically, adaptive holding control strategies can be categorized as optimization-based holding control and dynamic holding control.

Optimization-based holding control use control laws to minimize certain cost functions. Existing methods differ in different components of the optimization model.

(i) Scenarios: Sánchez-Martínez et al. [6] consider dynamic running times and passenger demands. Wu et al. [7] propose optimal holding control considering vehicle overtaking and distributed passenger boarding behavior. Sánchez-Martínez [8] et al. formulate event-driven holding control to adapt buses to the expected changes in running times and demand during events.

(ii) Objective functions: Barnett [3] and Zhao et al. [9] consider the passenger waiting time. Hall et al. [10], Delgado et al. [11] take transfer time into account when measuring the waiting time. Asgharzadeh & Shafahi [12] minimize passengers' total travel time which consist of passengers' ordinary waiting time, extra waiting time when the bus is at full capacity, and the waiting time of onboard passengers.

(iii) Constraints: Delgado et al. [13] consider the vehicle capacity constraint; while Eberlein et al. [14] consider the safe headway constraint.

(iv) Solution algorithms: Cortés et al. [15] propose generic algorithms to expedite model solving process. Koffman [16] developed a simulation model to evaluate the control effect of various control methods and solved the optimal control scheme accordingly. Chen et al. [17] present a multiagent reinforcement-learning framework to optimization bus operations in real time. Yu et al. [18] develop a SVM model to predict bus travel time and dwell time and minimize passengers' waiting time with the improved holding strategy.

Optimization-based holding control may face computational issues when solving their complicated formulations for the optimal solutions. Furthermore, the agencies can only observed the performance (e.g., accuracy and reliability) after

the implementation of the control strategies and may lead to unforeseeable control effect.

Dynamic holding control strategies apply negative feedback control to make bus system self-regulating to reduce headway deviation or schedule deviation. By communicating with control center, buses can continuously collect real-time data about the deviation of headway and schedule of its own and surrounding vehicles and dynamically determine their holding time to create a negative force to counteract those deviations. The advantages of dynamic holding control include the following: (1) all control parameters can be calibrated offline with archived bus AVL data and smartcard data and do not need to be calibrated in real time. (2) Each bus can dynamically generate its holding plan locally and efficiently without complex optimizations. (3) Performance measures like the schedule deviation, headway deviation, average waiting time and average in-vehicle travel time can be predicted beforehand.

Dynamic holding strategy is usually headway-based or schedule-based. Fu & Yang [19] investigate two different holding control models. The first model generates holding time based on the headway to the proceeding bus. The second model uses both preceding and following headway to determine the holding time. Daganzo [20] developed a forward-looking method to keep bus headways adhering to a predefined target headway. Holding time for buses is dynamically determined based on the deviation of their forward headways from targeted headway, where buses with smaller headways will be assigned with longer holding times, and vice versa. The method of convolution is introduced to simplify the modeling process, which is then widely used for the modeling of dynamic holding by other researchers. Daganzo & Pilachowski [21] further improved the control performance by proposing a two-way-looking control model. Bus cruising speed is continuously adjusted according to both forward headway and backward headway. Results showed that this two-way-looking method managed to maintain headway stability with faster bus travel speed than in forward-looking method. Bartholdi & Eisenstein [22] proposed a backward-looking control model that did not require predefined target headway and the information of passenger demands. This method has been proved to be effective in low demand by a field test. As pointed out by Xuan et al. [23], the above three headway-based methods can only maintain buses with stable headways, but do not ensure the adhesion to bus schedules. Xuan et al. [23] proposed a general holding control model and proved that all previous dynamic control methods are special forms of this one. In order to simplify the formulation and calibration, a one-parameter version called "simple control strategy" was developed. The "simple control strategy" can maintain both headway adherence and schedule adherence with relative smaller slack times than the previous methods. Liang et al. [24] developed a zero-slack version of two-way-looking control model. Simulation results shown that this control model further reduced passengers average travel time compared with Daganzo & Pilachowski [21]'s method. Zhang & Lo [25] propose a two-way-looking self-equalizing control method for both deterministic and stochastic running times. It is proved that the proposed control method keeps bus headway self-equalized under deterministic travel time and reduces the variance of headway to a certain value when travel time is stochastic.

Other researchers further investigate the above-mentioned dynamic holding methods to make them suitable for specific scenarios. Argote-Cabanero et al. [26] generalized Xuan et al.'s [23] "simple control strategy" into multiline systems by using both dynamic holding control and en-route driver guidance. The proposed control strategy applies to bus systems that mix with headway-based and schedule-based bus lines. Control effects are tested by both simulation and field study. Nesheli & Ceder [27] use real-time operational tactics to increase the actual occurrence of synchronized transfers and therefore reduce the magnitude and uncertainty of passengers' travel time. Estrada et al. [28] present dynamic cruising speed control methods and signal priority timing strategies to regulate bus operations. The proposed methods are based on the basic control logics of Daganzo [20]'s holding control strategy, but further consider the vehicle capacity constraints.

Studies are also focused on solving some key problems in implementing holding controls, such as the method of choosing control stops, trip time and optimal slack. Eberlein et al. [14] develop a deterministic quadratic program in rolling horizon scheme to formulate dynamic holding, and then test the holding model by simulations. Results suggest that holding buses at the first stop is of the highest efficiency, since even dispatching headways help to slow down the growth of headway variation along the route. Oort et al. [29] study how the choice of trip time, location and amount of control stops affect the reliability and efficiency of long-headway bus services. Simulation shows that a good combination of optimal trip time value and well selected control points can significantly reduce the additional travel time, where the optimal trip time usually ranges from 30-60 percentile value. In addition, bus route with two control stops achieves better control effects than that of one, but further increasing the number of control stops do not significantly improve the effect. Fu & Yang [19] develop both forward-looking and two-way-looking holding strategies. With simulation analysis, they find that usually two control stops are need to ensure better control effect, one at the terminal and the other at a high-demand stop near the middle of the route. By using $D/G/c$ queue model, Zhao et al. [30] generate a theoretical model that address the optimization problems of slack time for a schedule-based bus route with 1 bus and 1 control stop. They also present approximation algorithm for more general situations. Simulation shows that the proposed model can well describe how different value of slack time affects passengers' waiting time and delay.

Table 1 concludes recent works on the topic of holding control strategies. Recent research mainly carries out studies about holding control from three aspects: First, further improving control effect with new control logic, like forward-looking control, backward-looking control etc.; second, using holing control to tackle specific problems, like holding in multilines, holding to synchronize transfers; third, optimizing some key parameters about bus holding, like the optimization of slack time, trip time, number, and

location of control stops etc. It is well known that there are tradeoffs between the reliability and efficiency of bus operation when implementing dynamic holding control. Although large slack time, multicontrol stops and strong control coefficient helps to maintain bus operation of high reliability, they may seriously reduce the system's efficiency. Recent research mainly studies the optimization of slack time, number and location control stops and other parameters by simulations or empirical analyses. In addition, rare studies consider about the planning issue (e.g., determining bus schedule and dispatching headway when the fleet size and bus capacity are limited) in their holding strategies. This paper aims to integrate holding control and schedule planning together to achieve better control effects. The contributions of this paper are as follows: First, planning factors like fleet size, bus capacity, dispatching frequency and layover time are considered, which is critical for field operations. Second, mathematical formulas are generated to describe how control strength, slack time and the choice of control stop affect the reliability and efficiency of dynamic holdings. Third, formulas are proposed to quantify passengers' ordinary waiting time, extra waiting time and in-vehicle travel time. Fourth, an optimization model is presented to solve optimal control coefficients, slack time and the number and location of control stops.

2. Notation

The frequently used parameters in this work are listed in Table 2.

3. Modeling of Dynamic Holding Control

The general control model proposed by Xuan et al. [23] is widely studied to solve bus bunching problems. This paper adopts the basic control logic of Xuan et al. [23]'s work to formulate dynamic holding control method, and further improves the control efficiency by reducing the slack time needed. First, for the convenience of discussion, Xuan et al. [23]'s method is briefly introduced as background. Then mathematical analysis is conducted to discuss the reliability of dynamic holding in heterogeneous situations. Last, the concept of Equivalent Control Parameter (ECP) is proposed to quantitatively measure the impact of small slack time on the reliability and efficiency of dynamic holding.

3.1. Background. Equations (1) ~ (3) are often used to formulate bus motions (e.g., Daganzo [20], Xuan et al. [23], etc.).

$$t_{n,s+1} = t_{n,s} + \beta_s H + d_s + c_s \quad (1)$$

$$t_{n,s} = t_{n-1,s} + H \quad (2)$$

$$a_{n,s+1} = a_{n,s} + \beta_s h_{n,s} + D_{n,s} + c_s + \gamma_{n,s+1} \quad (3)$$

Equations (1) and (2) represent the scheduled bus motion, and (3) describes real bus motion. Combining (1), (2), and (3),

TABLE 1: Summary of Recent Dynamic Holding Strategies.

Authors	Approach	Highlight
Fu & Yang [19]	FWL+TWL	The optimal number and location of control points, and the optimal control strength are studied with simulations.
Daganzo [20]	FWL	The method of convolution is introduced to simplify the modeling process.
Daganzo & Pilachowski [21]	TWL	A cruising speed control method is proposed with the two-way-looking control logic.
Bartholdi & Eisenstein [22]	BWL	The proposed holding strategy does not require headway or schedule information.
Xuan et al. [23]	SB	A general holding control model is generated which represents a family of different control methods.
Liang et al. [24]	TWL	A zero-slack version of holding strategy is proposed.
Zhang & Lo [25]	TWL	A self-equalizing holding strategy with two-way-looking control logic is proposed.
Argote-Cabanero et al. [26]	SB	Holding control is generalized into multi bus lines.
Nesheli & Ceder [27]	FWL	Methods like holding control, boarding-limit control, and stop-skipping control are used to synchronize transfers for multi bus lines.
Estrada et al. [28]	TWL	Cruising speed control and signal priority control methods are proposed based on two-way-looking control logic.
Eberlein et al. [14]	—	The optimal location of control stop is analyzed by simulations.
Oort et al. [29]	—	Illustrating how the choice of trip time, location and amount of control stops affect the reliability and efficiency of long-headway bus services.
Zhao et al. [30]	SB	Mathematical analysis is carried out to address the optimization problem of slack time.

Notes: 'FWL' stands for 'Forward-Looking,' 'BWL' stands for 'Backward-looking,' 'TWL' stands for 'Two-way-looking,' and 'SB' stands for 'Schedule-based.'

TABLE 2: Summary of primary parameters.

Parameter	Definition
n	Bus number
s	Stop number
N	The amount of buses
S	The amount of stops
$t_{n,s}$	The scheduled arrival time of bus n at stop s
$a_{n,s}$	The actual arrival time of bus n at stop s
$\varepsilon_{n,s}$	The deviation from scheduled arrival time of bus n at stop s. $\varepsilon_{n,s} = a_{n,s} - t_{n,s}$
$h_{n,s}$	The headway between bus n and bus $(n-1)$ at stop s. $h_{n,s} = a_{n,s} - a_{n-1,s}$
c_s	The average cruising time between stop s and stop $(s+1)$.
$\gamma_{n,s+1}$	The random noise in the cruising time of bus n between stop s and stop $(s+1)$
$D_{n,s}$	The holding time applied to bus n at stop s
d_s	The amount of slack time inserted in the schedule at stop s
λ_s	The passenger arriving rate at stop s
T_b	The passenger boarding rate
β_s	A dimensionless measure for the demand rate at stop s. $\beta_s = T_b \lambda_s$
H	The target headway

the deviation from scheduled arrival time can be formulated as follows:

$$\varepsilon_{n,s+1} = \varepsilon_{n,s} + \beta_s \left(\varepsilon_{n,s} - \varepsilon_{n-1,s} \right) + \gamma_{n,s+1} + \left(D_{n,s} - d_s \right) \quad (4)$$

Xuan et al. [23] generate the holding time $D_{n,s}$ as a linear function of the schedule deviation of all buses at stop s:

$$D_{n,s}$$
$$= \max \left(0, d_s - \left[(1 + \beta_s) \varepsilon_{n,s} - \beta_s \varepsilon_{n-1,s} \right) + \sum_i f_i \varepsilon_{n-i,s} \right) \quad (5)$$

Assuming the slack time d_s is large enough, (4) can be simplified as (6):

$$\varepsilon_{n,s+1} = \sum_i f_i \varepsilon_{n-i,s} + \gamma_{n,s+1} \quad (6)$$

With numerical studies, Xuan et al. [23] notice that the control efficiency mainly depends on coefficient f_0. Therefore, a so called "simple control strategy" is generated as (7) and (8). The relationship of schedule variance between two consecutive stops can be expressed as (9). Note that parameter f_s represents the coefficient f_0 designated to bus stop s. For ease of expression, f_s rather than $f_{0,s}$ is used.

$$D_{n,s} = \max \left(0, d_s - \left[(1 + \beta_s) \varepsilon_{n,s} - \beta_s \varepsilon_{n-1,s} \right] + f_s \varepsilon_{n,s} \right) \quad (7)$$

$$\varepsilon_{n,s+1} = f_s \varepsilon_{n,s} + \gamma_{n,s+1} \quad (8)$$

$$\sigma_{\varepsilon,s+1}{}^2 = f_s{}^2 \sigma_{\varepsilon,s}{}^2 + \sigma_{s+1}{}^2 \quad (9)$$

So far, slack time d_s is assumed to be a large enough number. However, a large value of slack time leads to long average dwell time at stop. Xuan et al. [23] let $d_s = 3\sigma_{D,s}$ to guarantee a 99.87% confidence level of positive holding time, where $\sigma_{D,s}$ represents the deviation of holding time.

$$\sigma_{D,s}^2 = (1 + \beta_s - f_s)^2 \sigma_{\varepsilon,s}{}^2 + \beta_s{}^2 \sigma_{\varepsilon,s}{}^2 \quad (10)$$

3.2. Reliability Analysis. Xuan et al. [23] prove the "simple control strategy" to be a reliable control method under homogeneous circumstance, where inputs like c_s, $\gamma_{n,s+1}$, β_s are identical for all bus stops and road segments. We generalize their conclusions into heterogeneous situations.

Let us expand the right-hand side of (8) iteratively as follows:

$$\varepsilon_{n,s+1} = f_s \varepsilon_{n,s} + \gamma_{n,s+1} = f_s \left(f_{s-1} \varepsilon_{n,s-1} + \gamma_{n,s} \right) + \gamma_{n,s+1}$$
$$= \left(\prod_{k=1}^{s} f_k \right) \cdot \varepsilon_{n,1} + \sum_{i=2}^{s} \prod_{j=i}^{s} \left(f_j \cdot \gamma_{n,i} \right) + \gamma_{n,s+1} \quad (11)$$

Based on (11), the expected value of schedule deviation $\varepsilon_{n,s+1}$ can be formulated as $E(\varepsilon_{n,s+1}) = \left(\prod_{k=1}^{s} f_k \right) \cdot \varepsilon_{n,1}$. Its variance is as follows:

$$\operatorname{var} \left(\varepsilon_{n,s+1} \right) = \left(\prod_{k=1}^{s} f_k{}^2 \right) \cdot \operatorname{var} \left(\varepsilon_{n,1} \right)$$
$$+ \sum_{i=2}^{s} \left(\prod_{j=i}^{s} f_j{}^2 \right) \cdot \operatorname{var} \left(\gamma_{n,i} \right) + \operatorname{var} \left(\gamma_{n,s+1} \right) \quad (12)$$

Where $\varepsilon_{n,1}$ is the schedule deviation that can be measured as the difference between bus 's actual dispatching time and scheduled dispatching time at the first stop. Furthermore, $\operatorname{var}(\varepsilon_{n,1}) = 0$ since $\varepsilon_{n,1}$ is a deterministic value. Equation (12) can be rewritten as follows:

$$\operatorname{var} \left(\varepsilon_{n,s+1} \right) = \sum_{i=2}^{s} \left(\prod_{j=i}^{s} f_j{}^2 \right) \cdot \operatorname{var} \left(\gamma_{n,i} \right) + \operatorname{var} \left(\gamma_{n,s+1} \right) \quad (13)$$

Given that $\gamma_{n,s+1}$ follows the same distribution, i.e., $\operatorname{var}(\gamma_{1,s+1}) = \operatorname{var}(\gamma_{2,s+1}) = \dots = \operatorname{var}(\gamma_{n,s+1})$, we assign $\operatorname{var}(\varepsilon_{1,s+1}) = \operatorname{var}(\varepsilon_{2,s+1}) = \dots = \operatorname{var}(\varepsilon_{n,s+1}) = \sigma_{\varepsilon,s+1}{}^2$.

TABLE 3: Control effects in different conditions.

No.	Condition	ECP	Schedule deviation after control
1	$d_s - [(1 + \beta_s)\varepsilon_{n,s} - \beta_s\varepsilon_{n-1,s}] + f_s\varepsilon_{n,s} \geq 0$	f_s	$f_s\varepsilon_{n,s}$
2	$d_s - [(1 + \beta_s)\varepsilon_{n,s} - \beta_s\varepsilon_{n-1,s}] + f_s\varepsilon_{n,s} < 0$	$f_s' = -\dfrac{d_s - [(1 + \beta_s)\varepsilon_{n,s} - \beta_s\varepsilon_{n-1,s}]}{\varepsilon_{n,s}}$	$f_s'\varepsilon_{n,s}$

Furthermore, since $\varepsilon_{n,s+1}$ is the summation of many independent random variables, $\varepsilon_{n,s+1}$ can be assumed to be normally distributed according to the central limit theorem, i.e., $\varepsilon_{n,s+1} \sim N((\prod_{k=1}^{s} f_k) \cdot \varepsilon_{n,1}, \sigma_{\varepsilon,s+1}^2) \; \forall n$.

Let $F = \max(|f_1|, |f_2| \ldots |f_s|)$, and M be a large enough number to keep $M > \mathrm{var}(\gamma_{n,s+1}) \; \forall s$. Then we have:

$$\sigma_{\varepsilon,s+1}^2 = \sum_{i=2}^{s} \left(\prod_{j=i}^{s} f_j^2 \right) \cdot \mathrm{var}(\gamma_{n,i}) + \mathrm{var}(\gamma_{n,s+1})$$

$$(14)$$

$$< \sum_{i=0}^{s} F^{2i} \cdot M = \frac{1 - F^{2(s+1)}}{1 - F^2} \cdot M \approx \frac{M}{1 - F^2}$$

Equation (14) shows that the schedule variance $\mathrm{var}(\varepsilon_{n,s+1})$ is bounded when $|F| < 1$. Therefore, as long as we ensure $|f_s| < 1$ for each control stop s, buses will adhere to their schedules. It should be noted that regardless of whether the control parameter f_s is positive or negative, the reliability of the system remains the same as long as their absolute values are the same. However, (9) shows that when f_s is negative, a larger slack time is needed to ensure the effectiveness of the control. Therefore, a positive f_s is always selected in practice.

3.3. Optimization of Slack Time.

As aforementioned, Xuan et al. [23] let $d_s = 3\sigma_{D,s}$ to keep holding time positive. However, as to be shown in the case study, such magnitude of slack time is too large to keep bus operations of high efficiency.

To reduce slack time, the term $d_s - [(1 + \beta_s)\varepsilon_{n,s} - \beta_s\varepsilon_{n-1,s}] + f_s\varepsilon_{n,s}$ is allowed to be negative in some cases. This leads to $D_{n,s} = 0$ in (7). By introducing the concept Equivalent Control Parameter (ECP) in this paper, we redefine (9) to correlate slack time d_s with schedule variance $\sigma_{\varepsilon,s+1}^2$. Then the best slack time can be obtained by solving an optimization problem.

Table 3 shows the choice of ECP in different conditions. In condition 1, the predefined control parameter f_s does not lead to negative holding time. Thus the bus holding time can be determined by f_s. In condition 2, negative holding time occurs if the value of control parameter still sticks to f_s. In reality, however, the actual holding time will be 0 according to (5). In this case, if we replace the original control parameter f_s witt f_s', condition 2 can be converted to condition 1, satisfying $d_s - [(1 + \beta_s)\varepsilon_{n,s} - \beta_s\varepsilon_{n-1,s}] + f_s\varepsilon_{n,s} = 0$.

Statistics tell us that we can calculate the variance of variable X as $\mathrm{var}(X) = E(X^2) - E(X)$. Therefore, the variance of the schedule deviation with holding control can be expressed as follows:

$$\sigma_{\varepsilon,n}^{AC2} = \int_{-\infty}^{+\infty} \int_{-\infty}^{Bound} \left[(f_s\varepsilon_{n,s})^2 - f_s\varepsilon_{n,s} \right] \mathrm{f}(\varepsilon_{n,s})$$

$$\cdot \mathrm{f}(\varepsilon_{n-1,s}) \, \mathrm{d}(\varepsilon_{n,s}) \, \mathrm{d}(\varepsilon_{n-1,s})$$

$$+ \int_{-\infty}^{+\infty} \int_{Bound}^{+\infty} \left[(f_s'\varepsilon_{n,s})^2 - f_s'\varepsilon_{n,s} \right] \mathrm{f}(\varepsilon_{n,s})$$

$$\cdot \mathrm{f}(\varepsilon_{n-1,s}) \, \mathrm{d}(\varepsilon_{n,s}) \, \mathrm{d}(\varepsilon_{n-1,s})$$

$$f_s' = -\frac{\{d_s - [(1 + \beta_s)\varepsilon_{n,s} - \beta_s\varepsilon_{n-1,s}]\}}{\varepsilon_{n,s}}$$

$$Bound = \frac{[d_s - (1 + \beta_s - f_s)\varepsilon_{n,s}]}{\beta_s}$$

$$\varepsilon_{n,s} \sim N(0, \sigma_{\varepsilon,s}^2), \; \varepsilon_{n-1,s} \sim N(0, \sigma_{\varepsilon,s}^2)$$

where $\mathrm{f}(\varepsilon_{n,s})$ and $\mathrm{f}(\varepsilon_{n-1,s})$ are probability density functions

Then, (9) can be revised as follows:

$$\sigma_{\varepsilon,s+1}^2 = \sigma_{\varepsilon,n}^{AC2} + \sigma_{s+1}^2$$

$$(16)$$

Equations (15) and (16) correlate the magnitude of slack time to its corresponding impacts on the schedule variations. The next section will show how to optimize control coefficients based on these two s.

So far we assumed that all bus stops are control stops. At control stops, buses will be held for some extra time after passengers finish boarding to help buses meet their schedules. Too many control stops can significantly reduce the average bus travel speeds. This assumption can be relaxed. Equation (4) shows that the schedule deviations of different buses are interdependent without holding control. Bus's schedule deviation at stop $s + 1$ depends on the schedule deviations of both the bus itself and its preceding bus at stop s. If we implement control strategy at normal stops as the following, we can regain the independent feature.

$$D_{n,s} = \max(0, d_s + \beta_s\varepsilon_{n-1,s})$$

$$(17)$$

The schedule variance at stop $s + 1$ can then be expressed without terms related to the preceding bus $(n-1)$:

$$\sigma_{\varepsilon,s+1}^2 = (1 + \beta_s)^2 \sigma_{\varepsilon,s}^2 + \sigma_{s+1}^2$$

$$(18)$$

Based on trial simulations, the holding time shown in (17) is usually very small and can be ignored. Therefore, we can use s (16) and (18) to deduce the stop-by-stop schedule variances for every control stop and normal stop.

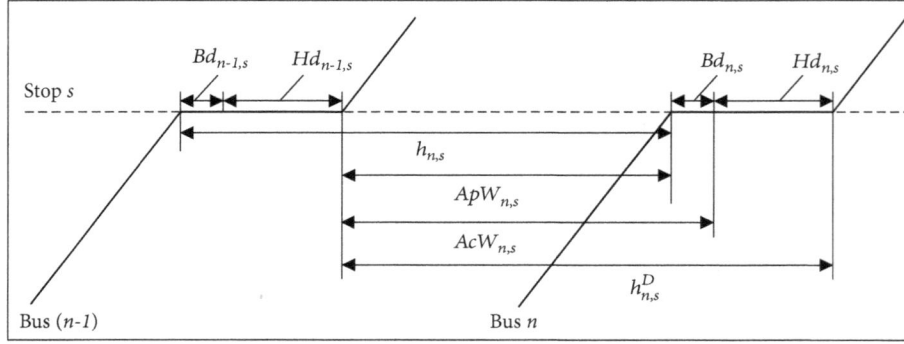

FIGURE 1: Illustration of passenger waiting time at stop.

4. Optimization of Bus Operation

Transit agencies need to provide reliable and efficient bus services to passengers with limited resources. In this section, we optimize the bus dispatch frequency and holding control methods to minimize passenger travel time when bus capacity and vehicle fleet size are limited.

4.1. Passenger Ordinary Waiting Time. When bus capacity is unlimited, all waiting passengers can hop on the first bus they meet, and we call their average waiting time as ordinary waiting time (OWT) in this circumstance.

Figure 1 depicts two consecutive buses dwellings at stop s. As shown, the bus dwelling time consists of the time for boarding ($Bd_{n,s}$) and holding ($Hd_{n,s}$) procedure. Existing researches assumed that passengers arrive within $h_{n,s}$ will board bus n. The OWT for all buses can be calculated as $\overline{T}_{Wait,s} = (H/2)(1 + \sigma_{h,s}^2/H^2)$ (Welding [2]), where $\sigma_{h,s}^2$ is the headway variance for all buses at stop s. Given $h_{n,s} = H + \varepsilon_{n,s} - \varepsilon_{n-1,s}$, $\sigma_{h,s}^2$ can be calculated as $\sigma_{h,s}^2 = 2\sigma_{\varepsilon,s}^2$.

In practice, only passengers who arrive during $AcW_{n,s}$ have to wait for boarding. $AcW_{n,s}$ starts from the departure time of bus $(n-1)$ and ends at the end of bus n's boarding procedure. Passengers who arrive during bus's holding procedure can directly get onboard without waiting. Furthermore, since the value of $Bd_{n,s}$ is much smaller than that of $AcW_{n,s}$, $AcW_{n,s}$ can be approximated as $ApW_{n,s}$. Then the OWT at stop s can be calculated as follows:

$$\overline{T}_{Wait,s}$$
$$= \frac{E(ApW_{n,s})}{2}\left[1 + \frac{var(ApW_{n,s})}{E(ApW_{n,s})^2}\right] \cdot \frac{E(ApW_{n,s})}{E(h_{n,s}^D)} \quad (19)$$

Where $E(ApW_{n,s})$ stands for the average value of $ApW_{n,s}$, $E(ApW_{n,s}) = (H - d_s)$ if we ignore the existence of boarding procedure. The average value of departure headway $h_{n,s}^D$ is $E(h_{n,s}^D) = H$. In (19), the weight of the OWT is assigned as $E(ApW_{n,s})/E(h_{n,s}^D)$ because among all passengers arriving during $h_{n,s}^D$, only passengers who arrive during $ApW_{n,s}$ have to wait to board. Since $ApW_{n,s}$ is decided by bus $(n-1)$'s leaving

time and bus n's arriving time, $var(ApW_{n,s}) = \sigma_{\varepsilon,n}^{AC2} + \sigma_{\varepsilon,s}^2$. Finally, (19) can be expressed as follows:

$$\overline{T}_{Wait,s} = \frac{H - d_s}{2}\left[1 + \frac{\sigma_{\varepsilon,n}^{AC2} + \sigma_{\varepsilon,s}^2}{(H - d_s)^2}\right] \cdot \frac{H - d_s}{H} \quad (20)$$

The OWT for passengers at all stops is just the weighted summation of OWT at each stop.

$$\overline{T}_{Wait} = \frac{\left(\sum_{i=1}^{S} \lambda_i \overline{T}_{Wait,i}\right)}{\left(\sum_{j=1}^{S} \lambda_j\right)} \quad (21)$$

4.2. Passenger extra waiting time. Due to the limit on bus capacity, some passengers may not be able to board the first bus they meet when the bus is overloaded, and therefore will experience an extra waiting time (EWT) for waiting the next bus. Supposing that all buses run along with schedules, bus's loads after leaving stop s are as follows:

$$p_{n,s}^t = p_{n,s-1}^t + H\lambda_s - \sum_{i=1}^{s} H\lambda_i l_{i,s} \quad (22)$$

Where λ_s is the average passenger arrival rate at stop s, and $l_{i,s}$ indicates the proportion of passengers who board at stop i and alight at stop s. However, when buses deviate from schedules, the actual passenger load becomes:

$$p_{n,s}^a = p_{n,s-1}^a + h_{n,s}\lambda_s - \sum_{i=1}^{s} h_{n,i}\lambda_i l_{i,s} \quad (23)$$

By combining (22) and (23), the passenger load deviation can be expressed as follows:

$$\varepsilon_{n,s}^p = \varepsilon_{n,s-1}^p + \lambda_s(\varepsilon_{n,s} - \varepsilon_{n-1,s}) - \sum_{i=1}^{s} \lambda_i l_{i,s}(\varepsilon_{n,i} - \varepsilon_{n-1,i}) \quad (24)$$

If we respectively replace $p_{n,s-1}^a$ and $\varepsilon_{n,s-1}^p$ in the right-hand side of (23) and (24) in an iterative way, we finally have the following:

$$p_{n,s}^a = \sum_{i=1}^{s} h_{n,i} \lambda_i \left(1 - \sum_{j=1}^{s} l_{ij} \right) \qquad (25)$$

$$\varepsilon_{n,s}^P = \sum_{i=1}^{s} \left(\varepsilon_{n,i} - \varepsilon_{n-1,i} \right) \lambda_i \left(1 - \sum_{j=1}^{s} l_{ij} \right) \qquad (26)$$

With (25), we have the average passenger load for all buses at stop s:

$$\overline{p_s} = H \cdot \sum_{i=1}^{s} \lambda_i \left(1 - \sum_{j=1}^{s} l_{ij} \right) \qquad (27)$$

Passenger loads variance can be calculated by (26):

$$\sigma_{\varepsilon^P,s}^2 = \sum_{i=1}^{s} 2\sigma_{\varepsilon,i}^2 \lambda_i^2 \left(1 - \sum_{j=1}^{s} l_{ij} \right)^2$$
$$= \sum_{i=1}^{s} \sigma_{h,i}^2 \lambda_i^2 \left(1 - \sum_{j=1}^{s} l_{ij} \right)^2 \qquad (28)$$

However, as illustrated in Figure 1, the number of boarding passengers is actually decided by the departure headway $h_{n,s}^D$ rather than the arriving headway $h_{n,s}$. Therefore, we replace the variance of arriving headway $\sigma_{h,i}^2$ by the variance of departure headway $\sigma_{lh,s}^2$ in (28), where we have $\sigma_{lh,s}^2 = 2\sigma_{\varepsilon,s}^{AC2}$ when stop s is a control stop, and $\sigma_{lh,s}^2 = 2(1 + \beta_s)^2 \sigma_{\varepsilon,s}^2$ if stop s is a normal stop. Then (28) can be revised as follows:

$$\sigma_{\varepsilon^P,s}^2 = \sum_{i=1}^{s} \sigma_{lh,i}^2 \lambda_i^2 \left(1 - \sum_{j=1}^{s} l_{ij} \right)^2$$

$$\sigma_{lh,s}^2 = 2\sigma_{\varepsilon,s}^{AC2} \quad \text{if stop } s \text{ is a control point}$$

$$\sigma_{lh,s}^2 = 2\left(1 + \beta_s \right)^2 \sigma_{\varepsilon,s}^2 \quad \text{if stop } s \text{ is a normal stop} \qquad (29)$$

Supposing that all buses are identical with capacity C_a, bus n's residual capacity at stop s will follow $C_{n,s}^R \sim N(C_a -$

$\overline{p_s}, \sigma_{\varepsilon^P,s}^2)$ $\forall n$. $C_{n,s}^R > 0$ means that there are still $C_{n,s}^R$ residual capacity left in bus n after it leaves stop s; while $C_{n,s}^R < 0$ indicates that bus n is fully loaded and $-C_{n,s}^R$ passengers will experience EWT for the next bus.

We consider $C_a - \overline{p_s} < 0$ as an unacceptable condition, because the number of waiting passengers will keep growing over time. Therefore we let the average EWT to be M_{ExWait} in this situation, where M_{ExWait} is a large enough number. When $C_a - \overline{p_s} \geq 0$, for those who have to wait for the following buses, they may manage to board the next bus $n+1$, or some of them may have to wait even longer if bus $n + 1$ is also fully loaded. Table 4 enumerates all possible cases.

Based on Table 4, we can calculate the average value of EWT.

In case 1, $-C_{n,s}^R$ passengers cannot board bus n. The average number of passengers who suffer the EWT in case 1 can be calculated as follows:

$$\overline{P_s^{case1}} = \int_{-\infty}^{0} -C_{n,s}^R f\left(C_{n,s}^R \right) d\left(C_{n,s}^R \right) \int_{-C_{n,s}^R}^{\infty} f\left(C_{n,s}^R \right) d\left(C_{n+1,s}^R \right) \qquad (30)$$

Suppose that passengers will experience an extra H waiting time for every one extra bus, the total EWT in case 1 becomes the following:

$$T_{s,ExWait}^{total,case1} = \int_{-\infty}^{0} -C_{n,s}^R H f\left(C_{n,s}^R \right) d\left(C_{n,s}^R \right) \cdot \int_{-C_{n,s}^R}^{\infty} f\left(C_{n+1,s}^R \right) d\left(C_{n+1,s}^R \right) \qquad (31)$$

Since the average number of passengers arriving between two consecutive buses is $\lambda_s H$, the average EWT in case 1 can be calculated as follows:

$$\overline{T_{s,ExWait}^{case1}} = \frac{T_{s,ExWait}^{total,case1}}{(\lambda_s H)}$$
$$= \frac{\int_{-\infty}^{0} -C_{n,s}^R H f\left(C_{n,s}^R \right) d\left(C_{n,s}^R \right) \int_{-C_{n,s}^R}^{\infty} f\left(C_{n+1,s}^R \right) d\left(C_{n+1,s}^R \right)}{(\lambda_s H)} \qquad (32)$$

Similarly, we have the following s for case 2.1 and case 2.2.

$$\overline{P_s^{case2.1}} = \frac{\int_{-\infty}^{0} -C_{n,s}^R f\left(C_{n,s}^R \right) d\left(C_{n,s}^R \right) \int_{-\infty}^{0} f\left(C_{n+1,s}^R \right) d\left(C_{n+1,s}^R \right) \int_{-C_{n,s}^R}^{\infty} f\left(C_{n+2,s}^R \right) d\left(C_{n+2,s}^R \right)}{(\lambda_s H)} \qquad (33)$$

$$\overline{T_{s,ExWait}^{case2.1}} = \frac{\int_{-\infty}^{0} -C_{n,s}^R \cdot 2H f\left(C_{n,s}^R \right) d\left(C_{n,s}^R \right) \int_{-\infty}^{0} f\left(C_{n+1,s}^R \right) d\left(C_{n+1,s}^R \right) \int_{-C_{n,s}^R}^{\infty} f\left(C_{n+2,s}^R \right) d\left(C_{n+2,s}^R \right)}{(\lambda_s H)} \qquad (34)$$

$$\overline{P_s^{case2.2}} = \frac{\int_{-\infty}^{0} -C_{n,s}^R f\left(C_{n,s}^R \right) d\left(C_{n,s}^R \right) \int_{0}^{C_{n,s}^R} f\left(C_{n+1,s}^R \right) d\left(C_{n+1,s}^R \right) \int_{-(C_{n,s}^R + C_{n+1,s}^R)}^{\infty} f\left(C_{n+2,s}^R \right) d\left(C_{n+2,s}^R \right)}{(\lambda_s H)} \qquad (35)$$

TABLE 4: All possible cases of EWT.

Case Depiction	Case 1	Case 2		Case 3
	Bus $n+1$ can take all the passengers who cannot board bus n.	Only part of the passengers who cannot board bus n are able to board bus $n+1$, and the remaining passengers have to wait bus $n+2$ for boarding.		When some of the passengers who cannot board bus n have to wait for more than 2 buses for boarding.
		Case 2.1	**Case 2.2**	
		Bus $n+1$ is already full before collecting those passengers and therefore all passengers have to wait for bus $n+2$.	Bus $n+1$ can collect some of the passengers, and all the other passengers have to wait for bus $n+2$.	
Condition	(1) $C_{n,s}^R < 0$ (2) $-C_{n,s}^R \le C_{n+1,s}^R$	(1) $C_{n,s}^R < 0$ (2) $C_{n+1,s}^R \le 0$ (3) $-C_{n,s}^R \le C_{n+2,s}^R$	(1) $C_{n,s}^R < 0$ (2) $0 < C_{n+1,s}^R < -C_{n,s}^R$ (3) $-C_{n,s}^R \le C_{n+1,s}^R + C_{n+2,s}^R$	(1) $C_{n,s}^R < 0$ (2) $-C_{n,s}^R > C_{n+1,s}^R + C_{n+2,s}^R$
Average Number of Passengers	$\overline{P}_s^{case1}$ Equation (30)	$\overline{P}_s^{case2.1}$ Equation (33)	$\overline{P}_s^{case2.2}$ Equation (35)	$\overline{P}_s^{case3}$ Equation (37)
Average Extra Waiting Time	$\overline{T}_{s,ExWait}^{case1}$ Equation (32)	$\overline{T}_{s,ExWait}^{case2.1}$ Equation (34)	$\overline{T}_{s,ExWait}^{case2.2}$ Equation (36)	$\overline{T}_{s,ExWait}^{case3}$ Equation (37)

$$\overline{T}_{s,ExWait}^{case2.2}$$
$$= \frac{\int_{-\infty}^{0} \left[C_{n+1,s}^R H - 2H \left(C_{n,s}^R + C_{n+1,s}^R \right) \right] f\left(C_{n,s}^R\right) d\left(C_{n,s}^R\right) \int_{0}^{C_{n,s}^R} f\left(C_{n+1,s}^R\right) d\left(C_{n+1,s}^R\right) \int_{-(C_{n,s}^R + C_{n+1,s}^R)}^{\infty} f\left(C_{n+2,s}^R\right) d\left(C_{n+2,s}^R\right)}{(\lambda_s H)} \quad (36)$$

Case 3 is the situation when some passengers have to wait for more than three buses for boarding, and we suppose that all passengers who suffer case 3 will experience a $3H$ EWT. Then the average EWT in this case can be calculated as follows:

$$\overline{T}_{s,ExWait}^{case3} = \frac{3H \overline{P}_s^{case3}}{(\lambda_s H)}$$

$$\overline{P}_s^{case3} = \int_{-\infty}^{0} -C_{n,s}^R f\left(C_{n,s}^R\right) d\left(C_{n,s}^R\right) - \overline{P}_s^{case1}$$
$$- \overline{P}_s^{case2.1} - \overline{P}_s^{case2.2} \quad (37)$$

It should be noted that all cases above are mutually exclusive events. Therefore the average EWT for passengers at all stops can be formulated as follows:

$$\overline{T}_{ExWait} = \frac{\sum_{i=1}^{S-1} \lambda_i \overline{T}_{i,ExWait}}{\sum_{j=1}^{S-1} \lambda_j}$$

$$s.t. \begin{cases} \overline{T}_{s,ExWait} = \left(\overline{T}_{s,ExWait}^{case1} + \overline{T}_{s,ExWait}^{case2.1} + \overline{T}_{s,ExWait}^{case2.2} + \overline{T}_{s,ExWait}^{case3} \right) & \text{if } C_a - \overline{p}_s > 0 \\ \overline{T}_{s,ExWait} = M_{ExWait} & \text{if } C_a - \overline{p}_s < 0 \end{cases} \quad (38)$$

4.3. Passenger In-Vehicle Travel Time (IvTT). With (3), we can get the bus travel time from stop s to stop $s+1$:

$$T_{n,s} = a_{n,s+1} - a_{n,s} = \beta_s h_{n,s} + D_{n,s} + c_s + \gamma_{n,s+1} \quad (39)$$

The average value of $T_{n,s}$ is as follows:

$$\overline{T}_s = \beta_s H + d_s + c_s \quad (40)$$

Consider the scenario that passengers travel from stop s to stop $s+m$ as an example for illustration. If we assume

all waiting passengers at stop s can get onboard immediately after the bus arrival. The average IvTT for those passengers can be calculated as $\sum_{k=s}^{s+m} \overline{T}_i = \sum_{k=s}^{s+m}(\beta_k H + d_k + c_k)$. The average IvTT for passengers at all stops is as follows:

$$\overline{T}_{inveh} == \frac{\sum_{i=1}^{S-1} \sum_{j=i+1}^{S} \sum_{k=i}^{j-1} (\beta_k H + d_k + c_k)}{\sum_{m=1}^{S} \lambda_m} \quad (41)$$

In reality, the waiting passengers cannot get onboard immediately, but keep on boarding during the boarding and holding procedure as shown in Figure 1. Therefore the average IvTT they spend at stop s is less than $\beta_s H + d_s$. We approximate the average value of IvTT at stop s as half of the average bus dwelling time, i.e., $(\beta_s H + d_s)/2$. Then (39) can be revised as follows:

$$\overline{T}_{inveh} == \frac{\sum_{i=1}^{S-1}\sum_{j=i+1}^{S}\sum_{k=i}^{j-1}\lambda_i l_{i,j}\left(\beta_k H + d_k + c_k\right) - \sum_{p=1}^{S-1}\lambda_p\left(\beta_p H + d_p\right)/2}{\sum_{m=1}^{S}\lambda_m} \qquad (42)$$

4.4. Fleet Size Limit. When all buses run according to schedules, the total travel time cost for one bus run is as follows:

$$T_{total} = \sum_{i=1}^{S-1}\left(\beta_i H + d_i + c_i\right) + L \qquad (43)$$

Where L is the layover time to recover buses from schedule deviations and to provide breaks for drivers. The mean value of T_{total} is $\overline{T}_{total} = \sum_{i=1}^{S-1}(\beta_i H + d_i + c_i) + L$. Supposing that the bus fleet size is N_{fleet}, we can calculate the target headway as follows:

$$H = \frac{\left[\sum_{i=1}^{S-1}\left(\beta_i H + d_i + c_i\right) + L\right]}{N_{fleet}} \qquad (44)$$

4.5. Optimization Model. Tradeoff exists when implementing dynamic holding strategy. Although bus holding has the potential to reduce passenger waiting time by lessening bus headway deviation (Welding [2]), sometimes excessive holding increases passengers' total travel time because of the following reasons. First, the added slack time increases buses' average dwelling time, thereby leading to long in-vehicle travel time for onboard passengers. Second, as illustrated by (44), bus dispatching headway increases with the value of slack time when fleet size is limited. As explained by Welding [2], passenger ordinary waiting time increases with average headway (i.e., dispatching headway) and decreases with headway deviation. When the benefit (i.e., reduction of headway deviation) of dynamic holding cannot compensate for the loss (i.e., increment of dispatching headway) caused by it, passenger ordinary waiting time will increase. Moreover, large dispatching headway also increase the likelihood of extra waiting time suffered by waiting passengers, especially if the supply of bus service is less than passenger demand.

The proposed optimization fully considers the positive and negative effect of dynamic holding on bus operation, and minimize average passenger travel time by solving the optimal locations of control stop, and the corresponding slack time and control parameter designated for each control stop. Limited bus capacity and fleet size are considered in this paper. As shown in function (45), passenger travel time consists of ordinary waiting time, extra waiting time and in-vehicle travel time as discussed before. Studies show that the perceived waiting time by passengers are usually much longer than the actual waiting time due to factors like adverse waiting environment, waiting anxiety, etc. (Psarros et al. [31],

Mishalani et al. [32]). Therefore, waiting penalty ζ_{wait} is added to the objective function to measure this perceiving difference. In this paper, we set the value of ζ_{wait} to be 2.1 as recommended in Transit Capacity and Quality of Service Manual [33].

$$\min \overline{T}_{total} = \zeta_{wait}\left(\overline{T}_{Wait} + \overline{T}_{ExWait}\right) + \overline{T}_{InVeh} \qquad (45)$$

Some major constraints are summarized as follows:

$$H = \frac{\left[\sum_{i=1}^{S-1}\left(\beta_i H + d_i + c_i\right) + L_{min} + 3\sigma_{\varepsilon,S}\right]}{N_{fleet}} \qquad (46.a)$$

$$\sigma_{\varepsilon,s+1}{}^2 = \begin{cases} \sigma_{\varepsilon,n}^{AC^2} + \sigma_{s+1}{}^2 & \text{if } \eta_s = 1 \\ \left(1 + \beta_s\right)^2 \sigma_{\varepsilon,s}{}^2 + \sigma_{s+1}{}^2 & \text{if } \eta_s = 0 \end{cases} \qquad (46.b)$$

$$\sigma_{lh,s}^2 = \begin{cases} 2\sigma_{\varepsilon,s}^{AC^2} & \text{if } \eta_s = 1 \\ 2\left(1 + \beta_s\right)^2 \sigma_{\varepsilon,s}^2 & \text{if } \eta_s = 0 \end{cases} \qquad (46.c)$$

$$\eta_s = \begin{cases} 0 & \text{if stop } s \text{ is acontrol point} \\ 1 & \text{if stop } s \text{ is anormal stop} \end{cases} \qquad (46.d)$$

Where (46.a) is the formula to calculate dispatching headway, (46.b) depicts the propagation of schedule variance along the bus route, (46.c) further illustrates how to calculation schedule various for both control stops and normal stops, the parameter η_s in (46.d) indicates whether stop s is a control stop or not. Note that η_s, d_s and f_s $(s = 1, 2, \ldots, S)$ are decision variables that need to be solved, and all other variables such as $\sigma_{\varepsilon,s}^2$, $\sigma_{lh,s}^2$ and H etc. are determined with certain value of η_s, d_s and f_s $(s = 1, 2, \ldots, S)$.

4.6. Parameter Calibration. We show how to calibrate the average cruising time c_s and the variance of cruising time $\sigma_{s+1}{}^2$ between stop s and stop $(s+1)$ with AVL data and signal timing data. Suppose that there are $N_{s,seg}$ road segments and $N_{s,int}$ intersections between stop s and stop $(s + 1)$. The average travel time $\overline{T}_{s,i}^{seg}$ and variance of travel time $\sigma_{s,i}^{seg^2}$ $(i = 1 \ldots N_{s,seg})$ for each segment can be calculated by using AVL data. In terms of signal delay, we assume that buses will not queue up at intersections in this experiment, then the average delay $\overline{DE}_{s,j}^{int}$ and variance of delay $\sigma_{s,j}^{int^2}$ at intersections can respectively be calculated with (47) and (48), where $RT_{s,j}^{int}$ is the red time and $CL_{s,j}^{int}$ is the cycle length.

$$\overline{DE}_{s,j}^{int} = \frac{\left(RT_{s,j}^{int}\right)^2}{2CL_{s,j}^{int}} \tag{47}$$

$$\sigma_{s,j}^{int\,2} = \frac{\left(RT_{s,j}^{int}\right)^3}{3CL_{s,j}^{int}} - \left(DE_{s,j}^{int}\right)^2 \tag{48}$$

The average value and variance of cruising time are the summation of the values of all segments and intersections between stop s and stop$(s + 1)$, respectively.

$$c_s = \sum_{i=1}^{N_{s,seg}} \overline{T}_{s,i}^{seg} + \sum_{j=1}^{N_{s,int}} \overline{DE}_{s,j}^{int} \tag{49}$$

$$\sigma_{s+1}^{\ 2} = \sum_{i=1}^{N_{s,seg}} \sigma_{s,i}^{seg\,2} + \sum_{j=1}^{N_{s,int}} \sigma_{s,j}^{int\,2} \tag{50}$$

5. Case Study

5.1. Data Collection. The case study uses field data collected from the Bus Route 56 in Chengdu, China. This bus route is one of the busiest in Chengdu, which serves more than 50000 passengers per day. The route runs across the city in the north-south direction, and connects several universities, large-scale business areas, hospitals and residential districts. We select 14 major bus stops located in the downtown area of the city as shown in Figure 2. In the simulation, we assume that these 14 stops form a loop. AVL data and smartcard data are collected during morning rush hour (7:30-8:30) from 17th to 21st (Monday to Friday) October 2016. Signal timing information is collect on 18th October 2016. The appendix list all the data results.

1. **AVL Data:** Each piece of AVL data contains information including bus ID, route ID, GPS location, GPS speed and timestamp. Buses regularly send a piece of AVL data to the control center every ten seconds. With large amount of AVL data available, the mean value and variance of cruise time can be calibrated for each road segment by using (47) ∼ (50).

5.2. Simulation Framework. A simulation platform is developed to test the proposed control method. As shown in Figure 3, this platform mainly consists of five components: bus, terminal, intersection, bus stop and road segment.

(i) **Bus:** Figure 3 illustrates how buses travel along the bus route. First, buses depart the terminal and travel in the outbound direction. After reaching the farthest bus stop, buses turn around and start to travel in the inbound direction. Finally, buses return to the terminal, waiting there for the next round of operation. As we can see, there are 4 buses waiting at terminal, 2 buses travelling on road segments, 2 buses dwelling at

stops, 2 buses waiting at intersections. Therefore the fleet size is 10.

(ii) **Terminal:** Each waiting bus at the terminal will be assigned a bar to record its status of layover time. The whole length of the bar represents the required layover time as aforementioned in (43). The green part of the bar indicates the elapsed time, and the grey part is the remaining layover time. Agency can release a bus from the terminal only when the following two conditions are met: First, the time since the last departure exceeds the predetermined bus headway. Second, there is (are) bus(es) at the terminal whose status bar is all green.

(iii) **Intersection:** All intersections are signalized and operate in fixed-time. Dedicated bus lanes are employed to avoid bus queuing problem at intersections. The key signal control parameters include cycle length, phase plan, offset and splits (splits are the portion of time allocated to each phase at an intersection).

(iv) **Bus Stop:** Passengers arrive at the bus stop at a fixed arrival rate. After a bus arrive at the stop, the waiting passengers start to board the bus at a fixed boarding rate until there are no more waiting passengers or the bus is fully loaded. The bus will leave the bus stop when both the boarding procedure and holding procedure are over.

(v) **Road Segment:** A road segment is a portion of the bus route which is separated by intersections and bus stops. Each road segment is assigned with two parameters: average cruising time of the road segment, and standard deviation of the cruising time.

Figure 4(a) shows a simulation result of how the route 56 performs under an uncontrolled situation. As shown, a total of 13 buses are assigned to this bus route, i.e., the fleet size is 13. At first, buses depart the terminal station in even headways. However, as time goes by, buses fail to maintain their headway to the target value when they travel in different speed and delay at intersections for different period of time. What's more, buses with larger headways serve more passengers and will lag further behind their preceding vehicle, and vice versa. Finally, buses bunch up together and move in pairs. More interestingly, when we observe the two rounds of operation of bus 1, it is easy to see that after bus 1 finish one round of operation, it has to take a layover time before depart for the next round.

Figure 4(b) shows the detailed view of the zoomed-in area in Figure 4(a). The horizontal lines at bus stops indicate buses' dwelling procedure, where longer line means longer dwelling time. Signal timings at intersections are represented by green and red horizontal lines. The green line and red line stand for green time and red time respectively. Figure 4(b) is a good example to show how bus bunching phenomenon occurs. Due to the large time gap between bus 13 and bus 1, many passengers are waiting at stop 8 and stop 9 for boarding, and therefore lead to an increase of dwelling time for bus 1. In contrast, the dwelling time of bus 2 is much shorter than that

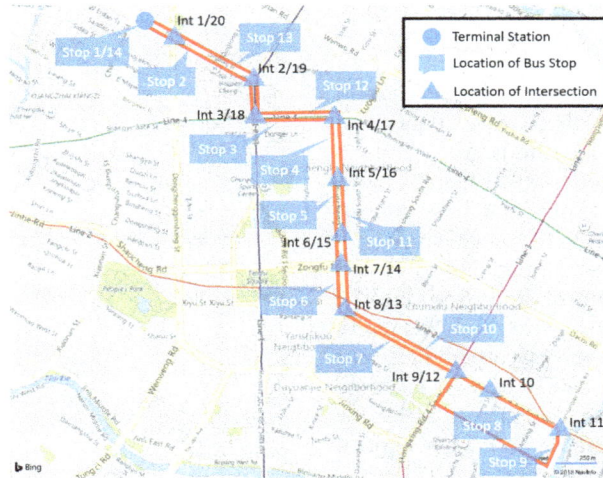

FIGURE 2: Map of bus route 56 in the Downton Area, Chengdu, China.

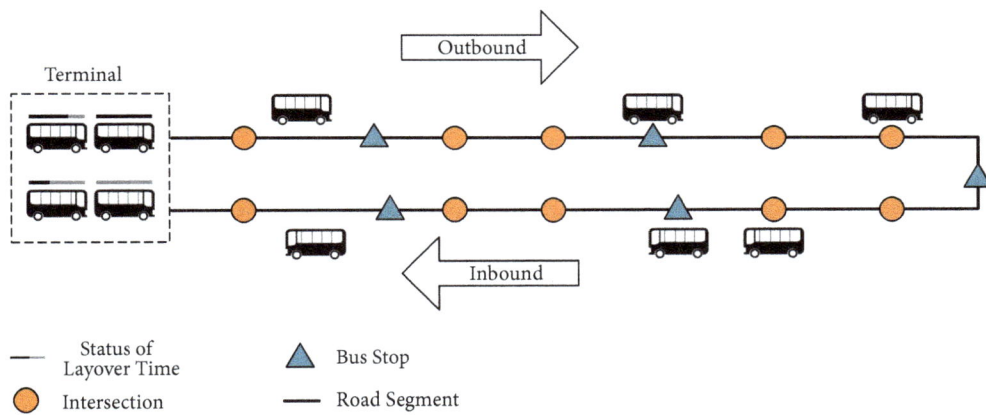

FIGURE 3: Illustration of the simulation platform.

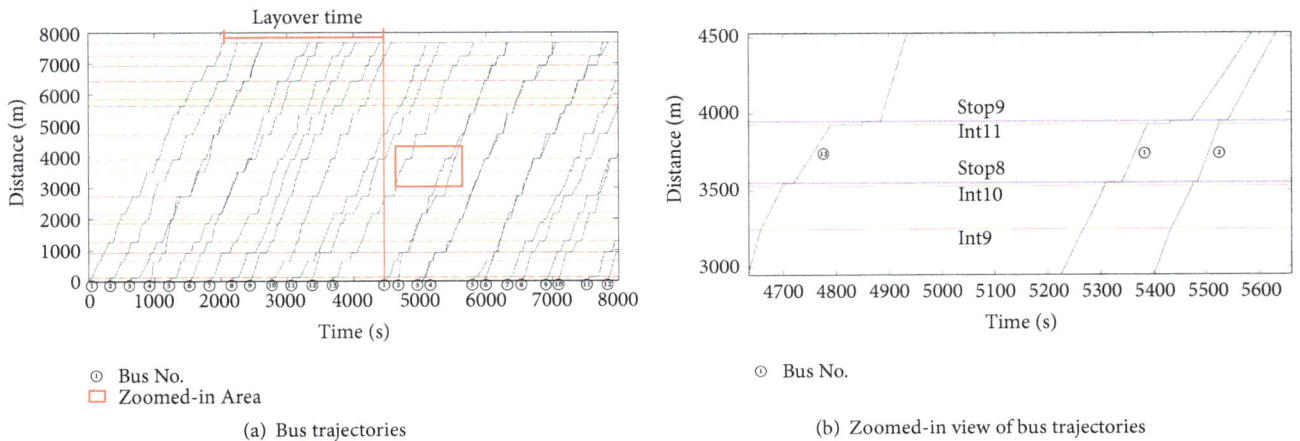

(a) Bus trajectories

(b) Zoomed-in view of bus trajectories

FIGURE 4: Illustration of the simulation results.

of bus 1, which is because only a small number of passengers arrive at stop 8 and stop 9 after bus 2 left. As a result, the headway between bus 1 and bus 2 will shrink over time until they finally bunch.

5.3. Simulation Result Analysis. In this section, the proposed simulation platform is calibrated with collected data from route 56 to evaluate different control methods.

As shown by (45) and (46.a), (46.b), (46.c), and (46.d), the proposed optimization model solve the best control scheme by choosing the prime location of control stops (η_s), and solving the optimal value of control parameters f_s and holding slack time d_s for each control stop. It should be noted that the proposed model is of very high flexibility which allows control stops to be unevenly distributed, and allows different control parameter and holding slack time for different control stops. However, to make it easier for comparison between our control model and other control methods, we simplify the optimization problem as follows:

1. Control stops are designated every 3 stops. Therefore, we have $\eta_s = 1$ ($s = 3, 6, 9, 12$), and $\eta_s = 0$ for the other stops.

2. The control parameters f_s at all control stops are of the same value, and the value of f_s ranges from 0.1 to 0.9 with a precision of 0.1;

3. We let $d_s = \alpha_s \sigma_{D,s}$, where α_s ranges from 0.1 to 3.0 with precision of 0.1. All control stops share the same value of α_s.

Figure 5 illustrates the predicted and simulated passenger average travel time (ATT) with different control parameter settings. The control coefficient f_s ranges from 0.1 to 0.9, and slack time coefficient α_s ranges from 0.1 to 3.0. The green surface represents the predicted ATT when bus overloading is minor events, i.e., $C_a - \overline{p_s} > 0 \; \forall s$. Both red and brown surface are ATT results generated by the simulations. It can be observed that the red surface matches well with the green surface with relative error less than 5%. We also use the simulations to generate the brown surface which represents significant overloading conditions when some bus stops satisfy $C_a - \overline{p_s} < 0$. The brown surface is outside of the green surface area. The value of ATT increases quickly when bus overloading become frequent events. Figure 5 illustrates that the proposed passenger-oriented performance measures can make good prediction on bus operational performance when buses are controlled under the proposed holding control strategy. Therefore, the optimization model shown in (43) and (44) can be a useful tool to optimize control parameters in order to minimize passenger average travel time.

Figure 5 also provides a comparison between the results from Xuan et al. [23]'s control method and the proposed control method in terms of passenger average travel time. Xuan et al. [23] set $\alpha_s = 3$ in their research to guarantee the predesigned holding control will always work in the valid range. However, their strategy requires very large slack time at control stops, which can lead to undesirable passenger travel costs. As shown in Figure 5, the minimum ATT that can be achieved by Xuan et al. [23] is 1256s when $\alpha_s = 3$ and $f_s = 0.9$.

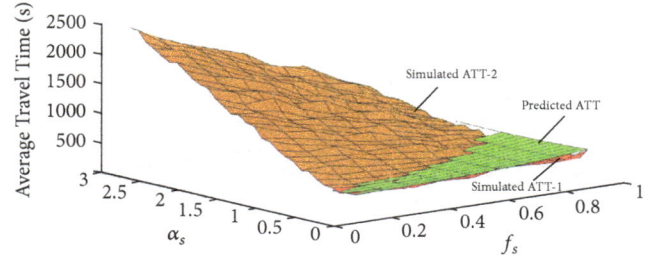

FIGURE 5: Predicted and simulated results of passengers average travel time.

By using the proposed control and optimization model, we can reduce ATT to 912s when we set $\alpha_s = 0.4$ and $f_s = 0.1$. This results in a 27.4% improvement compared to Xuan et al. [23]'s method.

We also compare our strategy with uncontrolled situations. Simulations are performed to find the best bus dispatching headway in uncontrolled cases. The bus operation reaches its best performance when dispatch headway is 345s, and the ATT is 1031s correspondingly. This means that the proposed method also outstrips uncontrolled case in terms of ATT by 11.5%. Figure 6 compares bus trajectories under controlled and uncontrolled cases. It is easy to see that under controlled case, buses travel in more uniform headways, and no bunching occurs in this case. However, the bus line suffers serious bunching problem when no holding control is implemented as shown in Figure 6(b).

Table 5 takes the terminal bus stop as a check point to see how passengers' travel time vary under different control methods. For example, when the proposed control method is implemented, the average travel time for passengers travelling from stop 9 to terminal stop will be 1335 seconds, and the standard deviation of their travel time is 293 seconds. It is irrational at the first glance when seeing that travelling from stop 12 to terminal takes much more time than those passengers who wait at stop 9. The reason is that buses are nearly fully loaded after leaving stop 11, and some of the waiting passengers at stop 12 may suffer an extra waiting time. It is easy to see the proposed method significantly reduce the mean value and SD value of passengers' travel time compared with other two methods, which means it can provide mo re efficient and reliable bus services. Especially for bus stops (like stop 12) with high passenger demand and low capacity supply, the other two methods lead waiting passengers to suffer undesirable extra waiting time, and increase their travel uncertainty.

5.4. Sensitivity Analysis. In this section, we compare control methods under different levels of passenger arriving rate and cruising time deviation. The best uncontrolled case is set as the benchmark to evaluate different control methods. By simulating uncontrolled bus operation under different dispatching headways, the best uncontrolled case can be found when the passenger travel time reaches the minimal value.

TABLE 5: Comparisons of the Mean Value and Standard Deviation of Passenger Travel Time under Different Control Methods (Unit: seconds).

| Origin Stop | Control Methods | | | | | |
| | The Proposed Optimization Model | | Xuan et al. [23] | | Non Control | |
	Mean	S.D.	Mean	S.D.	Mean	S.D.
Stop 9	1335	293	1400	304	1490	426
Stop 10	1199	289	1443	367	1420	459
Stop 11	973	330	1060	313	1082	420
Stop 12	1032	384	3960	858	1378	537
Stop 13	707	348	733	304	893	445

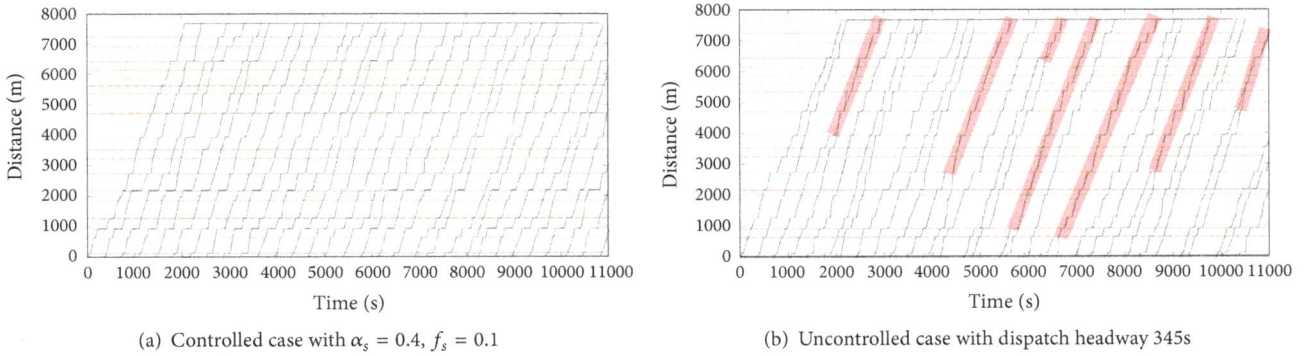

(a) Controlled case with $\alpha_s = 0.4$, $f_s = 0.1$

(b) Uncontrolled case with dispatch headway 345s

FIGURE 6: Bus trajectories under the proposed holding control and uncontrolled situation.

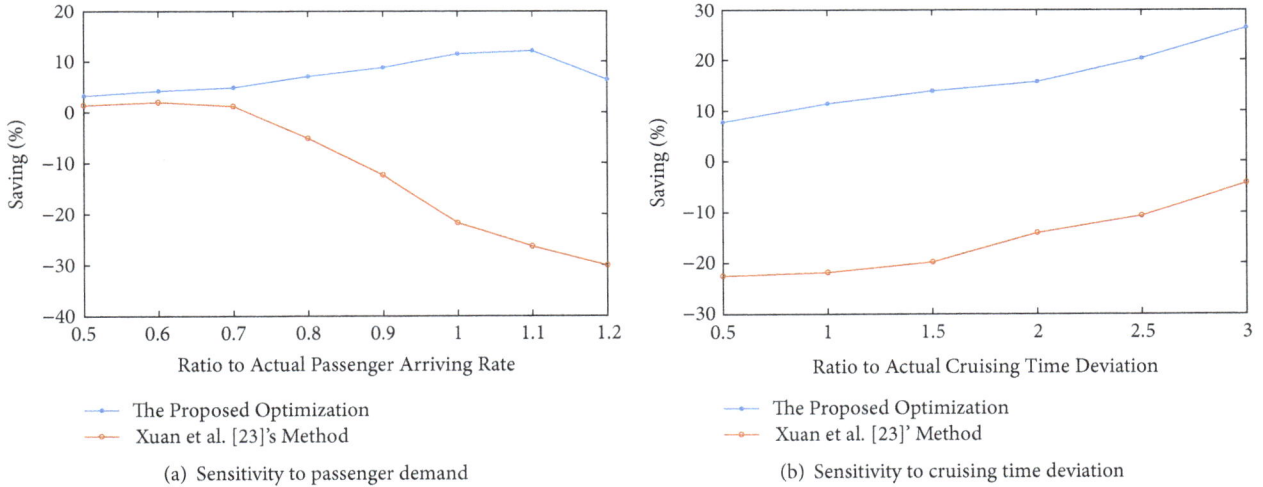

(a) Sensitivity to passenger demand

(b) Sensitivity to cruising time deviation

FIGURE 7: Saving of passenger travel time under different passenger demand and cruising time deviation.

Figure 7(a) presents the performance of different holding control methods when the ratio of simulated passenger arriving rate to actual passenger arriving rate ranges from 0.5 to 1.2. The vertical axis indicates the saving of passenger travel time compared with uncontrolled case. As shown, the proposed optimization outstrips uncontrolled case under all situations. When the ratio is in the range of 0.5 to 1.1, the saved passenger travel time increases with the increment

of passenger demand. The reason is that higher passenger demand leads to larger headway variance and more serious uneven bus load problem in uncontrolled case, which subsequently increases the ordinary waiting time and extra waiting time for passengers. In contrast, the proposed optimization helps to maintain bus headways to the target value, and therefore ensures smaller passenger travel time. However, there is a performance drop when the value of demand ratio reaches 1.2. This result clearly shows the tradeoff between reliability and efficiency when implementing holding control. Even though bus holding helps to improve the reliability of bus operation and alleviate bus bunching problem, the added slack time will increase the time cost per round of bus operation, which means less amount passengers can be served in a unit of time compared with uncontrolled case. When passenger demand is too high, oversaturated loading may even occur in controlled case, and such situation will counteract the benefits of the high reliability provided by holding control. It should be noted that route 56 is already a very heavy-loaded bus route, so that the ratio value 1.2 can rarely occurs in field operation. Due to large slack time required, Xuan et al. (23)'s holding strategy leads to oversaturated situation when the passenger demand ratio is higher than 0.7, and its control performance will drop quickly afterwards. Figure 7(b) evaluate control performances under different levels of cruising time deviation. Not surprisingly, the control performance of both methods will increase when cruising time deviation increase. That's because holding control results in much more stable and reliable bus operations compared with uncontrolled case under high randomness of traffic condition.

6. Conclusions and Future Work

Holding control strategy is an effective way to alleviate bus bunching phenomenon and improve the reliability of bus service, however too much slack time may significantly reduce the operational efficiency and lead to undesirable passenger travel cost and bus overloading. This paper proposes an integrated modeling of schedule planning and dynamic holding control. The proposed approach considers both planning factors (e.g., fleet size, bus capacity, dispatching frequency and layover time) and control factors (control coefficients, slack time, number and location of control stops) when modeling. An optimization model is presented to solve best control strategies. The proposed methods are tested by simulations. All parameters in the simulation environment are calibrated with field data. Simulation shows that the proposed optimization model can precisely predict the control effects of dynamic holding under different control parameters, and therefore can be used to optimize bus operation. Result shows that the proposed optimization achieves a 27.4% improvement compared with Xuan et al. [23]'s method under actual case. Sensitivity analysis further validate the proposed model under different levels of passenger demands and cruising time deviations. A summary of findings is listed as follows:

(1) We relaxed Xuan et al. [23]'s "simple control strategy" by reducing slack time required at control stops. Slack time is correlated with buses schedule deviation by introducing equivalent holding control parameters. We prove that buses can adhere to schedule even with quite small slack time.

(2) We propose performance measures from passenger perspective to precisely predict the control impact. The passenger travel cost consists of the waiting time, extra waiting time and in-vehicle travel time. We advance Welding [2]'s method by taking the holding procedure into consideration and eliminated the overestimation by assuming bus only collects passengers who arrive during the headway. Passenger loads on different buses are not even due to bus deviations from schedules. Some passengers will not be able to board the first bus they meet when that bus is fully loaded, and will therefore suffer extra waiting time. We enumerate all possible cases, and propose a performance measure which can theoretically predict passengers extra waiting time for any specific holding control strategies.

(3) As shown in case study, even though holding control may provide more reliable bus services than uncontrolled operations, it may increase passenger travel cost due to a large slack time. Benefiting from performance measures proposed in this paper, we formulate an optimization model by combining the schedule planning and holding control into an integrated procedure. The resulting model allows us to minimize passenger travel time with limited operating resources and layover time.

Although the research has reached its aims, there are still limitations that need to be solved in future studies. First, passenger arriving rate at each bus stop is treated as a fixed value in this work, which does not satisfy the real situation in some bus lines of high arrival randomness. Second, the proposed optimization is a non-convex integer programming, and it is time-consuming to solve optimal control strategy for some large-scale bus lines. Algorithms need to be proposed in future works which can solve the optimization more efficiently and accurately.

Appendix

Data Inputs for Case Study

Data inputs shown in Table 6 are calibrated by AVL data, smartcard data, and signal timing information. AVL data and smartcard data are collected during morning rush hour (7:30-8:30) from 17th to 21st (Monday to Friday) October 2016. More than 10,000 pieces of AVL data and more than 5,000 pieces of smartcard data can be collected each day during the rush hour. Signal timing information is collected on 18th October 2016.

Data in the second and third column of Table 6 indicates the average travel time and STD of travel time between nodes (stop/intersection). For example, the travel time from stop 1 to intersection 1 follows distribution of mean value 18 seconds and STD 9.47 seconds. Data in the fourth column represents passenger arriving rate at each stop. For instance, the passenger arriving rate at stop 1 is 0.045 person/second. Data in the fifth and sixth column is signal timing information. For example, the cycle length of the first intersection is 187 seconds, and the phase length for bus movement is 63 seconds.

TABLE 6: Data Inputs for Case Study.

(a)

Fleet Size	Bus Capacity	Layover Time	Passenger Boarding Rate
13 veh	80 prs/veh	40 min	1 s/prs

(b)

Node	Average Travel Time (s)	STD of Travel Time (s)	Passenger Arrival Rate (prs/s)	Green Time (s)	Cycle Length (s)
Stop 1			0.045		
Int 1	18 (from Stop 1 to Int 1)	9.47		63	187
Stop 2	19	6.29	0.059		
Int 2	56	26.62		63	179
Int 3	39	17.99		33	192
Stop 3	14	4.14	0.056		
Int 4	27	13.92		80	186
Stop 4	17	6.74	0.029		
Int 5	61	24.83		85	120
Stop 5	5	1.15	0.038		
Int 6	19	4.21		136	182
Int 7	26	6.01		48	182
Stop 6	12	3.88	0.024		
Int 8	69	25.52		36	192
Stop 7	19	7.18	0.021		
Int 9	53	23.88		88	194
Int 10	40	6.49		105	194
Stop 8	7	2.04	0.050		
Int 11	50	10.76		94	194
Stop 9	21	6.24	0.081		
Int 12	92	35.08		70	194
Stop 10	12	2.81	0.063		
Int 13	82	28.19		84	192
Int 14	37	12.58		81	182
Int 15	32	11.29		136	182
Stop 11	11	4.58	0.042		
Int 16	30	16.61		85	120
Int 17	41	13.29		48	186
Stop 12	15	4.42	0.113		
Int 18	55	20.28		68	192
Int 19	45	10.21		63	179
Stop 13	13	4.02	0.065		
Int 20	48	20.14		63	187
Stop 14 (Terminal)	12	3.22	0		

Note: Int stands for Intersection.

Based on smartcard data analysis, passengers average travel distance is approximately 3 bus stops. Therefore, we set passenger alighting rate as [0.1, 0.15, 0.5, 0.15, 0.1], where the vector indicates the percentage of passengers of different travel distance. For example, 10% passengers' travel distance is 1 bus stop; 15% passenger travel distance is 2 stops. In addition, when bus arrives at terminal, all boarding passengers have to alight. Therefore, all passengers who board at Stop 13 have to alight at the terminal.

Disclosure

Permission has been obtained for use of copyrighted material from other sources.

Conflicts of Interest

The authors declare that the received fund did not lead to any conflicts of interest regarding the publication of this paper.

Acknowledgments

This research is supported by Science & Technology Department of Sichuan Province, China (No. 2017JY0072). The authors would acknowledge Chengdu Public Transport Group Company for providing AVL data and smartcard data for this research.

References

[1] G. F. Newell, "Dispatching Policies for a Transportation Route," *Transportation Science*, vol. 5, no. 1, pp. 91–105, 1971.

[2] P. Welding, *The Instability of Close Interval Service. Operational Research Quarterly*, vol. 8, no. 3, pp. 133–148, 1957.

[3] A. Barnett, "On controlling randomness in transit operations," *Transportation Science*, vol. 8, no. 2, pp. 102–116, 1974.

[4] M. D. Rossetti and T. Turitto, "Comparing static and dynamic threshold based control strategies," *Transportation Research Part A: Policy and Practice*, vol. 32A, no. 8, pp. 607–620, 1998.

[5] S. Hanaoka and F. M. Qadir, "Passengers' perceptions and effects of bus-holding strategy using automatic vehicle location technology," *Journal of Advanced Transportation*, vol. 43, no. 3, pp. 301–319, 2009.

[6] G. E. Sánchez-Martínez, H. N. Koutsopoulos, and N. H. M. Wilson, "Real-time holding control for high-frequency transit with dynamics," *Transportation Research Part B: Methodological*, vol. 83, pp. 1–19, 2016.

[7] W. Wu, R. Liu, and W. Jin, "Modelling bus bunching and holding control with vehicle overtaking and distributed passenger boarding behaviour," *Transportation Research Part B: Methodological*, vol. 104, pp. 175–197, 2017.

[8] G. E. Sánchez-Martínez, H. N. Koutsopoulos, and N. H. Wilson, "Event-driven holding control for high-frequency transit," *Transportation Research Record*, vol. 2535, pp. 65–72, 2015.

[9] J. Zhao, S. Bukkapatnam, and M. M. Dessouky, "Distributed architecture for real-time coordination of bus holding in transit networks," *IEEE Transactions on Intelligent Transportation Systems*, vol. 4, no. 1, pp. 43–51, 2003.

[10] R. Hall, M. Dessouky, and Q. Lu, "Optimal holding times at transfer stations," *Computers & Industrial Engineering*, vol. 40, no. 4, pp. 379–397, 2001.

[11] F. Delgado, N. Contreras, and J. C. Munoz, "Holding for transfers," in *Transportation Research Board 92nd Annual Meeting*, vol. 2013.

[12] M. Asgharzadeh and Y. Shafahi, "Real-time bus-holding control strategy to reduce passenger waiting time," *Transportation Research Record*, vol. 2647, p. 16, 2017.

[13] F. Delgado, J. C. Muñoz, R. Giesen, and A. Cipriano, "Real-time control of buses in a transit corridor based on vehicle holding and boarding limits," *Transportation Research Record*, vol. 2090, pp. 59–67, 2009.

[14] X. J. Eberlein, N. H. M. Wilson, and D. Bernstein, "The holding problem with real-time information available," *Transportation Science*, vol. 35, no. 1, pp. 1–18, 2001.

[15] C. E. Cortés, D. Sáez, F. Milla, A. Núñez, and M. Riquelme, "Hybrid predictive control for real-time optimization of public transport systems' operations based on evolutionary multi-objective optimization," *Transportation Research Part C: Emerging Technologies*, vol. 18, no. 5, pp. 757–769, 2010.

[16] D. Koffman, "A simulation study of alternative real-time bus headway control strategies," in *Joint Automatic Control Conference*, vol. 14, pp. 441–446, 1977.

[17] C. X. Chen, W. Y. Chen, and Z. Y. Chen, "A Multi-Agent Reinforcement Learning approach for bus holding control strategies," *Advances in Transportation Studies*, 2015.

[18] B. Yu, J.-B. Yao, and Z.-Z. Yang, "An improved headway-based holding strategy for bus transit," *Transportation Planning and Technology*, vol. 33, no. 3, pp. 329–341, 2010.

[19] Z.-M. Zhang, X.-W. Guo, and R.-Q. Du, "Analysis of hydration heat-induced stresses and cracks in massive concrete walls," *Journal of Hohai University*, vol. 30, no. 5, pp. 12–16, 2002.

[20] C. F. Daganzo, "A headway-based approach to eliminate bus bunching: Systematic analysis and comparisons," *Transportation Research Part B: Methodological*, vol. 43, no. 10, pp. 913–921, 2009.

[21] C. F. Daganzo and J. Pilachowski, "Reducing bunching with bus-to-bus cooperation," *Transportation Research Part B: Methodological*, vol. 45, no. 1, pp. 267–277, 2011.

[22] J. J. Bartholdi and D. D. Eisenstein, "A self-coördinating bus route to resist bus bunching," *Transportation Research Part B: Methodological*, vol. 46, no. 4, pp. 481–491, 2012.

[23] Y. Xuan, J. Argote, and C. F. Daganzo, "Dynamic bus holding strategies for schedule reliability: Optimal linear control and performance analysis," *Transportation Research Part B: Methodological*, vol. 45, no. 10, pp. 1831–1845, 2012.

[24] S. Liang, S. Zhao, C. Lu, and M. Ma, "A self-adaptive method to equalize headways: Numerical analysis and comparison," *Transportation Research Part B: Methodological*, vol. 87, pp. 33–43, 2016.

[25] S. Zhang and H. K. Lo, "Two-way-looking self-equalizing headway control for bus operations," *Transportation Research Part B: Methodological*, vol. 110, pp. 280–301, 2018.

[26] J. Argote-Cabanero, C. F. Daganzo, and J. W. Lynn, "Dynamic control of complex transit systems," *Transportation Research Part B: Methodological*, vol. 81, pp. 146–160, 2015.

[27] M. M. Nesheli and A. Ceder, "Use of real-time operational tactics to synchronize transfers in headway-based public transport service," *Transportation Research Record*, vol. 2539, pp. 103–112, 2016.

[28] M. Estrada, J. Mensión, J. M. Aymamí, and L. Torres, "Bus control strategies in corridors with signalized intersections," *Transportation Research Part C: Emerging Technologies*, vol. 71, pp. 500–520, 2016.

[29] N. van Oort, J. W. Boterman, and R. van Nes, "The impact of scheduling on service reliability: Trip-time determination and

holding points in long-headway services," *Public Transport*, vol. 4, no. 1, pp. 39–56, 2012.

[30] J. Zhao, M. Dessouky, and S. Bukkapatnam, "Optimal slack time for schedule-based transit operations," *Transportation Science*, vol. 40, no. 4, pp. 529–539, 2006.

[31] I. Psarros, K. Kepaptsoglou, and M. G. Karlaftis, "An empirical investigation of passenger wait time perceptions using hazard-based duration models," *Journal of Public Transportation*, vol. 14, no. 3, pp. 109–122, 2011.

[32] R. Mishalani, M. McCord, and J. Wirtz, "Passenger wait time perceptions at bus stops: Empirical results and impact on evaluating real-time bus arrival information," *Journal of Public Transportation*, vol. 9, no. 2, pp. 89–106, 2006.

[33] *Transit Capacity and Quality of Service Manual*, Transportation Research Board, 2013.

17

Application of Finite Mixture of Logistic Regression for Heterogeneous Merging Behavior Analysis

Gen Li (iD)

School of Transportation, Southeast University, Nanjing 210096, China

Correspondence should be addressed to Gen Li; gilg4226307@aliyun.com

Guest Editor: Lele Zhang

A finite mixture of logistic regression model (FMLR) was applied to analyze the heterogeneity within the merging driver population. This model can automatically provide useful hidden information about the characteristics of the driver population. EM algorithm and Newton-Raphson algorithm were used to estimate the parameters. To accomplish the objective of this study, the FMLR model was applied to a trajectory dataset extracted from the NGSIM dataset and a 2-component FMLR model was identified. The important findings can be summarized as follows: The studied drivers can be classified into two components. One is called Risk-Rejecting Drivers. These drivers are consistent with previous studies and primarily merge in as soon as possible and have a distinct preference for the large gaps. The other is the Risk-Taking Drivers that are much less sensitive to the gap size and pay more attention to surrounding traffic conditions such as the speed of front vehicle in the auxiliary lane and lead space gap between the merging vehicle and its leading vehicles in the auxiliary lane. Risk-Taking Drivers use the auxiliary lane to get to the further downstream or less congested area of the main lane. The proposed model can also produce more precise predicting accuracy than logistic regression model.

1. Introduction

Congestion has become one of the most serious economic and social problems and has drawn great attention from the public, transportation research scientists, transportation managers, and so on. Understanding the causes and mechanism of traffic congestion can help traffic managers formulate targeted policies to make better use of the existing transportation infrastructures.

Merging areas are the bottleneck of freeway. Merging behavior is one of the typical mandatory lane changes when vehicles have to move from an on-ramp to the main road. It has been claimed in some studies that merging behavior at merging areas affects traffic operations and may trigger traffic congestions and breakdowns [1, 2]. Thus it is important to analyze the merging behaviors to help understand the mechanism of traffic jams to some extent from a microscopic viewpoint and build more accurate traffic simulation models.

Recently, driver heterogeneity has drawn great attention in microscopic traffic flow studies. Several studies investigated the driver heterogeneity during car-following process

[3–6]. Accommodating heterogeneity within the driver population is important in building microscopic traffic models. To investigate the heterogeneity in merging behaviors, a finite mixture of logistic regression (FMLR) model was proposed in this paper. This model can incorporate the unobserved heterogeneity and automatically segments the merging drivers into different homogeneous populations. More specifically, this paper aims to achieve the following objectives:

(i) Prove the existence of heterogeneity among merging drivers.

(ii) Identify different driving styles and attitudes during merging process.

(iii) Model the merging behavior more accurately.

The present study is organized as follows. The next section will provide a critical review on the existing relevant literature followed by Section 3, which describes the NGSIM data used in this paper. Section 4 gives the methodology to build FMLR model. Results and discussions are presented in Section 5.

Finally, the conclusions and future work are presented in Section 6.

2. Literature Review

Several methods have been adopted to model merging behavior, among which gap acceptance theory was the most widely used method [8–13]. The most important assumption in gap acceptance theory was that a driver makes a lane change when both the lead and lag gaps in the target lane are larger than the so-called critical gap. The critical gap is determined by the characteristics of the drivers, traffic conditions, and so on [14]. Gap acceptance models were initially built to estimate the capacity of unsignalized intersections. Different distributions of critical gaps were assumed in various studies [15–17]. Gipps [18] first used the gap acceptance theory to propose a comprehensive framework of lane-changing model. Gipps's framework has been widely used in several merge models [19, 20] and microscopic traffic simulation software [21, 22]. Different definitions of critical gap were used in these models and software.

Gap acceptance theory was often criticized as its basic assumption is often inconsistent with the real world observation because some lane change behaviors occurred when only the lead or lag gap or even none of them are larger than the critical gap [14, 23, 24]. To overcome this deficiency, discrete choice models such as binary logit model were used by some researchers [14, 25–27]. Built upon a series of studies [9, 10, 28], a framework for merging behavior with latent plans was introduced by Choudhury et al. [29]. Normal merge, merge with courtesy, and forced merge were considered in this framework. However, Marczak et al. [14] pointed out that in this framework only accepted gaps were considered and rejected gaps were ignored; and some of the estimated coefficients in the model were not significant.

Traffic behaviors are always uncertain and variable and heterogeneity cannot be ignored in traffic studies. Some studies investigated the heterogeneity among the macroscopic traffic flow [30, 31]. Others studied the heterogeneity in car following behaviors from microscopic viewpoint by deriving the joint distribution of model coefficients depending on an empirical basis [4, 5, 32–34]. However, only a few studies were found to investigate driver heterogeneity in lane changing models. A two-step clustering analysis was proposed by Li and Sun [35] to analyze heterogeneity of the merging maneuvers. However, this study ignored the heterogeneity during gap selection and decision process. An empirical analysis conducted by Daamen et al. [23] showed that different merging strategies might be adopted by different drivers under different traffic conditions. It has been pointed out by Keyvan-Ekbatani et al. [36] that different strategies might be used during gap selection process; however the sample size was too small to perform statistically relevant tests and build merging model.

Thus, a FMLR model was introduced in this paper to model the gap selection behaviors during merging process and investigate the heterogeneity among merging drivers. The FMLR model takes the advantage of two techniques: clustering and regression analysis. The model naturally incorporates

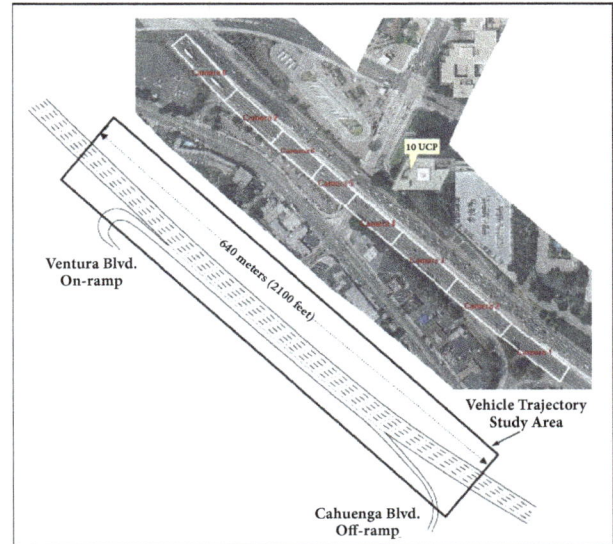

FIGURE 1: The section of US 101 [7].

the unobserved heterogeneity into logistic regression model and automatically segments the drivers into different homogeneous populations. The proposed FMLR model can explain the different strategies in merging behaviors.

3. Data Preparation

The NGISM dataset has been widely used for traffic flow and traffic simulation studies and proved to have high accuracy. Thus, in this paper, the vehicle trajectory data in NGSIM dataset collected on a segment of southbound U.S. Highway 101 (Hollywood Freeway) in Los Angeles, CA, are chosen [37]. Figure 1 shows the site for U.S. Highway 101. This US-101 section is 640 meters long and has five main lanes and one auxiliary lane. The vehicle trajectories were collected from 7:50 a.m. to 8:35 a.m. on June 15, 2005. The road section was covered by eight cameras and the dataset was updated at a resolution of 10 frames per second [7]. The dataset has three data subsets, all of which were collected in 15 minutes.

In this study, we focus on the behavior of merging vehicles and only trajectory data in the weaving section were used. However, it has been pointed out that the original trajectory data contain some noise and errors, which are caused by the system errors and tracking errors [38–41]. Several methods have been proposed to filter the data [38–40] or re-extract the trajectory data [41]. Re-extracting can produce the most accurate data especially the acceleration data, which however would also make too much effort. In this paper, a smoothing method called sEMA developed by Thiemann et al. [38] is applied to reduce the noise and errors. The sEMA method is also adopted in other studies of merging behaviors and has been proved to be able to provide enough precision for lane change studies [42–44]. This data smoothing technique was applied as follows:

(1) The velocities and accelerations of vehicles are directly estimated from the longitudinal positions.

TABLE 1: Examples with the same global coordinates in the first and second subsets.

Data Point	Sub dataset 1				Sub dataset 2			
	Vehicle ID	Frame ID	Local x	Local y	Vehicle ID	Frame ID	Local x	Local y
1	33	424	54.612	1397.746	36	847	54.612	1438.019
2	33	429	54.687	1420.332	1070	4878	54.687	1460.518
2	63	290	67.936	514.811	1472	5857	67.936	550.085

TABLE 2: Examples with the same global coordinates in the first and third subsets.

Data Point	Sub dataset 1				Sub dataset 3			
	Vehicle ID	Frame ID	Local x	Local y	Vehicle ID	Frame ID	Local x	Local y
1	42	446	53.395	1449.048	1721	8609	53.395	1483.814
2	63	483	41.056	1494.004	1280	6744	41.056	1528.773
3	296	967	53.8340	1389.247	905	4719	53.834	1424.013

(2) The locations (both local lateral and longitudinal coordinates), velocities, and accelerations of vehicles are smoothed by the symmetric exponential moving average filter (sEMA) proposed by Thiemann et al. [38] to decrease measurement errors in the data. The smoothing times of sEMA method are set as the suggested values for the U.S. Highway 101 dataset in Thiemann et al. [38].

Although the random errors can be reduced by the smoothing process, there are still some errors in the data. Thus, the following heuristic rules are applied to filter the datasets:

(1) Filter out the trajectories when there are no putative leading vehicles or putative following vehicles on the adjacent main lane. Such trajectories are recorded at the beginning or ending of the video tape and cannot provide the interactions of merging vehicles with their surrounding vehicles.

(2) Filter out the trajectories when putative leading or putative following vehicle of a merging vehicle runs around the lane boundary (it keeps touching the lane boundary before lane change or turns back the original lane in about 1 second). These trajectories are always caused by the tracking errors.

After filtering, a searching process was conducted to check the consistency of the local coordinates and global coordinates. Linear regression was performed between local coordinates and global coordinates for each subdataset. Three linear relationships were obtained for each subset:

$$Local y_1 = 0.3209 global x_1 - 1.1326 global y_1 \quad (1)$$

$$Local y_2 = 0.3291 global x_2 - 1.1334 global y_2 \quad (2)$$

$$Local y_3 = 0.3209 global x_3 - 1.1333 global y_3 \quad (3)$$

R^2 of three linear relationships are 0.9996, 0.9997, and 0.9997, respectively. It means that the local y of three subsets in US-101 datasets are inconsistent with each other. We cannot find simple linear relationship between local x and global x in US-101 dataset. This could be caused by the specific

coordinate system used and the special geometric shape of the road sections. It also could be caused by measuring errors.

To further verify the inconsistency of the US-101 dataset, several data points that have the same global coordinates among the three subsets were searched and obtained. By checking the local coordinates (local x and local y), it was found that the three subsets of US-101 dataset are consistent in local x, but inconsistent in local y. Tables 1 and 2 show the examples having the same global coordinates in the first and second subsets and in the first and third subsets.

One can find that, for the points with the same global coordinates, the three subdatasets have the same local x, but different local y. In the local longitudinal coordinate, the upstream edge (0 m) in datasets 1 is at 12.275m in dataset 2 and 10.598 m in dataset 3. Thus, the three datasets must be unified by using the local coordinates of one of the three subsets.

At every instant when offered a new gap, a merging vehicle driver assesses traffic conditions to decide whether to accept the offered gap or not. One merging vehicle could only accept one gap but could reject several gaps. After data processing, trajectories of 374 merging vehicles consisting of 925 observations were extracted from the dataset. The explanatory variables that may affect a driver's merging decision used as candidates for analyzing the merging behavior model are shown in Table 3.

4. Methodology

4.1. Finite Mixture of Logistic Regression. The FMLR model is based on the idea that the observed data come from a population with several subpopulations or components [45, 46]. The overall population is modeled as a mixture of the groups using finite mixture models.

Let $\mathbf{X}$ and $\mathbf{Y}$ denote random vectors with N samples and each sample has M_n observations $(\mathbf{x}_i, \mathbf{y}_i)$ $(i = 1, \ldots M_n, n = 1, \ldots N)$. Here, the response vector $\mathbf{Y}$ has values in $\mathbb{R}^d$ and the explanatory vector $\mathbf{X}$ has values in $\mathbb{R}^p$. Then, a FMLR with K components has the form

$$h(\mathbf{y} \mid \mathbf{x}, \psi) = \sum_{k=1}^{K} \pi_k f(\mathbf{y} \mid \mathbf{x}, \boldsymbol{\theta}_k) \quad (4)$$

TABLE 3: Descriptions of the explanatory variables.

Variable	Descriptions
D_n^i(m)	The size of the ith offered gap of merging vehicle n
V_n^i(m/s)	The speed of merging vehicle n at i^{th} offered gap.
Y_n^i(m)	The longitudinal position of the merging vehicle n to the start of the auxiliary lane.
ΔV_{nPL}^i(m/s)	The speed difference between the putative leading vehicle and the merging vehicle n at offered gap i.
ΔV_{nPF}^i(m/s)	The speed difference between the putative following vehicle and the merging vehicle n at offered gap i.
δ	Existence of a lead vehicle in the merge lane. If there is a lead vehicle in the merge lane, $\delta = 1$; otherwise, $\delta = 0$.
ΔD_{nlg}^i(m)	Lead gap of merging vehicle n in the auxiliary lane at offered gap i.
V_{nLead}^i (m)	The speed of the leading vehicle in the auxiliary lane at offered gap i.
ΔV_{nLead}^i(m/s)	The speed difference between the leading vehicle in the auxiliary lane and the merging vehicle n at offered gap i.

$$\sum_{k=1}^{K} \pi_k = 1, \quad \pi_k > 0 \qquad (5)$$

where $h(\mathbf{y} \mid \mathbf{x}, \Psi)$ is the conditional density of $\mathbf{y}$ given $\mathbf{x}$ and θ_k, π_k is the mixing proportion, θ_k is the component-specific parameter vector for the density function f, and $\psi = (\pi_1, \ldots, \pi_K, \theta_1, \ldots \theta_K)$ is the vector of all parameters.

Several finite mixture models can be extended based on (4) and (5). For multivariate normal f and $\mathbf{x} \equiv \mathbf{1}$ we get a finite mixture of Gaussians without a regression part, also known as model-based clustering. If f is a univariate normal density with component-specific mean $\beta_k' \mathbf{x}$ and variance σ_k^2, we have $\theta_k = (\beta_k', \sigma_k^2)$, and (4) describes a finite mixture of linear regression, also called latent class linear regression model or cluster-wise regression [47]. If f is a member of the exponential family, we get a FMLR models [48, 49].

The analyst does not observe directly which component, $k = 1, \ldots, K$, generated observation $\mathbf{y}_i$. The model assumes that individuals are distributed heterogeneously with a discrete distribution within the population. In order to impose the constraints in (2), the mixing proportions are parameterized with a multinomial logit form [50, 51]:

$$\pi_k = \frac{\exp(\alpha_k)}{\sum_{k=1}^{K} \exp(\alpha_k)}, \quad \alpha_K = 0 \qquad (6)$$

The constraint on α_K is imposed because only $K - 1$ parameters are needed to specify. The last proportion is one minus the sum of the first $K - 1$.

If individual specific characteristics are provided, the mixing proportions are extended as [50, 51]

$$\pi_{ik} = \frac{\exp(\theta_k \mathbf{z}_i)}{\sum_{k=1}^{K} \exp(\theta_k \mathbf{z}_i)}, \quad \theta_K = 0 \qquad (7)$$

where θ_K is the vector of component-specific parameters and $\mathbf{z}_i$ is an optional set of individual-specific characteristics for observation i.

For the observed random sample, $(\mathbf{x}_i, \mathbf{y}_i)$ $(i = 1, \ldots N)$, the log likelihood function for ψ is given by

$$\log L(\psi) = \sum_{n=1}^{N} \sum_{i=1}^{M_n} \log h(\mathbf{y}_i \mid \mathbf{x}_n, \psi)$$
$$= \sum_{i=1}^{N} \sum_{i=1}^{M_n} \log \left(\sum_{k=1}^{K} \pi_k h(\mathbf{y}_i \mid \mathbf{x}_n, \theta_k) \right) \qquad (8)$$

The maximum likelihood (ML) estimate of ψ is given by an appropriate root of the likelihood equation,

$$\frac{\partial \log L(\psi)}{\partial \psi} = 0 \qquad (9)$$

The conditional probability that observation $(\mathbf{x}_i, \mathbf{y}_i)$ belongs to component j is given by

$$P(j \mid \mathbf{x}_i, \mathbf{y}_i, \psi) = \frac{\pi_j f(\mathbf{y}_i \mid \mathbf{x}_i, \theta_j)}{\sum_{k=1}^{K} \pi_k f(\mathbf{y}_i \mid \mathbf{x}_i, \theta_k)} \qquad (10)$$

The conditional probabilities can be used to segment data by assigning each observation to the component with maximum conditional probability [50, 51]. A probabilistic segmentation of the data into K components can be obtained in terms of the fitted conditional probabilities. In the FMLR model we consider the latent component-indicator variables $\hat{z}_n = \hat{z}_{n1}, \ldots, \hat{z}_{nK}, n = 1, \ldots, N$, to classify each single observation:

$$z_{nk} = \begin{cases} 1, & \text{if } y_n \text{ belongs to component } k \\ 0, & \text{otherwise} \end{cases} \qquad (11)$$

The estimator of z_{jk}, $\hat{z}_{nk}$ is

$$\hat{z}_{nk}$$
$$= \begin{cases} 1, & \text{if } \hat{\pi}_k(\mathbf{y}_n; \widehat{\Psi}) \geq \hat{\pi}_h(\mathbf{y}_n; \widehat{\Psi}), \ (h = 1, \ldots, K; h \neq k) \\ 0, & \text{otherwise} \end{cases} \qquad (12)$$

4.2. Model Parameter Estimation. Parameters of FMLR models can be efficiently estimated through the EM algorithm [52].

(1) Initialization Step. Start with an initial seed (guess) for the parameter $\hat{\psi}$ using the K-means clustering algorithm [53].

(2) E-Step. Estimate the conditional component probabilities, $\hat{\pi}_{ik}$, for each observation using (7) and derive the mixing proportions as

$$\hat{\pi}_k = \frac{1}{N}\sum_{i=1}^{N}\hat{\pi}_{ik} \quad (13)$$

(3) M-Step. Maximize the log-likelihood for each component separately using the conditional probabilities as weights:

$$\max_{\theta_k}\sum_{i=1}^{N}\hat{\pi}_{ik}\log\left(\mathbf{y}_i \mid \mathbf{x}_i, \theta_k\right) \quad (14)$$

The EM algorithm alternates between the expectation and the maximization steps until the likelihood improvement falls under a prespecified threshold or a maximum number of iterations are reached.

But the drawbacks of EM algorithm are its possible slow convergence rate and long processing time in computer. Thus, in this paper, Latent GOLD 5.0 is used to estimate the parameters. Latent GOLD 5.0 can take the advantages of both EM and Newton-Raphson algorithms. It first uses EM algorithm to get close to the final solution and then switches to Newton-Raphson to finish estimation [54].

The most important and difficult step in building FMLR model is to determine K, the number of components. Since K is not a parameter, hypotheses on K cannot be tested directly. BIC or AIC [50, 51, 55, 56] are generally used as criterion to determine K. In this study, we determined K based on BIC:

$$BIC_{model} = -2LL + \gamma\log\left(N\right) \quad (15)$$

where LL is the log-likelihood value, γ is the number of free parameters to be estimated, and N is the number of observations in the data. A lower BIC value indicates a better model.

5. Results and Discussion

5.1. Results. To select an optimal model, we apply the FMLR model having an increasing number of components from 1 to 4 to fit, and apply Bayesian Information Criterion (BIC) as the indicator to select the most appropriate number of components. Table 4 shows the BIC values of models for different number of components. It can be observed from Table 4 that the lowest BIC value occurs at $K = 2$. Hence, it is plausible to select $K = 2$ as a proper number of components.

To select the model variables, the forward-selection method is adopted in this paper. It starts with no variables in the model, tests the addition of each variable using Wald-statics, and adds the variable that gives the most statistically significant improvement of the fit. In this paper, variables will be added one by one until none produce a significant Wald-statistic in all components.

Table 5 shows the estimation results. For comparison, the result of logistic regression is also provided. In this paper, the

Table 4: BIC value of FMLR model.

The Number of Components	BIC Value
$K = 1$	790.2955
$K = 2$	773.3871
$K = 3$	808.3826
$K = 4$	843.3061

component mixing proportions are a set of fixed constants (see (6)), as no sociodemographic characteristics of drivers are available in this dataset. The proportion of merging vehicle drivers in each component as indicated by H value in Table 5 is 67.2% and 32.8%, respectively.

By using (10)-(12), 374 drivers are classified into two components. One is the larger component, comprising 298 drivers and 612 observations, and the other is the smaller component, containing 75 drivers and 314 observations. To better understand the classification results, the mean values and standard deviations of related attribute variables are shown in Table 6.

5.2. Discussion. As seen from significance levels of parameters of Component 1 in Table 5, ΔD_{nlg}^{i} and ΔV_{nLead}^{i} fail to be significant at the 99% level. These suggest that front vehicles in auxiliary lane do not alter drivers' merge decisions in this component. Another impressing characteristic of this component is that the drivers have a distinct preference for the larger gaps. The negative sign of V_n^i indicates that drivers in this component tend to decrease their speeds during merging process. Consistent with previous studies, the decrease of speed difference between merging vehicle and putative leading vehicle and a gap located further towards the end of the auxiliary lane also increase the probability of accepting the current gap.

It is interesting to find that the parameter of D_n^i in Component 2 is much smaller than that in Component 1, which means the drivers in Component 2 do not pursue larger gaps as drivers in Component 1. In addition, speed difference between merging vehicle and putative leading vehicle is still important during merging process. Different from Component 1, ΔD_{nlg}^{i} and V_{nLead}^{i} are considered by drivers in Component 2. The sign of the parameter for ΔD_{nlg}^{i} is positive, suggesting that space in the auxiliary lane also affects the merging behaviors of drivers in Component 2 and the merging vehicle has a high probability of accepting a gap when there is an adequate space in front of the merging vehicle. One interesting finding from Table 5 is that the sign of the parameter for V_{nLead}^{i} is negative, suggesting that drivers in Component 2 are more likely to delay merge when the leading vehicle moves too fast. One possible reason for this result might be that when the leading vehicles move faster in the auxiliary lane, the drivers are provided more space in the auxiliary lane and they are using the auxiliary lane to reach further downstream in the main lane.

As illustrated in Table 6, the related variables show obvious differences across the two components. The average numbers of rejected gaps of the two components are 1.05

TABLE 5: Model estimation results of FMLR model.

Variables	Logistic Regression	FMLR(K = 2)	
		component 1(0.672)	Component 2(0.328)
	Parameter	Parameter	Parameter
V_n^i	-	-0.1810*	0.1063*
ΔV_{nPL}^i	-0.40848*	-0.3903*	-0.2557*
D_n^i	.05490*	0.1895*	0.0158*
Y_n^i	.01345*	0.0109*	0.0105*
V_{nLead}^i	-.07111*	-0.0400	-0.0568*
ΔD_{nlg}^i	.01370*	0.0037	0.0113*
Constant	1.26281*	0.8417*	-1.7619*

Note: * means that the parameters are significant at 99% level.

TABLE 6: Mean values and standard deviations of related variables in each component.

Variables	Component 1		Component 2	
	Rejected Gaps (Standard Deviation)	Accepted Gaps (Standard Deviation)	Rejected Gaps (Standard Deviation)	Accepted Gaps (Standard Deviation)
V_n^i(m/s)	15.050 (3.196)	13.418 (3.107)	13.505 (2.852)	14.272 (3.466)
ΔV_{nPL}^i(m/s)	8.611 (3.825)	1.985 (2.766)	5.375 (3.698)	3.187 (2.908)
D_n^i(m)	10.068 (5.274)	33.14 (22.32)	17.468 (15.175)	27.09 (23.65)
V_{nLead}^i(m/s)	9.841 (7.114)	11.627 (6.671)	10.529 (6.420)	8.062 (8.100)
ΔD_{nlg}^i(m)	43.42 (47.04)	44.83 (42.99)	33.05 (35.06)	26.79 (35.46)
Merge Location(m)	41.66 (57.87)		108.58 (64.19)	
Number of Rejected Gaps	1.05		3.19	

and 3.19 and the average merge location is 41.66m and 108.58m, which indicates that drivers in Component 2 tend to choose gaps further downstream and rejected more gaps than drivers in Component 1. The average rejected gap of Component 2 (17.468 m) is much bigger than Component 1 (10.068m) while the average accepted gap of Component 2 (27.09 m) is much smaller than Component 1 (33.14 m), indicating the inconsistency of gap acceptance theories. One can also find that the drivers in Component 2 increase their speeds during their merging process from 13.505m/s to 14.272 m/s, while drivers in Component 1 decrease their speed from 15.050 m/s to 13.418m/s, and in Component 2, the speed difference between the putative leading vehicle and the merging vehicle for accepted gaps is 3.187m/s, which is much bigger than Component 1, both of which indicate that drivers in Component 2 are more aggressive than Component 1. It is interesting to find that the standard deviations of the speeds for rejected gaps and accepted gaps in Component 1 are similar, which is not the case in Component 2. And one can also find that the standard deviation of rejected gaps for Component 2 is much bigger than that in Component 1. These findings indicate that the merging process of drivers in Component 2 is much more complicated than drivers in Component 1.

Figure 2 shows the relation between the gap size and location for the rejected and accepted gaps in the two components. One can find that the accepted gaps of drivers in Component 1 are almost all located in the beginning

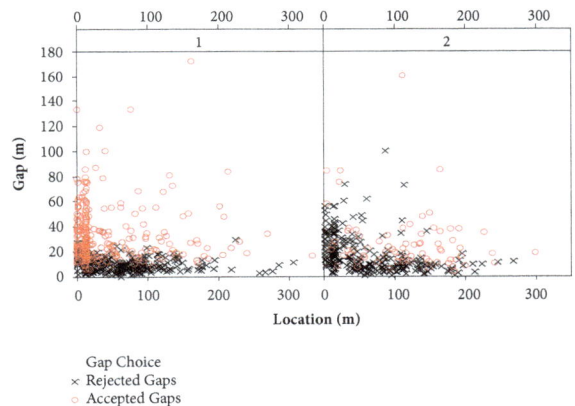

FIGURE 2: Relation between the gap size and location for the rejected and accepted gaps.

of the auxiliary lane while the accepted gaps of drivers in Component 2 are scattered along the lane. It is obvious that the rejected gaps of drivers in Component 2 are much larger than in Component 1 and are overlapped with the rejected gaps, while the overlapping area in Component 1 is much smaller.

Figure 3 shows the box plot of the reverse succession of offered gaps. The x-axis in Figure 3 is the reverse number of offered gaps before merging, in which 0 means the finally accepted gap and 1 means the last rejected gap before

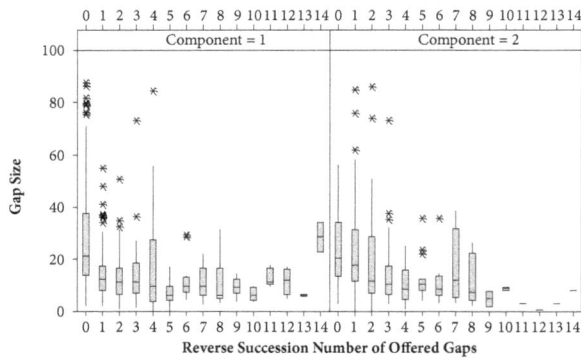

FIGURE 3: Box plot of the reverse succession of offered gaps.

TABLE 7: Comparison of estimated and observed values of logistic regression model.

Observed	Estimated		
	Reject	Accept	Total
Reject	**484.0**	68.0	552.0
Accept	92.0	**281.0**	373.0
Total	576.0	349.0	925.0

TABLE 8: Comparison of estimated and observed values of FMLR-2 model.

Observed	Estimated		
	Reject	Accept	Total
Reject	**510.0**	42.0	552.0
Accept	39.0	**334.0**	373.0
Total	547.0	378.0	925.0

merging. One can find that drivers in Component 2 might have several choices before merging, which indicates that drivers in Component 2 prefer to use the auxiliary lane to get further downstream.

Comparing the two components, drivers in Component 1 prefer larger gaps and lower speed difference, while drivers in Component 2 pay more attention to better surrounding traffic conditions and may sacrifice larger gaps to save travel time and get better traffic conditions. Thus, in this paper, Component 1 is named as Risk-Rejecting Drivers and Component 2 is named as Risk-Taking Drivers.

5.3. Accuracy of Developed Models. Tables 7 and 8 show the comparison of estimated and observed values of logistic regression model and 2-component mixture of logistic regression (FMLR-2) model. From theses tables, the proposed model improves the predicting accuracy from 82.70% to 91.24%. It can be concluded that the proposed model has better predictive power than logistic regression model.

6. Conclusions

To incorporate the unobserved heterogeneity into merge model, the present study builds a FMLR model which uses BIC to determine the proper number of mixing components

and performs parameter estimation by using Latent GOLD 5.0.

Given U.S. Highway 101 data, the identified optimal model is a 2-component mixture of logistic regression model, which means the drivers can be divided into two components characterized by the driving behavior heterogeneity. One is the Risk-Rejecting Drivers whose drivers are consistent with previous studies and primarily merge in as soon as possible. Drivers in this component have a distinct preference for the larger gaps. The decrease of speed difference between merging vehicle and putative leading vehicle and a gap located further towards the end of the auxiliary lane also increase the probability of accepting the current gap. Contrast to Component 1, Component 2 is constituted with the drivers that are much less sensitive to the gap size and have more emphasis on surrounding traffic conditions such as the speed of front vehicle in the auxiliary lane and space gap between the merging vehicle and its leading vehicles in the auxiliary lane. These drivers are using the auxiliary lane to get to the further downstream or less congested area of the main lane. Thus they are called Risk-Taking Drivers.

In addition, the proposed model can produce more precise predicting accuracy than logistic regression model.

However, more empirical studies are needed to apply this method to datasets in other sites with different demographics, climate, and geometric parameters in order to fully assess the effect of the factors affecting merging behaviors as well as fully understand the strengths and weaknesses of the proposed model.

Conflicts of Interest

The author declares that there are no conflicts of interest regarding the publication of this paper.

References

[1] L. Elefteriadou, R. P. Roess, and W. R. McShane, "Probabilistic nature of breakdown at freeway merge junctions," *Transportation Research Record*, vol. 1484, pp. 80–89, 1995.

[2] H. Yi and T. E. Mulinazzi, "Urban freeway on-ramps: Invasive influences on main-line operations," *Transportation Research Record*, no. 2023, pp. 112–119, 2007.

[3] S. P. Hoogendoorn and R. Hoogendoorn, "Generic calibration framework for joint estimation of car-following models by using microscopic data," *Transportation Research Record*, no. 2188, pp. 37–45, 2010.

[4] J. Kim and H. S. Mahmassani, "Correlated parameters in driving behavior models," *Transportation Research Record*, vol. 2249, pp. 62–77, 2011.

[5] S. Ossen and S. P. Hoogendoorn, "Heterogeneity in car-following behavior: theory and empirics," *Transportation Research Part C: Emerging Technologies*, vol. 19, no. 2, pp. 182–195, 2011.

[6] I. Kim, T. Kim, and K. Sohn, "Identifying driver heterogeneity in car-following based on a random coefficient model," *Transportation Research Part C: Emerging Technologies*, vol. 36, pp. 34–44, 2013.

[7] Cambridge Systematics, *NGSIM US 101 Data Analysis: Summary Report*, Prepared for Federal Highway Administration, 2005.

[8] Q. Yang and H. N. Koutsopoulos, "A microscopic traffic simulator for evaluation of dynamic traffic management systems," *Transportation Research Part C: Emerging Technologies*, vol. 4, no. 3, pp. 113–129, 1996.

[9] K. I. Ahmed, *Modeling drivers' acceleration and lane changing behavior*, Massachusetts Institute of Technology, 1999.

[10] G. Lee, *Modeling gap acceptance at freeway merges*, Massachusetts Institute of Technology, 2006.

[11] T. Toledo, H. N. Koutsopoulos, and M. Ben-Akiva, "Integrated driving behavior modeling," *Transportation Research Part C: Emerging Technologies*, vol. 15, no. 2, pp. 96–112, 2007.

[12] M. A. Ahammed, Y. Hassan, and T. A. Sayed, "Modeling driver behavior and safety on freeway merging areas," *Journal of Transportation Engineering*, vol. 134, no. 9, pp. 370–377, 2008.

[13] T. Toledo, H. N. Koutsopoulos, and M. Ben-Akiva, "Estimation of an integrated driving behavior model," *Transportation Research Part C: Emerging Technologies*, vol. 17, no. 4, pp. 365–380, 2009.

[14] F. Marczak, W. Daamen, and C. Buisson, "Merging behaviour: Empirical comparison between two sites and new theory development," *Transportation Research Part C: Emerging Technologies*, vol. 36, pp. 530–546, 2013.

[15] D. C. Gazis, R. Herman, and R. W. Rothery, "Nonlinear follow-the-leader models of traffic flow," *Operations Research*, vol. 9, no. 4, pp. 545–567, 1961.

[16] A. J. Miller, *Nine estimators of gap-acceptance parameters*, Publication of: Traffic Flow and Transportation, 1971.

[17] H. Mahmassani and Y. Sheffi, "Using gap sequences to estimate gap acceptance functions," *Transportation Research Part B: Methodological*, vol. 15, no. 3, pp. 143–148, 1981.

[18] P. G. Gipps, "A model for the structure of lane-changing decisions," *Transportation Research Part B: Methodological*, vol. 20, no. 5, pp. 403–414, 1986.

[19] P. Hidas, "Modelling lane changing and merging in microscopic traffic simulation," *Transportation Research Part C: Emerging Technologies*, vol. 10, no. 5-6, pp. 351–371, 2002.

[20] J. Wang, "A simulation model for motorway merging behaviour," *Transportation and traffic theory*, vol. 16, pp. 281–301, 2005.

[21] SiAS, *S-Paramics 2005 Reference Manual*, SIAS Ltd Edinburgh, 2005.

[22] PTV-Vision, *VISSIM 5.30-05 user manual*, 2011.

[23] W. Daamen, S. P. Hoogendoorn, and M. Loot, "Empirical analysis of merging behavior at freeway on-ramp," *Transportation Research Records: Journal of Transportation Research Board*, no. 2188, pp. 108–118, 2010.

[24] T. D. CHU, *A Study on Merging Behavior at Urban Expressway Merging Sections*, Nagoya University, 2014.

[25] H. Kita, "Effects of merging lane length on the merging behavior at expressway on-ramps," *Transportation and Traffic Theory*, pp. 37–51, 1993.

[26] H. Kita, "A merging-giveway interaction model of cars in a merging section: a game theoretic analysis," *Transportation Research Part A: Policy and Practice*, vol. 33, no. 3-4, pp. 305–312, 1999.

[27] J. Weng and Q. Meng, "Modeling speed-flow relationship and merging behavior in work zone merging areas," *Transportation Research Part C: Emerging Technologies*, vol. 19, no. 6, pp. 985–996, 2011.

[28] A. Rao, *Modeling Anticipatory Driving Behavior [MS Thesis]*, Department of Civil and Environmental Engineering, MIT, 2006.

[29] C. F. Choudhury, M. E. Ben-Akiva, T. Toledo, G. Lee, and A. Rao, "Modeling cooperative lane changing and forced merging behavior," in *Proceedings of the 86th Annual Meeting of the Transportation Research Board*, C., Ed., Washington, DC, 2007.

[30] L. Sun and J. Zhou, "Development of multiregime speed-density relationships by cluster analysis," *Transportation Research Record*, no. 1934, pp. 64–71, 2005.

[31] Y. Pan and L. Sun, "Characterizing heterogeneity in vehicular traffic speed using two-step cluster analysis," *Journal of Southeast University (English Edition)*, vol. 28, no. 4, pp. 480–484, 2012.

[32] S. Ossen and S. P. Hoogendoorn, "Car-following behavior analysis from microscopic trajectory data," *Transportation Research Record*, no. 1934, pp. 13–21, 2005.

[33] S. Ossen, S. P. Hoogendoorn, and B. G. H. Gorte, "Interdriver differences in car-following a vehicle trajectory-based study," *Transportation Research Record*, vol. 1965, no. 1, pp. 121–129, 2006.

[34] X. Ma and I. Andréasson, "Statistical analysis of driver behavior data in different regimes of the car-following stage," *Transportation Research Record*, no. 2018, pp. 87–96, 2007.

[35] G. Li and L. Sun, "Characterizing Heterogeneity in Drivers Merging Maneuvers Using Two-Step Cluster Analysis," *Journal of Advanced Transportation*, 2018.

[36] M. Keyvan-Ekbatani, V. L. Knoop, and W. Daamen, "Categorization of the lane change decision process on freeways," *Transportation Research Part C: Emerging Technologies*, vol. 69, pp. 515–526, 2016.

[37] V. Alexiadis, J. Colyar, J. Halkias, R. Hranac, and G. McHale, "The next generation simulation program," *ITE Journal*, vol. 74, no. 8, pp. 22–26, 2004.

[38] C. Thiemann, M. Treiber, and A. Kesting, "Estimating acceleration and lane-changing dynamics from next generation simulation trajectory data," *Transportation Research Record*, vol. 2088, pp. 90–101, 2008.

[39] V. Punzo, M. T. Borzacchiello, and B. Ciuffo, "On the assessment of vehicle trajectory data accuracy and application to the Next Generation SIMulation (NGSIM) program data," *Transportation Research Part C: Emerging Technologies*, vol. 19, no. 6, pp. 1243–1262, 2011.

[40] M. Montanino and V. Punzo, "Trajectory data reconstruction and simulation-based validation against macroscopic traffic patterns," *Transportation Research Part B: Methodological*, vol. 80, pp. 82–106, 2015.

[41] B. Coifman and L. Li, "A critical evaluation of the Next Generation Simulation (NGSIM) vehicle trajectory dataset," *Transportation Research Part B: Methodological*, vol. 105, pp. 362–377, 2017.

[42] Q. Wang, Z. Li, and L. Li, "Investigation of discretionary lane-change characteristics using next-generation simulation data sets," *Journal of Intelligent Transportation Systems*, vol. 18, pp. 246–253, 2014.

[43] X. Wan, P. J. Jin, F. Yang, and B. Ran, "Merging preparation behavior of drivers: How they choose and approach their merge

positions at a congested weaving area," *Journal of Transportation Engineering*, vol. 142, no. 9, 2016.

[44] X. Wan, P. J. Jin, H. Gu, X. Chen, and B. Ran, "Modeling Freeway Merging in a Weaving Section as a Sequential Decision-Making Process," *Journal of Transportation Engineering, Part A: Systems*, vol. 143, Article ID 05017002, 2017.

[45] G. McLachlan and D. Peel, *Finite mixture models , willey series in probability and statistics*, John Wiley & Sons, New York, USA, 2000.

[46] C. Fraley and A. E. Raftery, "Model-based clustering, discriminant analysis, and density estimation," *Journal of the American Statistical Association*, vol. 97, no. 458, pp. 611–631, 2002.

[47] W. S. DeSarbo and W. L. Cron, "A maximum likelihood methodology for clusterwise linear regression," *Journal of Classification*, vol. 5, no. 2, pp. 249–282, 1988.

[48] M. Wedel and W. S. DeSarbo, "A mixture likelihood approach for generalized linear models," *Journal of Classification*, vol. 12, no. 1, pp. 21–55, 1995.

[49] M. Wedel and W. A. Kamakura, *Market segmentation: Conceptual and methodological foundations*, Springer Science & Business Media, 2012.

[50] W. H. Greene and D. A. Hensher, "A latent class model for discrete choice analysis: contrasts with mixed logit," *Transportation Research Part B: Methodological*, vol. 37, no. 8, pp. 681–698, 2003.

[51] W. H. Greene, *Interpreting estimated parameters and measuring individual heterogeneity in random coefficient models*, 2004, Interpreting estimated parameters and measuring individual heterogeneity in random coefficient models.

[52] A. P. Dempster, N. M. Laird, and D. B. Rubin, "Maximum likelihood from incomplete data via the EM algorithm," *Journal of the Royal Statistical Society Series: B*, vol. 39, no. 1, pp. 1–38, 1977.

[53] G. McLachlan and T. Krishnan, *The EM algorithm and extensions*, John Wiley & Sons, 2007.

[54] J. K. Vermunt and J. Magidson, *Latent GOLD 5.0 upgrade manual*, Statistical Innovations Inc, Belmont, MA, 2013.

[55] H. Akaike, "A new look at the statistical model identification," *IEEE Transactions on Automatic Control*, vol. 19, pp. 716–723, 1974.

[56] G. Schwarz, "Estimating the dimension of a model," *The Annals of Statistics*, vol. 6, no. 2, pp. 461–464, 1978.

Permissions

List of Contributors

Zhiguang Liu
Department of Civil Engineering, Nagoya University, Nagoya 464-8603, Japan

Tomio Miwa
Institute of Materials and Systems for Sustainability, Nagoya University, Nagoya 464-8603, Japan

Weiliang Zeng
School of Automation, Guangdong University of Technology, Guangzhou, Guangdong 510006, China

Michael G. H. Bell
Institute of Transport and Logistics Studies, Business School,The University of Sydney, Sydney, NSW, Australia

Takayuki Morikawa
Institute of Innovation for Future Society, Nagoya University, Nagoya 464-8603, Japan

De Zhao
Jiangsu Key Laboratory of Urban ITS, Southeast University, Si Pai Lou No. 2, Nanjing 210096, China

De Zhao
Department of Civil and Environmental Engineering, National University of Singapore, Engineering Drive 2, E1A 08-20, Singapore 117576

Wei Wang and Yanjie Ji
School of Transportation, Southeast University, Si Pai Lou No. 2, Nanjing 210096, China

Ghim Ping Ong
Department of Civil and Environmental Engineering, National University of Singapore, Engineering Drive 2, E1A 07-03, Singapore 117576

Rick Wolbertus
Transport and Logistics Group, Department of Engineering Systems and Services, Faculty of Technology, Policy and Management, Delft University of Technology, Jaffalaan 5, 2628 BX Delft, Netherlands

Rick Wolbertus
Department of Urban Technology, Faculty of Technology, Amsterdam University of Applied Sciences,Weesperzijde 190, 1097 DZ Amsterdam, Netherlands

Bas Gerzon
PitPoint Clean Fuels, Gelderlandhaven 4, 3433 PG Nieuwegein, Netherlands

Qingfeng Zhou, Donghui Dai and Yaowu Wang
Harbin Institute of Technology Shenzhen Graduate School, Shenzhen, Guangdong 518055, China

Jianshuang Fan
Zhejiang University of Technology, Hangzhou, Zhejiang 310014, China

Quan Liang, Jiancheng Weng and Jian Rong
Beijing Key Laboratory of Traffic Engineering, Beijing University of Technology, Beijing 100124, China

Wei Zhou
Ministry of Transport of the People's Republic of China, Beijing 100736, China

Selene Baez Santamaria
Computer Science Department, Vrije Universiteit Amsterdam, 1081 HV Amsterdam, Netherlands

Jianming Ma
Texas Department of Transportation, Austin, TX 78717, USA

Raj Bridgelall and Pan Lu
College of Business, North Dakota State University, Fargo, North Dakota 58108, USA

Denver D. Tolliver
UGPTI, North Dakota State University, Fargo, North Dakota 58108, USA

Tai Xu
University of Modern Sciences, Dubai, UAE

Marian Brzeziński, Magdalena Kijek, Paulina Owczarek, Katarzyna Głodowska, Jarosław Zelkowski and Piotr Bartosiak
Military University of Technology, Faculty of Logistics, Institute of Logistics, Warsaw 00908, Poland

Weitiao Wu and Wenzhou Jin
South China University of Technology,Wushan Road, Tianhe District, Guangzhou 510641, China

Ronghui Liu
University of Leeds, Institute for Transport Studies, Leeds LS2 9JT, UK

Shi-chao Sun
College of Transportation Engineering, Dalian Maritime University, Dalian, 116026, China

Wei Zhang, Ruichun He, Changxi Ma and Mingxia Gao
School of Traffic and Transportation, Lanzhou Jiaotong University, Lanzhou 730070, China

Qi Ouyang, Yongbo Lv, Yuan Ren, Jihui Ma and Jing Li
School of Traffic and Transportation, Beijing Jiaotong University, Beijing 100044, China

Jian Zhang
Associate Professor in School of Transportation, Southeast University, 2 Si Pai Lou, Nanjing 210096, China

Zhibin Li
Nanjing University of Science and Technology, Nanjing 210094, China

Zhibin Li and Wei Wang
Professor in School of Transportation, Southeast University, 2 Si Pai Lou, Nanjing 210096, China

Fangwei Zhang
Assistant Professor in Shanghai Maritime University, 1550 Haigang Ave, Shanghai 201306, China

Yong Qi
Professor in School of Computer Science & Engineering, Nanjing University of Science and Technology, Nanjing 210094, China

Wenzhu Zhou
Associate Professor in School of Architecture, Southeast University, 2 Si Pai Lou, Nanjing 210096, China

Yong Wang
Associate Professor in School of Economics and Management, Chongqing Jiaotong University, 66 Xuefu Road, Nan'an District, Chongqing 400074, China

De Zhao
Postdoctoral Researcher in Department of Civil and Environmental Engineering, National University of Singapore, 1 Engineering Drive 2, E1A 07-03, Singapore 117576

Yi Luo and Dalin Qian
School of Traffic and Transportation, Beijing Jiaotong University, Beijing, China
MOE Key Laboratory for Urban Transportation Complex Systems Theory and Technology, Beijing Jiaotong University, Beijing, China

Agostino Nuzzolo and Antonio Comi
Department of Enterprise Engineering, University of Rome Tor Vergata, Via del Politecnico 1, 00133 Rome, Italy

Shixiong Jiang, Wei Guan and Liu Yang
MOE Key Laboratory of Urban Transportation Complex SystemTheory and Technology, Beijing Jiaotong University, Beijing 100044, China

Zhengbing He
College of Metropolitan Transportation, Beijing University of Technology, Beijing, China

Shuozhi Liu and Xia Luo
School of Transportation and Logistics, Southwest Jiaotong University, 111 Second Ring Road Beiyiduan, Chengdu 610031, China

Peter J. Jin
Department of Civil and Environmental Engineering, Rutgers, the State University of New Jersey, CoRE 613, 96 Frelinghuysen Road, Piscataway, NJ 08854-8018, USA

Gen Li
School of Transportation, Southeast University, Nanjing 210096, China

Index

www.ingramcontent.com/pod-product-compliance
Lightning Source LLC
Chambersburg PA
CBHW082023190326
41458CB00010B/3257